Robert Brown

The Countries of the World

Being a popular description of the various continents, islands, rivers, seas, and peoples of the globe. Vol. 2

Robert Brown

The Countries of the World
Being a popular description of the various continents, islands, rivers, seas, and peoples of the globe. Vol. 2

ISBN/EAN: 9783337301637

Printed in Europe, USA, Canada, Australia, Japan

Cover: Foto ©Andreas Hilbeck / pixelio.de

More available books at **www.hansebooks.com**

THE
COUNTRIES OF THE WORLD:

BEING

A POPULAR DESCRIPTION OF THE VARIOUS CONTINENTS, ISLANDS, RIVERS,
SEAS, AND PEOPLES OF THE GLOBE.

BY

ROBERT BROWN, M.A.

PH.D., F.L.S., F.R.G.S.

Author of "The Races of Mankind," etc. etc.

CASSELL, PETTER, GALPIN & CO.:
LONDON, PARIS & NEW YORK.

CONTENTS.

	PAGE
THE UNITED STATES:—	
THE INDUSTRIES AND THE MEN OF THE PACIFIC SLOPE	1
MINING	2
SOCIAL LIFE	3
"THE HONEST MINER"	6
MAIL DAY IN THE WEST	14
THE LANGUAGE OF THE PACIFIC SLOPE	24
"PUTTING THROUGH THE WINTER"	31
THE ROCKY MOUNTAIN STATES AND TERRITORIES:—	
NEW MEXICO	42
ARIZONA: THE COLORADO AND ITS CAÑONS	47
THE GREAT BASIN	58
UTAH	62
NEVADA	65
MONTANA	66
COLORADO	66
DAKOTA	72
THE PRAIRIES WEST OF THE ROCKY MOUNTAINS	74, 75
THE PRAIRIES EAST OF THE ROCKY MOUNTAINS	74, 79
PRAIRIE AND FOREST: THEIR SANITARY ASPECTS	84
WYOMING	88
THE "WONDERLAND" OF AMERICA	91
THE MISSISSIPPI BASIN	100
NEBRASKA	107
MINNESOTA	108
IOWA	111
KANSAS	112
WISCONSIN	114
ILLINOIS	115
MISSOURI	116
ARKANSAS	123
THE INDIAN TERRITORY	126
LOUISIANA	130
TEXAS	138
MARYLAND	150
VIRGINIA	155
NORTH CAROLINA	167
SOUTH CAROLINA	170
GEORGIA	171
FLORIDA	173

	PAGE
THE UNITED STATES (continued):—	
ALABAMA	180
MISSISSIPPI	183
TENNESSEE	185
KENTUCKY	190
INDIANA	194
MICHIGAN	195
OHIO	199
PENNSYLVANIA	208
NEW YORK	214
NEW JERSEY AND DELAWARE	216
NEW ENGLAND	219
CONNECTICUT	220
RHODE ISLAND	222
MASSACHUSETTS	223
VERMONT	227
NEW HAMPSHIRE	230
MAINE	233
MEXICO:—	
ANAHUAC, AZTEC, AND SPANISH	234
COLONIAL MEXICO	240
MEXICO INDEPENDENT	243
ITS PHYSICAL GEOGRAPHY AND RESOURCES	247
THE CLIMATE	251
VEGETABLE PRODUCTS	252
ANIMAL PRODUCTS	255
ITS MEN AND MANNERS	256
THE PRIESTS AND THE CHURCH	258
INDIANS	262
MESTIZOES	272
THE MINERS	277
THE RANCHEROS	280
THE CATTLE-BREEDERS AND HERDSMEN	282
THE TOWNS OF OLD SPAIN AND NEW SPAIN	286
LIFE ON THE PLAZA	287
THE "COMMON PEOPLE"	298
CREOLES	298
THE PROLETARIANS	299
ITS COMMERCE	302
ITS PROSPECTS	304
THE WEST INDIES:—	
A GENERAL SKETCH	304
JAMAICA	307
HAYTI	315
CUBA	319

LIST OF ILLUSTRATIONS.

	PAGE
A Mining Camp on a North-west American River (Leech River. *From an Original Photograph*)	*Frontispiece.*
Mining "Cement" by the Hydraulic Process	5
Sacramento Street, San Francisco	9
A Miner's Cabin by the American River, California (*From an Original Photograph*)	13
View of "The Gorge," Victoria Harbour, Vancouver Island (*From an Original Photograph*)	17
View of the Western Suburbs of Victoria, Vancouver Island (*From an Original Photograph*)	20
Shouswap Indians, British Columbia	21
Stage Coach Starting from a Railway Station in Western America	25
A Summer Encampment in a North Pacific Forest (*From an Original Photograph*)	29
View of Moore's Lake, Utah ... *To face page*	33
View of the Snoqualami Falls, Washington Territory (*From an Original Sketch*)	33
View of Sooke Lake, Vancouver Island (*From an Original Sketch*)	37
Ute Indian Chief	40
Warrior of Blood Indian Tribe	41
A Prairie Dog "Town"	44
Snowy Range of the Sierra Madre, Rocky Mountains	45
The great Cañon of the Colorado: Raft Precipitated over a Cataract	49
The *Cereus giganteus*, or Monumental Cactus	53
View of Green River, Utah	56
View of the Valley of the Bubbling Waters, Utah	57
A Street in Salt Lake City, Utah	60
View of Salt Lake City, Utah	61
An Hotel in Salt Lake City, Utah	64
View of Clear Creek Cañon, Colorado *To face page*	65
Emigrant Train in Colorado (*From an Original Sketch*)	65
A Pack Train Ready for Loading (*From an Original Sketch*)	68
A Street in Denver, Colorado	69
The Colorado Beetle (*Doryphora decemlineata*)	72
A Dakota or Sioux Indian	73

	PAGE
Steamer on the Lower Missouri	76
A Blood Indian	77
View in the Valley of the Upper Mississippi	81
A Herd of Buffaloes (*Bison Americanus*) on the Prairie	85
Station of the Pacific Railway at Omaha	88
View of Yellowstone Lake, Wyoming	89
View of the Cliffs in the Grand Cañon of the Yellowstone, Wyoming	93
View of the Hot Springs on Gardiner's River, Wyoming	96
View of the Falls of St. Anthony, Minnesota. *To face page*	97
View of the Tower Falls and Column Mountain, Wyoming	97
View on the Upper Mississippi, Minnesota	101
Rafts of Trees on the Missouri	104
View of the Source of the Arkansas, one of the Tributaries of the Mississippi	105
View in Nebraska	108
Winnebago Indian	109
View of "The Pictured Rocks," Southern Shore of Lake Superior	113
View of St. Louis, Missouri	117
A Cotton Shoot	121
View of "The Tower Rock," Rock Island, on the Upper Mississippi	124
View of a *Levée* on the Lower Mississippi	125
A Cheyenne Indian Chief	128
Loading a Cotton Steamer ... *To face page*	129
Creek Indians	129
An Election Day in New Orleans	133
A Street in New Orleans	136
Picking Cotton	137
View on the Shore of the Gulf of Mexico	141
Catching Wild Horses on the Prairies with the Lasso	145
View on the Baltimore and Ohio Railroad (Maryland)	149
View of Jefferson's Rock, Cemetery Hill, Harper's Ferry, Maryland	152
The Bridge Across the Potomac at Harper's Ferry	153

LIST OF ILLUSTRATIONS.

	PAGE
View of Goshen Pass, Rockbridge County, Virginia	157
View of the Port of Richmond, Virginia	160
View of Balcony Falls, James' River, Virginia. *To face page*	161
View of the Alleghany Mountains, Virginia	164
View of Stiles Falls, Virginia	165
View of Grandfather Mountain, North Carolina	169
View near Morganton, North Carolina	172
View of Watauga Falls, Western North Carolina	173
A Negro Village in Georgia	176
Loading Cotton at Savannah	177
View on one of the Tributaries of the St. John River, Florida	181
View on the St. John River, Florida	184
Hauling in a Drum Fish (*Pogonias chromis*) off the Coast of Florida	185
Forest Scenery in Florida	189
Interior of the Mammoth Cave of Kentucky	192
View of Lake George, New York State. *To face page*	193
Exploring the Echo River, Mammoth Cave, Kentucky	193
View of "The Grand Chapel" Rocks, Lake Superior	197
An Ohio Farmhouse	201
View of the Juniata River, near Lewistown, Pennsylvania	205
View of "The Bridal Veil" Falls, Raymondskill River, Pennsylvania	209
View of the Vale of Wyoming, with Susquehanna River, Pennsylvania	212
View of the Falls of Niagara, Western Side	213
View of Lake Champlain, New York State	216
View of a Spur of the Blue Mountains, Delaware Water Gap, New Jersey	217
View of Salmon Brook, Granby, Connecticut	221
View of Boston, Massachusetts, from Bunker's Hill	224

	PAGE
View of Mount Mansfield, Vermont *To face page*	225
View on the Connecticut River, Massachusetts	225
View of Salem, Massachusetts	228
View on the Missisquoi River, Vermont	229
View in the White Mountains, New Hampshire	232
View on the Coast of Maine	233
View of Vera Cruz, on the Gulf of Mexico	237
View of the Pyramid of Cholula, near Puebla	240
Colossal Head Carved in Stone, in an Aztec Ruin at Izamal, in Yucatan	241
Echmea paniculata, a plant of tropical America (not Aloes, as in the text)	245
Tropical Climbers	249
A Lagoon in the *Tierras Calientes*	253
Gathering Vanilla (*Vanilla planifolia*)	256
The Cathedral of Mexico *To face page*	257
A Mexican Monk of Former Times	257
View of Puebla (from the East and the West)	261
An Aztec Ruin at Tuloom, in Yucatan	265
Mexican Indians of the *Tierras Calientes*	269
Mexican Mestizo Lady and Maid	273
Indians Dressing Ore for the Corralitos Smelting Works	277
Interior of Smelting Works at Chihuahua	281
Plaza of Guadalajara, in the State of Jalisco	284
An *Evangelista* and his Clients	285
View of the City of Mexico	288
Negro Huts, Jamaica *To face page*	289
Fountain and Aqueduct, City of Mexico	289
Mexican Serenos (Night Watchmen)	293
Cypress Grove of Chapultepec, near the City of Mexico	297
A Rural Kitchen in the *Tierras Calientes*	301
View of Charlotte Amalie, St. Thomas, West Indies	305
View of Newcastle, Jamaica	309
A Planter's House, Jamaica	313
View in Hayti	316
View of the City of San Domingo, Hayti	317
View in Cuba	319

THE COUNTRIES OF THE WORLD.

CHAPTER I.

IN the first Volume of this Work we have endeavoured to convey to the reader an idea of the scenery, and the aspects of the plant and animal life of the countries which lie on the Pacific Slope of the Rocky Mountains. Before commencing in our Second to describe the States and Territories of the Great Republic, which lie in and around the Rocky Mountain range itself, it may be well to speak somewhat more systematically of one or two of the industrial features, and especially of the men following the chief occupation of that region. Much of what I shall tell the reader may be already familiar to him from other sources. If such be the case, the writer will be sufficiently pleased, for he will have the gratification of knowing that the impressions he obtained from personal observation are so far accurate that they have struck others beside himself. Moreover, as some of what follows are pictures-memoranda of a life that can never more return to the "Wild Lone Land," he is the more anxious to describe it before it is looked upon as mere romance.

THE UNITED STATES: THE INDUSTRIES AND THE MEN OF THE PACIFIC SLOPE.

Taking then California as the best type of the Western Rocky Mountain slope, its resources seem almost endless. At one time gold was the only thing the State yielded; this, though useful as a stimulant to other industries, was not, however, in itself riches. Every ounce taken out of a country makes it so much the poorer, without really making the world either richer or better. But the wealth of California remained to a great extent within its bounds. The swarms of gold-diggers who rushed to it from all the world in many cases remained to cultivate the soil, to erect manufactories and towns, to plant vineyards, hew down the forests, export the timber, build ships, run steamers on the rivers and lakes, and in a hundred other ways aid in the development of the country's resources. Hence gold-mining, though still a prominent occupation of the Californians, is a resource which, if even it failed to-morrow, would scarcely injure the country; indeed, it would rather help it, for it would release a large number of men for other pursuits, and especially to engage in agriculture, for which the country is so admirably suited. We need not repeat the already well-worn tales of the enormous wheat yield of California, of the mile after mile under grain, of the cutting it by machinery, threshing on the field by machinery, and sacking by the same means all in one day. Neither is it necessary to tell of the enormous fruit yield, of the peaches, the pears, the apples, and the strawberries all the year round, or of the grapes, oranges, and other semi-tropical fruits of the South. Suffice it to say that in California are every manufacture and every industry which flourish in the rest of America, in addition to some which are more peculiarly its own.

MINING.

It was, however, gold-digging which gave this State—and indeed the whole coast—its original impetus, and gold-digging and gold winning will always be associated in the popular mind with this land of gallant men. Gold is mined in various ways. It originally all came from quartz veins, but by the crumbling down of the matrix it is now scattered through the earth, and more particularly in the sand and gravel of rivers which have washed it far away from its original home. It was in such localities that the gold was first washed, and to some extent it is still mined there. The varied apparatus for separating it from the *débris* it is mixed with is all constructed on the same principle, viz., by aid of water to wash the gold scales to the bottom, these being heavier, and leave the sand, earth, and stones at the top to be thrown away. If the gold be very "fine," *i.e.*, in very minute scales, it is caught by means of quicksilver mixed with the mass in process of washing (Vol. I., page 310). The implements in use, or which have been invented, are simply endless. A comparatively recent mode of washing the gold is, however, so interesting, that we may describe it, and have illustrated it by a figure (Vol. I., Plate X.). This is the hydraulic method. Its principle consists in letting water down from a considerable height, and throwing it under the pressure of its own weight against the "pay dirt," which is thus torn down, dissolved, and carried into the sluice below. This is effected by means of a strong hose, and is used not to wash the dirt, but to save digging with shovels, and to carry it to the sluice. "The hydraulic process is applied," writes Mr. Hittel, who has given us the best account of it, "only to claims when the dirt is deep, and when the water is abundant. If the dirt were shallow in the claim and its vicinity, the necessary head of water could not be obtained. Hydraulic claims are usually in hills. The water is led along on the hill at a height varying from fifty to two hundred feet above the bed rock to the claim at the end or the side of the hill, when the water, playing against the dirt, soon cuts a large hole, with perpendicular, or, at least, steep banks. At the top of the bank is a little reservoir, containing perhaps in it more than 200 gallons, into which the water runs constantly, and from which the hose extends down to the bottom of the claim. The hose is of heavy duck, sometimes double, sewn by machine. This hose when full is from five to ten inches in diameter, and will bear a perpendicular column of water fifty feet high; but a greater height will burst it. Now, as the force of the stream increases with the height of the water, it is a matter of great importance to have the hose as strong as possible; so for this purpose in some claims it is surrounded by iron bands, which are about two inches wide, and are connected with four ropes, which run perpendicularly down. The rings are about three inches apart. The 'crinoline' hose thus made is very flexible, and will support a column of water 150 or 200 feet in height. The pipe at the end of the hose is like the pipe of a fire-engine hose, though usually longer. Sometimes the pipe will be eight inches in diameter when it connects with the hose, and not more than two inches at the mouth, and the force with which the stream rushes from it is so great that it will kill a man instantaneously, and tear down a hill more rapidly than could a hundred men with shovels." This stream is directed against the bank with such force that soon the cliff is undermined, and the large mass of dirt tumbles down. (See page 5, which shows another phase of hydraulic mining, viz., when a number of streams of water are directed against the hard "cement" or conglomerated gravel.) This the shovel-men wash away into the sluice or timber-lined ditch, when they commence at the bottom of the bank again, and so on. The gold in the "dirt" is caught in riffles, or cross-bars, placed here and there in the sluice, or impeded in other ways; and then at leisure—generally once a week—cleaned up by the company to which it belongs. The mud from such a hydraulic claim forms long, dreary, grassless flats, very characteristic of such localities, and the quantity washed into the Californian rivers by the operations of the gold miners is now getting so plentiful that these streams are no longer clear as once they were, but roll along milky in hue, and in some cases are absolutely getting shoaled up. Quartz mining is also an important industry on the Pacific. In this case the rock containing the gold is directly mined and crushed in mills or by means of the "arastra," a piece of rock which is drawn by means of mules or horses round in a trough over the broken pieces of quartz, and thus, as in the cases of the mill, though more slowly and cheap, grinding it up into powder. This is aided by the addition of water, while the gold is caught by means of mercury added from time to time to the creamy liquid. This "amalgam" is then heated in retorts, the quicksilver driven off in fumes, and condensed again in water, while the spongy mass of gold remains behind.

Cinnabar, or the ore of mercury, is also extensively mined in California, chiefly at New Almaden, a little west of San José. At one time the mines yielded from 2,500,000 to 3,500,000 lbs. of quicksilver in a year; but the production has now fallen to about 1,000,000 lbs. per annum. Copper is found, though at present no mines are wrought. Zinc, tin, lead, and iron are in the same category. The coal is of tertiary and cretaceous age, and poor, though it is mined to the extent of about 175,000 tons per annum. Borax was exported in 1873 to the extent of 100,000 dollars; but silver mining, though extensively followed in the neighbouring region—more particularly in Nevada State—has never been very successful in California, except at Cerro Gordo, where the ore is chiefly galena. These mines yielded in 1875 about a million dollars of metal, six-tenths of which was silver. Altogether in the latter year the worth of the silver mined in California was 3,000,000 dollars, in Nevada, 28,600,000 dollars, while in California 17,000,000 dollars, and in Nevada 12,000,000 dollars worth of gold were mined. In 1876 the total value of the gold and silver obtained in California, Nevada, Utah, Colorado, and the other states and countries of the United States, was about 100,000,000 dollars.

SOCIAL LIFE.

Perhaps no country has ever had its social life so frequently described as California. To this region came the choice spirits of every nation, and also the reckless desperadoes of the world. All nations amalgamated together, and founded here almost a new race. The wild excitement of the gold winning, which for years exercised supreme control over the manners and thoughts of the people, has also had its influence in forming society in these regions. The result is a people in many respects unique, though rapidly getting moulded down to that uniform level to which civilisation, railways, telegraphs, and newspapers tone all mankind. A new element has comparatively recently been added in the Chinese; and independently of all questions connected with their effect on the labour of the country, there cannot be a doubt but that in time these Mongol hordes will exercise a wonderful influence on the life of the Western United States. The Indians are rapidly getting exterminated, but the Chinese are not. On the contrary, they arrive in greater numbers than ever, and flourish to an extent strange to them in their native land. The whole population of the State was, in 1870, 582,031, an increase of fifty-three per cent. since the census of 1860; but as San Francisco had in the former year only 149,473 people, and has now, it is claimed, a population of little less than 250,000, we may judge from this fact that the State is much more populous than it was seven years ago. However, it must be remembered that fully one-half of the people live within an area of 4,000 miles from the chief town (San Francisco). The capital (Sacramento) had in 1870 16,283 people; but no other town had anything like so numerous a show of citizens. The wandering character of the population has a prejudicial effect on its morality, while the greed of gold, and the all-potent effect it had in the development of the country, have exercised an evil effect on commercial morality. The scarcity of women, and the want of home-life over a great portion of the country, is a vicious element in Californian life, and indeed exercises a mischievous influence on social existence all over the Pacific slope. There are generally few respectable women living in the mines: with a result that can be easily imagined. The first ornament of the female mind is too often absent in California, and the same love of gain which is developed in the men shows itself in the women, though in an infinitely more fatal manner. I state this on the authority of the greatest historian of the State, otherwise I should hesitate at so sweeping a conclusion from merely a few passing visits. Divorces are very common, and unions unrecognised by the law scarcely less so. The recklessness of life which ever characterised the State still exists to some extent, though in San Francisco and other large cities quiet people can live as securely, and indeed enjoy as polished society, as in almost any other town in America. Extravagance of living is common; but, on the other hand, no man is compelled to live in a style beyond his means merely to "keep up his position in society." Never was there a people among whom the stranger could feel more at home—never a more "sociable" race. In general society there is no asking as to what family the new arrival belongs, whether even he is wealthy. The main facts desired to be ascertained about him before he is welcomed are whether he is well educated, pleasant, and entertaining. There is a liberal tone in all classes of society, and an almost cosmopolitan sympathy with any eccentricity in thought, in manners, or in religion. The people love to be amused, and will pay for it. Accordingly, California is the El Dorado of all actors, singers, and showmen generally, and indeed of every one who has anything to dispose of. Society has been turned upside-down here; the rich people have once been poor, those poor formerly are now wealthy; hence the tolerance, the freedom, and the slight *soupçon* of roughness which prevail everywhere,

MINING "CEMENT" BY THE HYDRAULIC PROCESS.

even among people whose surroundings would lead us, in other countries, to expect something else. But if we described Californian life fully a volume would be too small, and the result such as would please no one. The gold-digger is probably as little changed as any one. Indeed, he is only now settled down into a marked feature in social life. Formerly anybody and everybody were gold-diggers, but now mining is a profession; a trade or a labour we cannot call it, for to mine for gold is never degrading to any man, who might think he was lowering himself were he to drive a team or plough a field. In this respect, it must be confessed, Californian society has a little changed. Still, though you can find men of all classes working at mechanical toil, it is now rare to find men highly educated and well behaved, finding any necessity for so doing. They have either left the State in despair, or gravitated into their proper position. I met, however, in British Columbia, in early times, a good deal of this. But the senior wrangler who found his accomplishments as an oarsman the means of earning his living as a boatman in San Francisco Bay seems to have departed from that locality; though, indeed, pages could be easily filled with similar tales, not more apocryphal, and some indeed which I could vouch for. Let us, however, as our space is limited, say a little about the gold-digger and his ways; and the sketch equally applies to the miner all over the gold regions of North America.

"The Honest Miner."

One autumn a few years ago I struck into the wild mountain region of Southern Oregon, just north of the Californian boundary line. I had not gone far on the trail before I overtook a stalwart, grey-shirted, knee-booted individual. He had a pair of scarlet blankets strapped on his back, and as he trudged along, for want of better company, he held an animated conversation with himself, an oath being every now and then very innocently, no doubt, introduced when the merits of the case seemed to demand it. He was an old gold-digger returning to his favourite "creek." He had been on one of the usual digger wild-goose chases after some fancied El Dorado at a distance, but was returning disappointed to the place where he had mined for many a year. Every locality was familiar to him. As we walked together over the mountains, or by the banks of the creek or stream, down in the wooded valley, my companion would point out to me, with a half-regretful pride, where "big strikes" had been made in former times. Pointing to a ruined log cabin, out of which a cayote wolf rushed, he assured me that the owner had washed some forty thousand dollars out of a patch twenty or thirty yards in extent. Cañon Creek, the locality in question—it must be familiar to many of my readers—is a fit specimen of many similar localities all over the Pacific coast. It was one of those "dead cities," the weird associations of which have been introduced into a well-known work. In Tuolumne County, California, are very many such. "We lived," writes a graphic and humorous writer, though in this case the description is soberly in earnest, "in a small cabin on a verdant hill-side, and there were but few other cabins in view over the wide expanse of hill and forest. Yet a flourishing city of two or three thousand people had occupied this grassy, dead solitude during the flush times of twelve or fifteen years before, and where our cabin stood had once been the heart of the teeming hive, the centre of the city. When the mines gave out the town fell into decay, and in a few years wholly disappeared—streets, dwellings, shops, everything—and left no sign. The grassy slopes were as green and smooth and desolate of life as if they had never been disturbed. The mere handful of miners had seen the town spring up, spread, grow, and flourish in its pride; and they had seen it sicken and die, and pass away like a dream. With it their hopes had died, and their zest of life. They had long ago resigned themselves to their exile, and ceased to correspond with their distant friends, or turn longing eyes towards their early homes. They had accepted banishment, forgotten the world, and had been forgotten of the world. They were far from telegraphs and railroads, and they stood, as it were, in a living grave, dead to the events that stirred the globe's great populations, dead to the common interests of men, isolated and outcast from brotherhood with their kind. It was the most singular, and almost the most touching and melancholy exile that fancy can imagine. One of my associates in this locality for two or three months was a man who had received a university education; but now for eighteen years he had decayed there by inches, a bearded, rough-clad, clay-stained miner, and at times, amid his sighing and soliloquisings, he unconsciously interjected vaguely-remembered Latin and Greek sentences, dead and musty tongues, meet vehicles for the thoughts of one whose dreams were all of the past, whose life was a failure; a tired man, burdened with the present, and indifferent to the future; a man without ties, hopes, interests, waiting for rest, and the end." Cañon Creek was such a place as that so graphically described in this extract. It had been, I was told, once a "bully old diggin'," but the stream having been pretty well washed out, the miners had decamped to parts unknown, leaving no address behind them. Like the Arabs, "they folded their tents, and silently moved away." Here was a half-ruined building, choked up with weeds, bearing record that it had been once the "El Dorado Saloon"—in other words, a gambling hell, or worse—and around it were a few cabins. This had been the town site, and the projectors no doubt imagined that it was to be "the right smart chance of a city." However, fate had decided otherwise, and the only traces of former greatness to be seen were piles of washed stones and gravel, and long trenches, and half-ruined ditches, which gave the spot the appearance of a place where some great engineering operations had been left half finished. Here and there a solitary Chinese slunk about, intent on his own business, and, if my companion was to be believed, in pursuit of stray cats. As we turned a corner of the rough trail we suddenly emerged in front of the store. By the door were sitting half a dozen of the old *habitués* of the creek, lazily talking. My friend was delighted. "There they are," he cried, "boatin' about, chawin' baccy, jest as nat'ral as anything!" He seemed to be a popular man among them. As his friend—friendships are quickly made in the West—I was received with vociferations of welcome, and the choice of half a dozen shanties to "spread a blanket in." In this way I saw a good deal of the honest miner of Cañon Creek, and learned not a little of his ways of life and thought, in this lonely little dell in the Siskiyou mountains. Of course, we have all read about the miner in California, British Columbia, or Australia; about his extravagance, his boisterousness, and his conduct generally; and we are too apt to think of him only as the roystering blade in the palmy days of 1849 or 1853, when gold could be had for the picking up. The typical miner in 1877 is a very different man from that of 1849, even though he be the same individual. No longer do you, as a rule, see the many fine-looking handsome fellows of the early days of California twenty or twenty-five years ago. They were all young men then, but hardship has told upon them; for, in many cases, they have pursued, with varying luck, that business of gold-digging ever since. The 'forty-niners are the "blue blood" of the coast, but they are proverbially poor. Accordingly, these men, among whom I associated on Cañon Creek, were very different from our usual notion of the gold-miner, but were yet at the same time very characteristic types of what is well known on the Rocky Mountain slopes as the "honest miner." He differs in many respects from the settler of late years. Enter his cabin, and there is always indubitable evidence of that miserable life of single blessedness with which Bret Harte has familiarised his readers. The gold-digger is almost universally unmarried. The rough blanket-spread cot; the axe-hewn table, with its scanty array of crockery; the old battered stove, or fireplace, built of clay and stones; the inevitable sack of flour, half sack of potatoes, and junk of pork; the old clothes and old boots, and a few books and newspapers, go far in making out the extent of the miner's worldly possessions. A little patch of cultivated ground, enclosed by old "sluice-box" lumber, is sometimes an accompaniment, as well as a dog, a cat, or a few fowls. The inhabitant of this cabin is often rough, grey, and grizzly. He came out twenty years ago, and his residence has, with few exceptions, always been on the gulch where we now find him. Probably it rejoices in the euphonious name of Horse-beef Bar, Bull-dog Forks, Jackass Gulch, Rough and Ready, Rag Town, Puppy Town, Love-letter Camp, Jim Crow Cañon, Gospel Gulch, Gouge Eye, Shirt-tail Bar, Bogus Thunder, or Ground Hog's Glory—all veritable localities. By one of these names his home may or may not be found on the surveyor-general's map, but at all events it goes by no other. He "does his trading" at a store at Diggerville. Credit he calls "jaw-bone," and talks about "running his face" for "grub," but sometimes this is objected to by the storekeeper, as the gulch is not "paying" well, and behind the counter you may see a mule's "jaw-bone" significantly suspended, and below the words "played out!" Here the honest miner purchases a few pounds of flour, a little tea, coffee, and brown sugar, and as much whisky as he can buy. He can tell where all the rich spots have been in the rivers, bars, gulches, and flats; but that was in the glorious, wicked, cutting, shouting, fortune-making times of yore. He can't tell where there are any rich spots now. He is certain there is a rich quartz ledge in the mountain yonder, and, if he could get water on the flat, he is sure it would pay good wages. Excess of fortune spoiled him in '49. Economy is a myth with him, and he cheerfully entertains half a dozen friends, though his magazine of provisions, as well as of money, be in an advanced state of exhaustion. His supper cooked, he thinks of home—that is, the home of twenty years ago. In reality he has no home. Mentally, he sees the faces of his youth, fresh and blooming; but they are getting old and withered now. He sees the peach orchard and the farmhouse, from which he wandered, a young rover, when first the news of golden California burst upon the astonished ears of the world. That home is now in the hands of strangers. Were he to "go East," as he calls it, he would find himself a stranger in a strange land. He thinks he'll go back "some time or other." Fortune occasionally favours him a trifle more than usual; and then he may make a trip to "the Bay," as he calls San Francisco. He stops—or he did in my day—at the "What Cheer House." He may be seen there by hundreds. Poor fellow! He came here to enjoy himself, but he doesn't well know how. The novelty of the city wears off in a day or two. Without occupation, his routine of life broken, he becomes a victim to a disease for which the French alone could have invented a name—*ennui*. At night he may go to the theatre; but by day he sits in rows in the hall of the hotel, crowds the entrance, and sometimes blocks up the street. If he have money enough, and be so inclined, he may "go on the spludge," and possibly get drunk; but that with this class of miner is not very likely. His face wears an expression of

SACRAMENTO STREET, SAN FRANCISCO.

wild bewilderment and intense weariness. Unaccustomed to the hurry and bustle of the city, he "collides" frequently with the denizens of the metropolis. The spruce, fashionably-dressed, frizzle-headed clerks who flit by excite in him feelings of contempt and indignation. For all "airs" and "frills" he has an honest disdain. The swarms of youthful females in the streets astonish, delight, and tantalise him. They are something so new to him. There are few on Jackass Gulch, and they would be better away. When he knew "Frisco" it was not much more than a collection of cotton tents on some sand-hills. Now it is a fine city of 250,000 inhabitants. Females were almost unknown, and the announcement by a steamboat proprietor of "four lady passengers to-night" was quite enough to ensure a crowded patronage for his vessel. But the digger of the auriferous soil often leaves the city with the knowledge that the world has gone far ahead of him during his lonely residence in the mountains. He had far better not have come. In Diggerville he is somebody. In San Francisco he is lost among the crowd, or at best is only a "rusty old miner," those who thus contemptuously talk of him forgetting that he and such as he were the founders, and are yet, to a great extent, the stronghold of California.

I have spoken of the deserted villages, and the melancholy wrecks they have become. "You may see," writes the same author I have already quoted, and I can vouch for the accuracy of his statement, "such disfigurements far and wide over California, and in some places where only meadows and forests are visible, not a living creature, not a house, no stick or stone, or remnant of a ruin, and not a sound, nor even a whisper, to disturb the Sabbath stillness. You will find it hard to believe that there stood at one time a fiercely-flourishing little city of 2,000 or 3,000 souls, with its newspapers, fire company, brass band, volunteer militia, banks, hotels, noisy Fourth of July processions and speeches, gambling hells, crammed with tobacco-smoke, profanity, and rough-bearded men of all nations and colours, with tables heaped with gold-dust, sufficient for the revenues of a German principality, streets crowded and rife with business, town lots worth 400 dollars a front foot, labour, laughter, music, dancing, swearing, fighting, shooting—*everything* that delights and adorns existence; all the appointments and appurtenances of a prosperous and promising young city; and *now* nothing is left of all but a hopeless, homeless solitude. The men are gone, the houses have vanished, even the *name* of the place is forgotten. In no other land in modern times have towns so absolutely died and disappeared as in the old mining regions of California. It was a driving, vigorous, restless population in these days. It was a *curious* population. It was the only population of the kind that the world has ever seen gathered together, and it is not likely that the world will ever see it again. For, observe, it was an assemblage of 200,000 young men—not simpering, dainty, kid-gloved weaklings, but stalwart, muscular, dauntless young braves, brimful of youth and energy, and royally endowed with every attribute that goes to make up a peerless and magnificent manhood, the very pick and choice of the world. No women, no children, no grey and stooping veterans, none but erect, bright-eyed, quick-moving, strong-handed young giants, the strongest population, the finest population, and the most gallant host that ever trooped down the startled solitudes of an unpeopled land. And where are they now? Scattered to the end of the earth, or prematurely old and decrepit, or shot or stabbed in street affrays, or dead of disappointed hopes and broken hearts— all gone, or nearly all, victims devoted upon the altar of the golden calf, the noblest holocaust that ever wafted its sacrificial incense heavenward. It is pitiful to think upon." It will be easy to amplify this text by a thousand instances and strange tales. But the subject is unpleasant. It is too saddening. Let us turn from it.

I fancy that I do not really wrong the honest miner in saying he does not possess much religion. Yet if a clergyman by any chance come into his camp, he makes a point of attending "meeting" on much the same principle, and with feelings of about equal reverence, with which he would go to a dog-fight, or a tight-rope performance: because he looks upon it as "the right thing" to *patronise* "the affair." If the parson look on as he is washing for gold, he will ask him if he would like to "wash out a pan?" and as this invitation is usually accepted, the worthy fellow will contrive to slip in among the gravel a tolerable nugget, so that the washer may be nothing the worse for his clerical visit, custom in such cases providing that the contents of the pan go to the visitor. A friend of mine, whose lot it was to officiate as a clergyman among them at one time, used often to tell me that he had to ring a bell in the morning along the apology for a street, inviting his parishioners to divine worship, and that, finding nobody in church when he came in, he first looked into one gambling saloon or tavern, and then into another, inviting the company to come to church. "All right, parson," would be the good-natured reply; "we'll be there as soon as we've played out this hand for the whiskies. Jest be goin' ahead, and we'll be along for the preachin'!"

I have spoken of the miner's propensity to "take a drink." This taking of "drinks" is one of his characteristics. No bargain can be made, or any other matter of business or sociality settled, without the indispensable "drinks." The same clerical friend, whose experience I have just related, was shocked on his first arrival among the miners at being asked to "stand drinks," after he had received a very liberal subscription towards the building of his church. Two mining companies that I know something about threw dice to determine which of them should treat the "whole creek" to champagne, and as that wine was sold at fifteen dollars per bottle, the cost to the loser may be guessed. In most mining localities it is looked upon as a cause of mortal offence to decline drinking with the first man who shouts, "Let's put in a blast, Cap.!" In some places it is quite a serious breach of etiquette not to ask all who are sitting round in the bar-room of a tavern, though total strangers, to "Step up and take a drink." Sometimes they do not require any invitation. An acquaintance of mine having had a long ride one day, dismounted at a tavern to take *more Americum* (some refreshment), when, to his utter astonishment, fourteen men who were sitting around stepped up, and "'lowed they would take sugar in thar'n." He paid for the fifteen "drinks," as it was in strict accordance with the custom of the country; but he took care not to go back to that hostelry again. The Australian gold-digger is in many respects different from the Californian, but still he evinces the same carelessness of money. It used to be the custom for these men to come down to some village after they had made a slight "pile," go each to his favourite public-house, and give the money into the landlord's hands, with the information that he "shouted" (or asked all and sundry to drink) until it was finished. Then the landlord at intervals would say, "Step up, boys, it's Jim Jenkins's shout!" Then they all wished Jim luck, until Jim's "shout" was out, and then he went back to his "gully," proud that he had "spent his money like a man." On one occasion a miner came down and handed his money over to the landlord; but, contrary to expectation, nobody would respond to his "shout." He had been a convict, and "lagged" for some grievous offence. The man was at his wit's end. At last he struck upon the brilliant expedient of engaging an idler at labourer's daily wages—eight shillings—to drink with him. And thus he got through his holiday.

No one can tell where a rich mine will be discovered, or where it will not. Even quartz mines, which require skill to diagnose, have been equally discovered by chance. Geologists were never more successful than other people in this search, or than those ignorant of "veins," or "dykes." A robber fired at a man standing with his back to a rock, but missed him. As the ball splintered the moss-grown quartz, the miner who was attacked saw specks of gold sparkle in the moonlight. It afterwards proved one of the richest mines in California. Two miners about to leave the country, and in despair, just to celebrate the event, "got on the splurge"— to use their own expression — the night before their intended departure. As they were coming home to their cabins, in mere foolishness they commenced rolling stones down a slope. One of these struck the point of a rock, which, on being examined, was found rich with specks of gold. This changed their plans, and they stayed and stayed to some purpose, for they afterwards became very wealthy men. The "honest miner" is far from being a domestic character. If he was making five dollars per diem "to the hand" at "Greaser's Camp," and heard that somebody else was making six at "Hellgate Cañon," in "Mountain Goat Gulch," the chances are that he would presently make his appearance at the new El Dorado. Now it was Gold Bluff to which all were rushing. That failed, but it did not dishearten the failures. They next rushed in thousands to Gold Lake. In 1855 the Kern River fever raged; and in 1858 came the "Fraser River rush," seizing the Californian miners like a frenzy. Few were very successful, and most of them came back—to use their own term—so "strapped" that eventually it became a matter of personal offence to ask a man if he had ever been to Fraser River. But he was not cured. In 1863 the infuriated searcher after gold was blocking all the mountain trails in the Sierra Nevada Mountains, and "Washoe" was the cry. In 1864 it was Blackfoot, and in 1866 I saw hundreds rushing through slush and snow for Big Bend, in British Columbia, declaring that Cariboo and Stickeen, all former "rushes," were not "a patch on it," and that, at all events, they would "see the elephant." This vagabond propensity will fasten on a man who allows himself to sit in front of a frying-pan and a bundle of blankets, on the ridge-pole of a sore-backed horse; and I verily believe that there are many men among the gold-diggers who, if their history were known, have travelled more and endured greater hardships than some whose names are famous in the annals of vagabondism, and whom the Geographical Society delights to honour. The true seeker after El Dorado does not stop at distance or difficulties. The Pacific gold coast miner does not care to be called, like the Australian, a "digger," this term in the former region being applied to, and associated with, the lowest race of native Indians. He likes to be called the "honest miner." That he is honest enough, as honesty goes in North-West America, nobody will deny to the profession as a whole, but still there is occasionally the "dishonest miner" too. We do not speak of the rascal who is caught stealing gold out of the "sluice-box," and gets lynched for his pains, but of the equally rascally individual who "salts" a claim before selling it—that is, he scatters a few pieces of gold through the gravel before the buyer comes to test it. In California some of the "claims" are wrought summer and winter. Indeed, the winter is there more favourable than the summer, because water is more plentiful. In the hot weather California "dries up;" but in British Columbia, near the Rocky Mountains, the frost causes work to be suspended. Then the claims are "laid over," and the great body of the miners come down to Victoria and other towns to pass the winter months, and to spend the money they have made during the summer. They also often try to dispose of some rather doubtful "claims" at this time, and one of the means adopted is to report "having struck a good prospect" just before leaving. It is remarkable, to say the least of it, how many good prospects are "struck" just about this time. The endless swindles connected with quartz companies are leaving painful records in the memory of certain gentlemen in the City of London and elsewhere, whose purses were

A MINER'S CABIN BY THE AMERICAN RIVER, CALIFORNIA.

larger than their foresight. Gold-mining will always be a staple industry of the Rocky Mountain slope, and the increased immigration and attention excited by the Pacific railroad will greatly increase the business; but the old miner will be "killed off." Large companies will work his "claims," and shoals of new faces will crowd the yet solitary valleys. Philistines who know not the old traditions, and have no sympathy with the old manners and the old men. He will himself meet them half-way, as the fur-traders and trappers have done, and will unconsciously lose many of his characteristics and peculiarities. He will get toned down to the duller routine of other workmen as his pursuit takes its place among the " industries."*

---

* See also the author's sketches of Rocky Mountain Men and Manners in Dickens's *All the Year Round*, *The Field*, *Field Quarterly Magazine*, &c., for 1869 et seq.

Mail Day in the West.

To describe the many curious phases of life which the Furthest West now presents, and did present even more markedly before civilisation trod so fast on the heels of barbarism, would be beyond the powers of the space at our disposal, agreeable as would the task be to the writer, and possibly interesting as it might prove to the reader. The trapper and the fur-trader, the voyageur and the gold-digger, we have already briefly sketched. To complete the picture of the typical men of the Rocky Mountains and Rocky Mountain slopes, we should require to devote some space to the frontier man "who is always moving west"— his simplicity and his shrewdness; his savage likings and rude ways, and yet his kindliness of heart, and his rough hospitality; to the "loafer" and the "bummer," disreputable specimen of the *genus homo*, neither peculiar to America, nor yet to the Pacific Coast section of it, though now, owing to the miner's lazy, good-natured tolerance of everything and of all men, found there in greater abundance than elsewhere; to the ruffian, "road agent," or desperado, who finds his uncertain home in the midst of the "society" of the men of all nations who resort to gold countries; but these, and other types, we must perforce pass over. I cannot, however, leave men and manners in this region without giving a brief sketch of some ones peculiar to the Pacific Slope, but which, year by year, are getting rarer, and will soon disappear. In describing the mail day and mail arrangements of the country, from California to British Columbia, an opportunity may be afforded for saying something about the many sharply-hewn individualities which the hard life of these regions has scattered through it, among the less characterised streams which older civilisations have poured into the provinces, territories, and states of Western America.

Bang! It is a dull sound, as of a cannon, which wakes us out of our sleep on four bearskins under a blue blanket and an old coat in our "shanty," or cabin, in Victoria, Vancouver Island. We are as yet new to the ways of the place, and rub our eyes wondering what it can all mean. While we are cogitating, we hear the scuttling of many feet along the wooden "side-walk;" and the companion who, for the time being, shares our mansion, rushes in, dressing as he makes for the door, and tells us to "hurry up," for the mail is in. "Hurrying up" means, in this case, jumping into some clothes and a pair of boots, and joining the people who are now running from hither and thither down the quiet streets towards the harbour. It is yet early morning, but half the population seem to be up, and all going one way. The hotels, and many private and business houses, are flying flags. You also notice that, though this is nominally a British town, fully one half of the colours are American. Our Transatlantic cousins are "great on bunting;" and on high days, holidays, and on steamer day, are in no way backward to display the "goose and gridiron" to the breeze. There is the mail steamer from San Francisco lying alongside the wooden wharf, blowing off steam, and already surrounded by draymen, black and white, all shouting, most of them swearing, and not a few of them with gold watches in their sleeved-waistcoat pockets. Remember that we are in El Dorado. The chief citizens are also down deep in conference—three deep— with the purser, who, cigar in mouth, is busy with invoices and bills of lading, while here is paterfamilias, much excited and very hot, seeing to the landing of his wife and family, whom at last, having prepared a new home for them, he has brought away from struggling, overstocked England. They look very happy, but wondrously bewildered, at the new scenes around. And yonder is a sweet English girl, who has come all the way from fair Devon to pine-clad Vancouver, to wed the Bideford lad who has been toiling in the mines all these years for her; and as we see that brave lass escorted by the happy lover, and the brother who has come out with her, to the Hotel de France, and thence to the little wooden church upon the hill, we feel certain that all the world looks bright to them, and all the mean-looking board houses gilded palaces. There are also idlers like ourselves, seeing if any acquaintances have come, and what "new chums" have arrived for our colonial society. Here seems to be a popular man, who has just come out of the steamer. Half a dozen young fellows are round him, and he is laughing and shaking hands. He seems an old colonist, who has been away on a visit, and has returned again. "Glad to see you, boys," we hear him saying. "mighty glad! Tell you what, the old country's not what we thought it, and I'm glad to be back from their small twopenny-halfpenny-wheelbarrow ways. I'm going to stick here, I tell you, and I guess you'd better all do the same!" The incredulous, sad-looking smile on some of these young English faces show that they don't half believe the enthusiastic returned colonist, and then we hear one say to the other, "Ah! it's all very well for Stephens, with his town lots and Cariboo claim; but I guess if I'd got his chance, you'd soon see the last of this child!" Nevertheless, they all go up and "take a drink" with the jubilant Stephens in old Ben Griffin's, at the Boomerang. There is already quite a brisk business going in that same way. "Ben's" seems to be the English house, and there the newly-arrived Briton may, while quenching his thirst, indulge in the new arrival's amusement of abusing the "Yankees" to his heart's content, without any fear of ulterior consequences. It seems apparently etiquette for the new arrivals and the old hands to go and "take a drink" before starting into the serious work of breakfast. Nobody has, however, much time for breakfast to-day. Even the lazy — the serenely lazy — Indians are now too excited to sit on the "side-walks" lazily watching the busy multitudes of pale-faced strangers. Even they are down at the wharf acting as porters to the different hotels, for omnibuses and cabs are as yet unknown to Victoria. We get clear of the drays, and trunk-laden aborigines, and go up to the post-office, a little wooden building which also does duty for the harbour-master's office; for the Postmaster-General, being at that time an old sea-captain, was made by an economical legislature to do duty also as captain of the port. The postman's knock, so familiar to us at home, is quite unknown on the Pacific Coast. There everybody goes to the post-office for his own letters. Accordingly, by the time we have reached that building, merchants and merchants' clerks, or men who have boxes in the office, for which they pay a round sum per annum, are rushing for their "mail matter," as it is called. The general public have, however, the advantage of no such aristocratic luxury, and are forming in line to await their turn at the office window. This arrangement, from long custom, has become familiar to the heterogeneous mob who are waiting outside. Noiselessly, and without any non-sense, each new comer takes his turn at the end of the single file, until it reaches in a long *queue* up Wharf Street, or away towards the Hudson's Bay warehouse. Now and then, indeed, some more than usually bumptious individual will attempt to step into the line out of his order; but he speedily becomes convinced of the little mistake he had made as he is politely but promptly handed back, until, to his astonishment, he lands at the end of the tail. If the mail arrives when the town is full of gold-diggers, it will sometimes be hours before the last of the human *queue* can reach the head of the line; and though he need never attempt to go out of his place, it lies quite within his purpose to effect this by love or money. It is not often that he attempts to do so by the former means; the latter comes more within the bounds of possibility. Accordingly, you are almost sure to see in the line now and then some tall, gaunt, grey-shirted fellow who, you are perfectly certain, expects no letter, and who is in no great hurry. At a wink the individual to whom a letter may be of importance buys him out, and takes his place in the rank. In San Francisco, in the old roystering, money-scattering days, ten dollars were often paid for this favour.

At last we are near the head of the line. There are still two before us, and we take our cue from them. Number one presents his head at the opening in the boarded window—"Bock, Hiram J."—this in a nasal accent. The Postmaster-General is assisted by the Deputy Postmaster-General—we are fond of titles in the colonies—and both rapidly turn over the piles of letters arranged in the pigeon-holes under the different letters of the alphabet. "Nothing;" and Hiram J. Bock, late of Martha's Vineyard, Massachusetts, turns away, and cuts a chew of tobacco to console him for his disappointment. Number two is absorbed in an exchange of compliments with an individual behind him who is kicking his shins, and requires a peremptory shout of "Next man!" to remind him that the "Honourable gentleman" at the window requests his order. He darts forward and shouts —not in the half-whispering tone of some modest individual, but after the manner of a free-born Briton—"Smith!" "What Smith?" "John Smith!" A laugh comes from the inside, as the Postmaster takes a peep at his man, and asks, "What John Smith?" A little altercation ensues, which seems settled to their mutual satisfaction, for John Smith lays down three bits—about eighteen pence—for some partially paid letters, and moves off with "his mail." We come next, and so the line goes on. The newspapers are not distributed to-day. There is no time for that. The "newspaper men" are, however, in the room behind, looking over the pile on the floor for their "exchanges," and I daresay not very particular in making use of any tempting periodical, even though the address on it should not be theirs. Perhaps it is only in the United States and in the British provinces on the Pacific that private individuals are allowed to compete with the Government in carrying letters. These are called "express companies," and one or other has an office in every place of any consequence all over the country. These express companies usually conduct a banking business, commission agency, and are also carriers of parcels—in fact, general factota. The chief of these is Wells, Fargo, and Company, or, as they are familiarly called, "Wells Fargo." In every large town Wells Fargo's office is one of the best situated and most substantial building. If anybody be in difficulty about getting anything to or from any place he goes to Wells Fargo. Nobody, so far as I am aware, ever saw either of the gentlemen so called; indeed, some profane

VIEW OF "THE GORGE," OR NARROWEST POINT OF THE WESTERN PROLONGATION OF VICTORIA HARBOUR, VANCOUVER ISLAND, WITH TREES OF *ABIES DOUGLASII*. (*From an original Photograph.*)

individuals will not hesitate to hint that they are of the nature of two Mrs. Harrises, but, at all events, everybody knows their representatives. Many people, Americans especially, have a most enduring belief in them, and prefer to send their letters by them rather than through the post-office, even though the former mode of conveyance is more expensive. All you have to do is to go to their office, buy one of their envelopes, put your letter in it, and then hand it over to them. You may be almost certain that it will be delivered to your correspondent more safely, and, probably, quicker than it would be if committed to the hands of the postal authorities. The agents of these express companies travel far and near, and often to places where there are no postal arrangements. Through the wildest parts of the country you will meet them in canoes, on horseback, in stage-coaches, all bound on the same errand, carrying treasure, parcels, and letters. The mining population could scarcely exist without them, and have a most unwavering faith in them, I believe in every case well founded, for I have known very few things committed to their hands go astray. Though, perhaps, the mail coming to Victoria from England by Wells Fargo will be small compared with that by the legitimate channels, that leaving by them will be nearly as large, and to San Francisco probably larger. We step up accordingly to Wells Fargo's office in Yates Street to see if there are any letters for us there. The large room is full of people. The agent is standing upon the counter with a pile of letters, alphabetically arranged, in his hands, shouting out the different names, and tossing them hither and thither with an adroit spin, learned by long practice, in the direction of the applicant. The only person at all idle there on this busy day seems to be the captain of the steamer, who is sitting quietly in the "express" office sanctum reading the colonial papers, and now and then nodding to his friends.

To-day you need never attempt to speak to any man on private business. Under ordinary circumstances anybody in Victoria must be unwontedly busy if he has not time to "take a drink," but to-day unless it be in the way of business, nobody has leisure even for that, or to do anything but read his letters, and write hasty answers to his correspondents. The provincial legislature by silent consent never think of meeting on mail day, and the Honourable John Jones must perforce postpone that great attempt to overthrow the government on the momentous subject of the Hog and Goat Bill, until he has written a smart note to Gingham, Cheatem, and Co., of Manchester, about the quality of that last invoice of long shirtings. Even Her Majesty's Courts of Justice must put off the trial of the Hydah Indian for murder until the mail is gone. The chief justice is, besides, too busy signing various legal papers to-day to attend to his ordinary judicial functions. The sheriff—and the functions of a colonial sheriff are more useful than ornamental—is very busy, for he knows, by long experience, that on every mail day a number of gentlemen who may be in pecuniary trouble are apt to give their creditors the slip, and bid farewell for ever to an ungrateful colony. This intention comes to the sharp ears of their anxious friends in the way of business, and instantly these gentlemen rush over to the chief justice, and swear that to their certain knowledge such is the intention of the individual aforesaid. A capias, or "cap'us," as it is familiarly known among those accustomed to it, is then issued, ordering and commanding the sheriff to see that So-and-so—*ne exeat regno* —does not, in a word, abscond, to the loss of his mourning creditors. In the course of the afternoon the sheriff and his myrmidons may be seen hanging about the steamer armed with these bits of paper, and then between wanted and wanter there is a fine trial of skill, and not unfrequently, by a display of careless nonchalance, the victim slips from under the watchful fingers of the legal functionary. On mail day bills become due, and as everybody has his bills to meet on that day, everybody expects everybody else to pay what he is due. But, of course, as always happens in these cases, the creditor frequently reckons without his host. Accordingly, the steamer gun is at times the signal for gentlemen who "have something out" to have a particular engagement in the country until the mail has gone. In San Francisco, before the steamer day was superseded by the Pacific railroad, this necessity of making up bills against steamer day gave rise to a peculiar set of men, who made a business of lending money "from steamer day to steamer day," the rate for the ten days or a fortnight being from one to two per cent., almost invariably also with "collateral security," that is, a deposit receipt for merchandise in a warehouse, or some such easily transferable document. This was rather exorbitant, even in a country where the ordinary interest on money, with good security, is eighteen per cent. per annum. But then at no time in California has it been looked upon as dishonourable either to lend or to borrow money at the highest rates of interest, and the "from steamer day to steamer day" money lenders grew rich.

The newspaper offices are full, for the different local journals issue a special edition for steamer day, containing a summary of the last ten days' or two weeks' news, and people are busily buying these at one shilling a copy, in the sanguine hope, cherished in spite of many failures, that they will reach the persons to whom they post them. As we pass down by the post-office again, most of the town people have received their letters, but the settlers from the immediately outlying districts have begun to come in. You notice that nearly all of these people, though roughly dressed, are yet quite a different style from our familiar agricultural labourer. Most of them have an air of intelligence, and several are even refined in appearance, manners, and language. For months these men have been shut out from all news from home. Some have just come down from the mines, and you can see by the look of them have been unsuccessful. Others are "putting through the winter" as best they can, hunting, working on farms, or living from hand to mouth until the snow clears off the mountains, and they can start off to try their luck in the gold-fields once more. Some, as they receive their letters, cram them into their pockets, and move away to some quiet place to read them, while others, all careless of the throng, move along Wharf Street and up Bastion Street, diligently perusing the long-expected missive. Another will comfortably sit down on the edge of the wooden "sidewalk," with his feet in the gutter, and, heedless of passers-by, peruse his correspondence from beginning to end. A nervously excited man will open his letter, peep into it, and then rush off to devour it in the quiet corner of some neighbouring "saloon," or public-house; and now and then you will see an anxious face, and notice often a tearful eye glancing at a deep black-edged envelope, which tells that one more link which bound the far-off colonist to the mother country has been severed.

As the hour for the steamer sailing approaches, the whole town gets into a fever of excitement; and when the whistle begins to blow in a spasmodically shrieking manner, and the black clouds of smoke which announce that steam is getting up, darken the air, you would suppose that the end of the world-colonial was approaching. Here a returning colonist, surrounded by a knot of envious friends, and with an air of pity on his face for us who are remaining, is rushing down to the steamer, or hastily taking the inevitable farewell "drink" before shaking the colonial dust off his colonial highlows, while every other man seems to be rushing with a letter for somebody to post in San Francisco, having been too late for the mail. At last the steamer begins to ease off, then the crowd give

VIEW OF THE WESTERN SUBURBS OF VICTORIA, VANCOUVER ISLAND. (From an Original Photograph.)

a cheer or two, which is returned with interest from the crowded deck of the steamer. Steady! and she is round the arbutus-covered point by the Indian village, the crowd turn off, and once more we Vancouverites are left to ourselves, and mail day is over for the next fortnight.

Only the day's work is not over for the journalists. A hasty dinner swallowed, the colonial sub-editors grind up the editorial scissors, and set to work to get up the summary of European news, while, perhaps, an idle friend may look over the exchanges, and make up their foreign correspondence in Paris, Vienna, or St. Petersburg, with a sufficiency of local colouring derived from experience, or from Murray's Guide Books. Next day the night's work will appear pretty well put together, with lists of the passengers, the imports and exports, the amount of gold-dust despatched by each banking-house, and a variety of notices headed " Personal." These note that " we are glad to see" that such and such a distinguished citizen " has again arrived home, and looking well after his visit to Europe," or that such and such another citizen, also distinguished after a fashion, had, to the regret of his numerous creditors, managed to elude the active and vigilant sheriff. Then follow a list of acknowledgments, " To the gentlemanly and obliging purser of the steamer *Brother Jonathan*, for late San Francisco papers," or to " our eminent citizen, Hon. Donald Macdonald, who has just arrived from a prolonged visit to Canada, for late Dominion

SHUSWAP INDIANS, BRITISH COLUMBIA.

exchanges." In a few days things settle down to their ordinary dead level. The gentlemen " wanted" get at their ease again, while the citizens who had " something out" return again to town, of course terribly shocked to hear that the mail had been in and gone. Until the great mail day arrives again the even tenour of the mercantile way is undisturbed, except by the arrival of a few local mails "up river," or 'long coast local steamers, smacks, and trading schooners, or by the still more primitive conveyance of an Indian canoe.

It is not difficult to remember when this latter method was the only postal convenience in the country; but that was when the Honourable the Hudson's Bay Company reigned supreme over these territories. At that time the only civilised spots through this immense tract of country were the forts of the great fur-traders. When the officer in charge of one of these lonely outposts wished to send a letter to another fort, he merely sought out an Indian, wrapped the letter well up in a piece of oil-cloth, and despatched him. There was a stated rate of remuneration, according to the season of the year, for this service, and this every Indian knew. Accordingly, the aboriginal courier might go the whole way and receive the whole reward, or—what was much more likely—he might not have his way all clear before him, and prudently stop as long as his skin was whole, or his scalp intact. In that case he sold the letter to another Indian for a share of the reward, proportionate to the distance yet to be gone over. To the Company it was quite the same, for eventually the letter reached its destination in safety, when the full payment was claimed by the contractor for the last division of the route traversed. In those days the one annual ship to Fort Vancouver took about twelve months on the passage from England. It halted to take in wine at Madeira, coffee at Rio, cattle at the Falklands, at Valparaiso to water, at the Sandwich Islands to trade tortoise-shell, and at San Francisco to bribe Don Castro, the Mexican governor—all before the gun from Astoria reported that she had passed the bar of the Columbia River. Then she went to Canton and sold her sea-otters to the mandarins, and bought nankeens, teas, and silks, and so made the round voyage. Accordingly, the Company adopted another method to send their letters and the rarer furs more quickly to England. Every year the fur brigade crossed the whole breadth of the American continent on foot, on horseback, in birch-bark canoes, and in batteaus to York Factory, in Hudson's Bay. The accountant of the Company then soldered up the papers in a shining tin box, which he strapped on his back, and—the observed of all observers (the tin box, not he)—Fort York, or Moose Factory, as the case might be, was reached in about five months from Fort Vancouver. These were the primitive days of the postal system in the Far West.

We thought we had got an immense advance when the pony express carried the letters by relays of fearless riders over the plains to the Furthest West railway station, and thence returning to Sacramento, in California. I can well remember the steamer lying puffing at the wharf at Sacramento until clatter! clatter! the last pony-rider, a rough, hardy, wiry fellow armed cap-à-pie, galloped on board with the mails from "St. Joe's" (Saint Joseph), in Missouri, apologising to the little knot that gathered round him by the funnel for keeping them waiting, by relating (as he lit a cigar) that "a mile or a mile and a half on this side of Brown's Hole, down by the sulphur spring, Joram Hicks, the pony-rider, had got his har lifted by Pah-utes, and it was 'nation time, boys, afore the stock could be got up, and the bags ketched from the critters and sent on!" And the speaker went off as unconcerned as if he were relating the most trivial incident in the world. But it was a very trivial incident in those days, for one who did business on the great prairies to lose his scalp between sunrise and sunset. Then came the stage-coaches—such as we have figured on page 25—and everybody thought the end of the world could not be far off, when you could, by travelling day and night, and getting the life almost shaken out of you amid a cloud of dust, go from Virginia City to Omaha in seven or eight days. Now you can go from New York to Sacramento, in luxurious carriages, in less than that time. Still the stage-coach is, and will long be, an institution in many parts of the West, both for passengers and mails. It will long remain the most civilised mode of reaching the remoter diggings until railways are general, which will be a long time. There was such "a stage" going between Portland, in Oregon, and Marysville, in California, within the last few years. In the winter it had to be ferried over swollen rivers on rafts, and often, when crossing prairies which, when flooded, look like great lakes, the passenger suddenly became aware that the coach was floating, and that if he didn't look sharp he'd be drowned like a rat in its hole. How long it took I should be afraid to risk even a conjecture. Rumours—probably unfounded—are extant of a too confiding person having died of old age on the road! Still, I can vouch that a journey by it used to be very amusing—after it was all over.

Even after the overland coach began to run, mail day was still an institution in San Francisco, only a limited mail being conveyed across the plains. Then, among other things, the Eastern (U.S.) papers always sold for a shilling (twenty-five cents, or two bits) a copy. Somehow or other the mail used to manage, nine times out of ten, to come in after dark, and often at very unseasonable hours. Often when coming home from some pleasant party, at an early hour in the morning, the quiet streets would resound with the yells of the newsboy (and a California newsboy is the sharpest of his race), "Pa—nama st'm'r in. New Yor—k Herald, Tri—bune, World. Here you are, sir! two bits!" and he was off again to intercept some other reveller on the opposite side of the street. You hurried home, lit your lamp, and prepared to extract what enjoyment there might be in the journal purchased, until, as you unfolded it, a sort of suspicion began to steal over you that there must be some mistake. The paper was a year old! Then, as you dashed it aside, recollection dawned that this was the 1st of April, and that in buying a paper you had yourself been sold!

In addition to all these methods, a good number of letters, in the most out-of-the-way places in the Far West, are carried by private hands. There are localities so remote, that even the stage-coach is not heard of, and the express-man does not find it worth his while to reach them. Letters from such places are, as might be expected from what we have learned of the lonely position of some of the fur-trading posts (Vol. I., page 194, &c.), few and far between. There are also distant knots of gold-diggers in some secluded mountain valley, washing the sands of a nameless stream, who might quietly slip out of men's memories, did they depend upon Government reaching them with their letters. Luckily, however, the miner is not much addicted to epistolary correspondence, and what he does find it necessary to indite he manages to get conveyed to the coast somehow or other. If you have anything like an extensive acquaintance in one of these gold-digging regions, you need not be astonished some autumn day to find your door in Victoria opened in a free-and-easy way, and a knee-booted, grey-shirted individual walk into the nearest apartment. Then, as he begins to search his pockets for something, he introduces himself: "Name of Brown, Mister? Guessed so. Mine's Job. 'Job,' sez he to me. 'Job, I calc'late I'll put off a line to Mister Brown down to Victory,' sez he to me. 'Easy know him,' sez he. 'Ain't much to look at,' sez he, "but——

of I ain't lost it!" With this the honest miner stands open-mouthed, until, finally, a brilliant idea strikes him, and, with a self-satisfied slap, he throws down on the table a greasy document, which he produces from the lining of his felt hat, and which you find has been about five months on the way, while the amateur postman has been "prospecting" in and about the Rocky Mountains enough to merit great things at the publisher's hands if only he could write as fluently and as well as he talks.

And so, by drops and by driblets, "mail matter" trickles in to moisten our letter-thirsty souls in the Far West, until—too quick for the man whose bill is then due, too slow for everybody else—the days roll along, and with a hurry and a scurry, a running hither and thither, mail day again comes round.

This was written some years ago. There have been changes since then; though to the country north of California it is, to a great extent, still applicable; while to nearly all of the more out-of-the-way parts of the country mail day will always bear much the same character as that which I have attempted to describe.

THE LANGUAGE OF THE PACIFIC SLOPE.

The reader need expect no philology in this section. He will hear nothing of the twenty or thirty languages or dialects spoken by the Indian tribes of the great region under description, nor one word of any of the several yet unknown and unrecorded languages existing, though soon to become extinct, among the aborigines. English is now the language universally spoken by all civilised men north of Mexico. Spanish is still spoken in Southern California, but only among the descendants of the Mexican inhabitants, though they all understand English. The French-Canadian voyageurs and their mongrel families still speak French, but they also understand English. And all those who have much, or anything whatever, for that matter, to do with the native tribes, understand the Chinook jargon, a kind of *lingua franca*, composed of fragments of many tongues, aboriginal and European, though based upon the language of the Indian tribe, called the Chinooks, who at one time dwelt at the mouth of the Columbia River, but of whom only fragments now exist on the Siletz Reservation, near the Oregon Coast. Spanish words have, however, got intermixed with the English spoken by the whites, while native Indian words are also frequent.

But what distinguishes the vernacular English of the Pacific from even that of the West generally is the number of "slang" words that corrupt the "well of English undefiled." "American"—as the Transatlantic Secretary of Embassy styled the tongue of Shakespeare and Milton as spoken by the Emperor of Russia—is rich in these unrecognised fragments of English. But the *argot* of the Pacific slope is to a great extent peculiar to itself, though every year less so, and smacks so much of the soil that, before understanding it, the student must learn much of the Pacific Coast, its people, its industries, and conditions of life. It may be, therefore, useful to devote a few paragraphs to this by-path of philology.

First, then, in a country where everybody works, it is natural that the idler should be contemptuously spoken of, though good-naturedly tolerated. In California the habitual

STAGE COACH STARTING FROM A RAILWAY STATION IN WESTERN AMERICA.

idler is called a "bummer." He is always well dressed, affable in conversation, and ready to "take a drink" with any one. He peculiarly affects a cut velvet waistcoat of gorgeous hue, "California diamonds" in his shirt-breast, a polished quartz seal hanging at his fob, possibly a gold buckle to his broad watch ribbon, and has an infinite acquaintance, whom he insists on introducing to you, which introduction generally results in their being asked to "take a drink." I have noticed that they always have an "interest in a quartz lead," and are "down at the Bay" to get up a company to work it. A "loafer" is not so bad as a "bummer," for though a loafer may become a nuisance by calling at unseasonable hours at places of business, yet this does not stamp him as endowed with this permanent quality of character, but only as being seized with temporary idleness. A "bummer," on the contrary, is a low, disreputable, lazy fellow, very equivalent to the familiar "sponge." The term is probably, according to Mr. Hittel, derived from the vulgar German words *bummelin* and *bummeler*, which are about equivalent to "loafer" and "loaf." Its origin has been attributed to *Boehmen*, the German name of Bohemia, a nation proverbial for the number of its sharpers and adventurers. In France the gipsies are called *Bohémiens*, because of their roving and worthless character. If there is anything worse than a "bummer" it is a "whisky bummer," a term I have heard more than once applied to the Pacific Coast reputation of some gentlemen who in after years, and in a more severe atmosphere, became famous statesmen and distinguished generals.

"On it," is a peculiar and expressive Pacific phrase. Endless tales hinge on the use of this phrase, and it is remarkable how, with the almost inexhaustible resources of Californian anecdote to draw on, you will hear the same wretched story related over and over again, until the very initiatory syllable of it being a precursal sign of what is about to be inflicted, sets the wearied traveller on his guard against the coming boredom. Men travel so much about on the Pacific slope that they soon acquire all the stock stories, and, of course, it is those that are most frequently repeated, and the most stupid of them, that seize the shallow sense of humour in the dull men, who are found even in these countries of sharp wits. A man who is "on it" is generally looked upon as a "scaly customer;" and I regret, for the honour of the legal profession, that an indignant litigant frequently finds it necessary to apply this term to his attorney.

A man may be "on the make" when he is keen after filthy lucre; "on the fight" when he is combatively inclined; "on the shoot" when ready to "back his opinion" with his revolver; "on the spludge" when under the influence, more or less, of alcoholic excitement, his driving furiously about town in a buggy, halting at every other "saloon," "standing" champagne "to the boys," smashing a mirror or two, and paying for them out of a leather bag of twenty-dollar pieces, or "slugs," "cleaning out the town" with his pistol, and generally "spending his money like a man"—or a donkey—if one could imagine so sagacious an animal acting in such a foolish way.

The occupation of the gold-digger has suggested to him numerous expressions to signify his desire for his friends to partake of vinous hospitality with him. He prays them "to put in a blast," and they in their turn, as they lift up their glasses, are civil enough to say to their host, "Here's to you, old man, and hopin' your pay dirt'll pan* out gay;" in other words, that the soil on which he is working may yield rich supplies of gold-dust and nuggets. At other times, the same hospitable request to partake of stimulants is indicated by the rather more generally-used expression of "take a smile," or the inviter is asked to "nominate his pisin," or, as somebody takes your arm at the corner of the main and the only street of Diggerville, you are requested in a cheerful tone of voice to "hist in a drop of pisin;" the two latter phrases being intended to express the popular opinion regarding the quality of the whisky, also vulgarly known as "chain-lightning," "mountain howitzer," "tangle leg," and "tarantula juice."

Spanish expressions naturally supply a large quotum of Pacific Coast, and especially Californian, phrases, which have, however, now got so perfectly incorporated into every-day language that they can hardly be classed as either slang or vulgarisms. For instance, the enclosure, or "corral," in which mules, horses, and cattle are confined, has supplied a verb and noun in common use. When a man is "cornered" in giving utterance to some untenable proposition, he is said to be "corraled;" when a horse-thief is caught he is "corraled," generally preparatory to his being hung; and a successful operator on the Stock Exchange will be described as having "corraled" all the shares in the Root-Hog-and-Die Quartz Mining Company. A farm is universally called a "ranch," derived from the Spanish "rancho," and a farmer a "rancher," from the Spanish "ranchero." An Indian village is also called a "ranch," from the Spanish word "rancheria," and when, as is the custom in the more out-of-the-way places, and was universally so in more primitive times, a young man lives in a "shanty" or cabin by himself, he is said to be "ranching it." When the writer first wintered in Victoria he was asked by a most accomplished lady if he "roomed at a hotel, or ranched it?" There are also numerous other Spanish words in common use—such as "aparejo," a Mexican pack-saddle; "adobe," a sun-dried brick; "arroyo," a brook, or the dry bed of a rivulet; "cañada," a deep ravine; "alforja," a bag, usually made of raw cow-hide, used for holding the articles to be carried by a pack-horse, &c. A muleteer who carries goods to the mines or elsewhere is always called a "packer;" and to "pack" anything means to carry it, whether on the speaker's own back or on a quadruped's it does not matter. "Cañon" we shall see by-and-by is a deep ravine worn by a river. To "cayote" is to dig a hole in a bank in search of gold, or to burrow like a cayote wolf. A late eminent Californian senator was familiarly known as "Cayote Joe," from his addiction to this method of mining before he betook himself to Washington. "Reata" is a raw-hide rope used for lassoing. "Rubric" has nothing whatever to do with matters ecclesiastical, but is merely the flourish with which Mexicans and native Californians end their signatures. The rubric is even a legal form of signature, and was the only way in which Francisco Pizarro could sign any document. To "rubricate" is to sign with a rubric, &c.† Numerous Indian words have also got incorporated in the

---

* A pan is a metal dish in which the digger washes out a test quantity of earth or gravel, and then judges his "claim" or mine by the result. In America a "claim" is said to be worth so many "cents" to the pan, just as in Australia it is talked of as yielding so many "pennyweights to the bucket." The pan is shaped like a wash-hand basin, and is used to knead bread in, or to wash the owner when he indulges in that luxury.

† Hittell's "California," page 385.

in the British possessions, where the natives are much more numerous than the whites, and on more friendly terms with the latter than in the United States. In some parts of the country, beside numerous Indian names, still retained for localities, many words in daily use are derived from some of the native dialects. A man talks of having no "*chickamen*," or money, on his person, and will indignantly inveigh the assumptions of the "*tyhees*," or great men, and even coins a noun, "tyheeism," to denote this arrogance of the tyhees. The lumbermen, who live close to the Indian villages, and see few other people beside the natives, use these words to such an extent that to a stranger their conversation is almost unintelligible. The frontier-men and frontier-women too, also use in their language Indian words, in most cases derived from the Chinook jargon. For instance, I have been asked, when approaching a border farmhouse, whether, "when *clattowaying* (going) along thar," I had "*nanuitched*" (seen) a *clagl* (black) cow?" for the questioner's "*tumtum*" (heart, temper) was "*cultus sikes*" (indifferently well pleased) at the "*hyou*" (plenty) trouble the beast was giving him.*

We should tire our reader's patience out, did we go over the various expressions used in the gold-digger's every-day life; how he takes a "square meal" when he comes to his inn, or how, when he gets into narrow circumstances, he is "dead broke," or "caves in," or "goes up a flume." He will promptly tell a bore to "dry up," a very expressive phrase, when one remembers how the Californian streams dry up during the summer. To show the application of some of these odd expressions, perhaps we cannot do better than to parody in miner English a portion of a proclamation of the Governor of British Columbia, anent some mining regulations, which lies before us. It may add some interest to this *jeu d'esprit* if the reader be told that it is based on a similar one, the authorship of which is usually credited to a late prime minister of the province, and an eminent law officer of the Crown! With this we may conclude our brief survey of the wide and fresh field of Western slang.

PROCLAMATION! HAVING THE FORCE OF LAW. YOU BET!

Whereas, a change in the mining laws is expedient. Be it enacted as follows:—
1. That all former proclamations are hereby repealed and "played out." Interpretation clause.—In the construction of this act the word "guv" shall mean the governor of this colony, and "commish" shall mean the gold commissioner for the time being. The words "fizzled," "played out," "pettered," "caved," and "gone up a flume," shall respectively mean, when applied to a mining claim, that the same is worthless; and when applied to an individual, that he is ruined, helpless, dead, or in debt, and the terms "dead broke" and "busted" shall, for the purposes of this act, be construed to mean the same thing. That the words "pile," "the dust," "the colour," and "bottom dollars," shall be construed to mean the current coin of this realm. The term "free-miner" shall mean every person entitled to mine. The term "on it" shall imply a willingness to buy, sell, or

---
* For a vocabulary of this rude linguistic medium of communication between the whites and the Indians see Smithsonian Institution's "Miscellaneous Collections," 1863 (No. 161), and "Dictionary of the Chinook Jargon" (McCormick, Portland, Oregon, 1865).

A SUMMER ENCAMPMENT IN A NORTH PACIFIC FOREST: COFFEE MOUNTAIN, NEAR ALBERNI, ON THE WESTERN SHORES OF VANCOUVER ISLAND (From an elegant Photograph.)

get drunk; and "on the make" shall mean a determination to make money, honestly, if you can; if you can't—make money; "on the sell" shall mean a willingness to sell, and "on the buy" a willingness to purchase. The term "you bet" shall be used to remove any doubt which may possibly exist in the mind of the individual addressed; and "you bet your life" shall be applied in the same way, but shall be more conclusive; and the term "you bet your boots" shall be equivalent to "you bet your life;" and the term "you bet your bottom dollar" shall, for the purposes of this act, mean "your life," or "your boots." The word "chain-lightning" shall mean very ardent spirits; and "mountain howitzer" shall mean liquor that kills at over one thousand yards; and "scorpion juice" and "tarantula-juice" shall be construed to mean "mountain howitzer," or "chain-lightning;" and "drinks for the crowd" shall mean any and all of the foregoing, for the persons present, but not any others. That "in a horn" shall be equivalent to the old classical term of "over the left;" and, for the purposes of this act, "in a horn" shall be equivalent to "in a hog's eye." These terms imply doubt, and shall be equivalent to "no you don't." That the term "vamoose the ranch" shall mean that the individual referred to has left for parts unknown; and "slope" shall be equivalent to "vamoose the ranch;" and "make tracks" shall, for the purposes of this act, be equally as expressive as the two foregoing terms. That the term "got the dead wood on him" shall not refer to any kind of timber whatever, dead or alive, but shall be used when one individual has obtained a fair or unfair advantage over another; and the term "got the bulge on him" shall be as strong as "getting the dead wood on him," and getting either the "bulge" or "dead wood on him," may result from "sloping," "making tracks," or "vamoosing the ranch." That "spotted," when applied to an individual, shall have no reference to the state of the skin of any white man, or any spot thereon, but shall mean that he is watched; and, when applied to mining, shall mean that the gold is scattered; and the term "biz" shall mean business. That "sock it to him" shall be equivalent to the old word "punish;" and "give him fits" shall be equivalent to "sock it to him;" provided also, that the word "fits" shall not include apoplexy or epilepsy. That "jawbone" shall mean credit, provided also that the size, shape, and contour of such "jawbone" shall not, for the purposes of this act, be material. That "nare a colour" shall be equivalent to "dead broke," and there shall be no difference between "nare a colour" and "nare a red." That the phrase "there's a heap of trouble on the old man's mind" shall mean that the individual referred to is either "gone up a flame," "pettered," or that he has "struck the bed-rock pitching" the wrong way; and a "young man" shall, for the purposes of this act, be an "old man," and the feminine gender shall be included in the masculine, and both in the neuter. That "bully for you," or "bully for him," shall mean a term of approval; and "good on your head," or "good on his head," shall mean the same thing. That the terms "old hoss," "doc," "judge," "col'nel," "cap," and "old hoss," are all equivalent, and the term "or any other man," shall have no definite meaning, and may be applied indiscriminately to all things. And "slum-gullion" shall mean clay; "pay dirt," dirt containing gold; and "good prospects" shall not mean a pleasing landscape, but plenty of "pay dirt;" and "wash-boulders," "wash-gravel," and "bed-rock pitching," shall mean indications of gold somewhere. That a "jumper" shall not mean a person who indulges in the active exercise of jumping, but shall mean a person who possesses himself of another man's claim because it is paying; and an invalid, or cripple, or woman, may be a "jumper." 2. It shall be lawful for the guv, you bet, to appoint one or more commishes, as he may think proper, to transact the biz of the mines of this colony. 3. That no jumper shall hereafter be allowed to indulge in that exercise, and if the commish shall find him on it, he shall have power to sock it to him, or fine him drinks for the crowd, you bet your life. 4. That all honest miners, who are on the buy, may purchase more than two claims from those who are on the sell, provided also that both parties may or may not be on the make. 5. That any honest miner who shall, after the passing of this act, allow any other miner to get the dead wood on him, shall, you bet your boots, upon complaint made to the commish that there's a heap of trouble on the old man's mind, be spotted a muggins, and be ordered in consequence to pay a fine of two ounces, or, in default of payment, catch fits, and the commish shall approve of the conduct of the one that's on it, by saying "bully for you," and may add at his pleasure, "or any other man." 6. Not finished, and therefore this act is to save time. Issued under our seal of Cariboo, this ninth day of May, and the tenth year of the mines. By the Guv's command, X. Y. Z., Boss of the Colonial office. God save the Queen, and good on her head!*

New phrases are always coming into use. For instance, of late years a California "wastrel" has been called a "hoodlem," the origin of the phrase being unknown to me.

"PUTTING THROUGH THE WINTER."

Perhaps I cannot better conclude this chapter, and these sketches of Pacific coastmen and life, than by saying a little about wintering on the shores on the North Pacific, or "putting through the winter," to use the familiar phrase.

When winter catches the traveller in Victoria, Vancouver Island—the western suburbs of which we have engraved on page 20—or anywhere thereabout, he must look out for dreary rain or sleet driving along the streets of the little town. This is the time for mud ankle deep on every road. It is a season when no man travels very far afield. Then all exploration is stopped, and every one endeavours as best he may to "put through the winter" in the manner most agreeable or suitable to his inclinations. Victoria is the general rendezvous, and now in November the streets are crowded with stalwart sunburnt fellows, attired in all the glory of new and very ill-fitting ready-made clothes, "loafing" around the corners of taverns and billiard saloons. A few years ago we— that is, the writer and some companions—were likewise forced to "put through the winter" in the same place. We took a small house, and fitted it up in a rough way, as is the custom among bachelors there. For our food we bargained with a French restaurant at so much a month. Our "cord wood" we got in; and, our books and household gods around us, with a small Indian boy to act as henchman, we were prepared for a siege of rain and snow. The small Indian was no less an individual than a prince

* *All the Year Round* (1871).

of the blood royal—a son of old "King Freezy," the drunken Indian chief of the Tsongeisth on the other side of the harbour. Victoria is wet and drizzly, and far from pleasant in the winter; but away in the Rocky Mountains country, in the northern forest-clad regions which slope down to the Pacific, winter means something even more stern and dismal. Snow covers the whole tract for a great portion of the winter months, and to the genial and more than Italian summer succeeds a cold almost Arctic in its intensity. Few men are abroad at this season, unless necessity compel them. The Indian keeps within his lodge, the trader and the hunter within the pickets of their forts, unless the latter come out to look at his traps, set at no great distance; and the gold-miner in Cariboo or Idaho fortifies himself to "put through the winter" as best he may. The ground is too hard to work profitably, and for five or six months these lie on their oars, until the spring sun melts away the winter blanket of snow. Up in these parts, by the time October's first snows are whitening the tops of the mountains, and the ice is beginning to form in the "prospect hole," at which they wash themselves of a morning, hundreds of sturdy honest fellows are very seriously cogitating about "putting through the winter." If they are in funds, or if they have little money, they "come down below," and at this season you may meet them by hundreds, trudging or riding over the mountain trails that lead down to the milder, wetter regions of the coast. If you ask the traveller whither he is bound, he will tell you to "put through the winter." If his "claim" has yielded well, he will pass the next few months in Victoria or San Francisco in state, generally spending all that he has toilsomely earned through the past summer. He exchanges his old indiarubber boots, his grey and not over clean woollen shirt and ragged trousers, for the spruce attire most affected by the "honest miner," as he delights to call himself, and lavishes his money in a way so foolish that only the honest miner would ever think of. For him the gambler sets his snares, in the shape of "small games" of monte, "seven up," and "cut-throat poker;" and the "saloon keeper" and all his satellites spread their nets to catch this reckless, foolish bird. If he be poor, he will take a little board "shanty," in partnership with some other "mates" in the same predicament, and will, probably, take unto himself some aboriginal damsel whose attractions have captivated him, and loafing about the street corners, disporting himself at the squaw dance house, and possibly in indulging in a "little game" also, he will pass the winter. If he be wholly penniless he still manages to get through. The vicissitudes of fortune are too common in a new mining country for anybody to be surprised at a man being, in Pacific parlance, "strapped," or "flat broke;" he may be poor to-day and rich to-morrow, and is treated accordingly. If he be known he has generally no very great difficulty in getting some restaurant keeper to trust him with a winter's board; a "dry goods man" to rig him out in a suit of clothes; and he is a friendless or unbefriended man indeed who cannot get somebody to lend him enough of coin to pay for his occasional morning "cocktail," or other little expenses which the honest miner looks upon as quite indispensable to his existence. It is really surprising with what coolness a man who has not a penny in the world will what he calls "run his face" for a winter's board, and how leniently the hotel keeper looks upon this little weakness of his. I fancy few in this country would care to trust anybody to the extent of £50, £60, or even £100, for a hotel bill; and yet that is done every winter

VIEW OF MOORE'S LAKE, UTAH.

to these miners on the other side of the Rocky Mountains. It is rarely that a miner, even if he be unknown, is refused a meal at a roadside house on the way to the mines.

VIEW OF THE SNOQUALAMI FALLS, WASHINGTON TERRITORY. (From an Original Sketch.) *

He looks upon it not as a favour, but as a right. I fully believe that the old miner thinks that he is entitled to a suit of clothes on credit for the asking anywhere; he has spent money in the country, he argues, and deserves something from it. Yet it is astonishing how seldom these tradesmen, who require to be so full of faith, lose by their customer. He may possibly not see him for years, but he generally pays him in the end.

* See Vol. I., p. 307.

If the miner be wholly friendless, creditless, and moneyless, he may work a little during the winter. He will saw wood, hunt for the game-shop and the butcher, through the wet winter, or do something or other; but still, that is about the last of his thoughts. This class of men seem to look upon it as a settled matter that they shall earn in the summer enough to pass the winter in some town jovially, and, whether rich or poor, are rarely depressed in spirit—a better time is coming.

"We're putting through the winter," an old acquaintance (a Government official) writes me, from near the Sierra Nevada Mountains, from "Carson City"—a little town much afflicted with such gentlemen—"just in the old style. We ain't got a cent—there isn't a cent in the treasury; if there was we couldn't draw; but I've got a good room, a dog that can sponge for itself, and *the boarding-house men hold out astonishingly!*" My correspondent is the type of a class—happy under the circumstances. I fancy if anybody in London attempted such systematic credit, he would be very apt to be looked upon as a suspicious character, and to be committed to the charge of X 0124, as an individual worthy of the notice of the nearest police magistrate. Happy is it, therefore, for the band of stalwart fellows who are pioneering in the West, and sometimes get beggared in the attempt, that North-West America is not England. But, in whatever way the gold-digger "puts through the winter," he is only too restless to be off again, and has generally considerable difficulty to keep intact enough of coin to form what he calls, in the gambling language only too familiar to him, "a stake," to pay his expenses "up," and enable him again to set to work at his claim by the time the Gold Commissioner declares it "jumpable."

About April the exodus commences from Victoria in the shape of an advance guard of the "hardest up" of the lot; and from that time until June you may daily meet long troops of them trudging through the snow, sleet, and mud, with a rusty frying-pan, a tin pot, and a pair of red, blue, or green blankets, containing their "possibles," on their backs, very dull and very tired, for a British Columbian road is a "mighty hard" one to travel. A joyful meeting at the end of their journey ensues with their old comrades, who have wintered on the creek (too poor, or too disinclined to come down); news is exchanged, parties made up to "prospect" for new diggings, and before another winter is come, and the snow on the Bald Mountain warns them to be getting under shelter, doughty deeds are done by brave men, who, though they never heard of the Geographical Society, are yet for all of that as great explorers as some at whose diluted narratives the ladies clap their hands in Burlington House. Strange tales of wonder do they tell that winter, as they sit round the warm stove or blazing log-fire in their cabins, of their summer's deeds—stories which would make the fortune of some small travellers who seek to "make a point" in their books: how they have come, by the banks of some nameless stream, on some poor fellow's skeleton, with his empty provision-bag by his side, or a tin pot, on which is rudely inscribed that he was dying of starvation; or a bullet through that fleshless skull tells how this weary wanderer met his end. It is then that we learn that there are other methods of passing the winter than the reckless devil-may-care way we witnessed on the coast. Happy is he who lives to tell the tale. I have before me a letter written a few years ago, and in it the writer relates how he passed the winter in a log-cabin in the Rocky Mountains:—"I started, as you know, for Idaho; but, when I got to the Columbia River, I saw people from the Blackfoot country, and they gave me such glowing accounts of it that I concluded I would put out for the favoured spot. I fell in with a party of men of my own sort, and three of us bought ponies from the Indians, and loaded them with provisions and camp kit, and started out with high hopes of fortune, as I, at least, had done twenty times before. I have seen the rough sides of backwood life, and have endured many hardships, but this trip was like a dark night to a small nigger! We were forty-eight days on the trail from Walla Walla to this place. We walked every foot of the way, and it either rained or snowed every day and night. The half of the tramp was intensely cold, and the ears and fingers of my two comrades froze badly. I am living now in a cabin, built with timber, within four miles of the summit of the Rocky Mountains. The inside of it is not elegant, but it is warm, and that is the main thing. The snow is I don't know how many feet deep, and there is not the slightest chance of working at mining until April. It is very cold. This is called a pleasant day; but I am, nevertheless, sitting as snug by the fire as I can without burning, and the ink freezes in my pen. I have to keep the door partly open for light, and it feels as if the butt end of the North Pole was punching me in the back. Frozen hands and feet are common as flea-bites in California, and two of my neighbours have died from cold. My chum is off on snow-shoes to bring home from the trading place a hand-sled load of provisions; but the snow is commencing to drive, and I am getting afraid for him. Provisions are to be got, but are not very cheap. For instance, flour is (cheap) 34 dollars (£7 1s. 8d.), per cwt.; beef, 22 cents (11d.) per lb., by the quarter; bacon, 75 cents per lb. (3s. 1½d.); sugar, 80 cents (3s. 4d.); syrup, 78 cents (3s. 3d.) per gallon; beans, 60 cents (2s. 6d.) per lb. I bought quite a stock of coffee, and I have traded it off for beef, which is more filling at the money. If I don't 'winter kill,' as some of my neighbours have done, I intend to do some tall wrestling in the coming spring and summer, and get out before another winter. One like this will do for me—you bet!"

Bad enough is it if he stays at home, but the man who passes his life in the Far West is apt to be of a restless turn of mind, and wearies of remaining in his cabin. At all events, woe betide the forlorn explorers who are caught in the mountains in these snows. Such a fate was Fremont's, in the Sierra Nevada of California; Marcy's, in the Rocky Mountains, and "Mountaineer Perry's," who journeyed on snow-shoes over the Tête Jaune Cache Pass to Jasper's House, a solitary post of the Hudson's Bay Company.[*] The winter traveller in British Columbia may get snowed up, as were the Canadians who, in 1863, attempted to pass that way, and whose remains were found next summer on the upper waters of Fraser River shattered and torn, with marks which told that what Indians and wild beasts had failed to effect, the living, maddened by famine, had done to the dead: they had resorted to the last resource of men maddened by hunger—cannibalism. The hunter maybe caught in the snow, or the trapper shut up in his little camp, seeing daily his small

---

[*] This is the same individual who crossed "the plains" trundling his worldly effects before him in a wheelbarrow.

stock of provisions lessening and the storm still blowing without. Sad is the lot of such, but still sadder even is that awful tale which yet lingers in California—of the trains of emigrants who, on their way to California, were caught, with their wagons and cattle, in the snows of the Sierras in the winter of 1847. They built themselves little huts, and as long as their provisions held out they maintained hope. Soon the snow covered their cattle, and they knew not where to find them. They then lived on hides, though some refused to touch them. Little parties attempted to reach the settlements and bring necessaries, but failed; and then scenes ensued which pen would fail to describe. There were women and children in that band; but women forgot their womanhood, and children's youth and mother's love were no protection to them. At last the terrible news reached the settlements, and help was brought them. It was too late for many, and those who were saved would, in their shame, have preferred death .... The log cabins which had been the scene of such revolting acts were razed to the ground, and all tangible traces of such a tragedy erased. Many years have now elapsed since this happened, but as late as 1865 I was pointed out a survivor who had taken part in it. It seemed to me, as I shuddered at the remembrance of the story, that his face yet bore a sullen aspect—a Cain-like mark by which all men might know him. The story related is one of the most terrible in the annals of California, and all the more horrible in that it is strictly true, and has been more than once almost paralleled. It nevertheless deserves to be recorded as an episode in the earlier life of the young state—as a horrible concomitant of the *auri sacra fames*.

About the southern end of Vancouver Island wintering is scarcely such dreary work as in the far-off wilds. The winters are usually very wet and "muggy;" but little snow falls, and even then it only lies for a few days. The cold is not severe, and most of the ice used so extensively during the summer months is brought from Harrison Lake, in British Columbia. The winter of 1863-64 was unprecedently moist. Rain fell in an almost continual pour, so that none of the hundreds of idle men who were "putting through the winter" in Victoria could go very far afield. Still, now and then, tired of the prison-life of that little town, where every face was familiar, almost *ad nauseam*, we would wander out by the road nearly knee-deep in mud, or into the dripping woods, seeing what was to be seen, or away into the forest-paths or byways, where the "dead-broke miners" were felling trees, or heaping up cord-wood for the market. For this privilege in the vicinity of towns they generally pay a small royalty to the owners of the land. Here and there in other quiet places Chinese would be also working away at the same business. They would generally be in partnerships of half a dozen to a dozen, and long before we came in sight of them we could hear their noisy chattering resounding through the woods, or we would overtake their industrious compatriot, the Chinese pedlar, trotting along with two bamboo woven covered hampers dangling at either end of a bamboo pole, suspended on one shoulder.

Most of the younger men among the gold-diggers would remain in town all winter hanging about the bar or billiard rooms, "ranching it" in "shanties" in twos and threes in a more or less decidedly bachelor life. But others of a sporting proclivity, or so scant of coin as to be impelled to leave town, would go in small parties and establish

VIEW OF SOOKE LAKE, VANCOUVER ISLAND.
(From an Original Sketch by Mr. Frederick Whymper.)

a kind of Agapemone camps beside some lake, such as the Sooke (p. 37), and out of which the Sooke River flows into the bay of the same name, near which deer abounded, or by the mouth of the Fraser, San Juan, Salmon, or other stream, where the wild geese and ducks collect in countless numbers in the winter. There they would remain for weeks shooting for the Victoria market. Such a winter camp is by no means so pleasant as the summer one we have already described (p. 29). Still, the winter hunting-lodge is not without its charms to men of simple tastes. The deer come down to the lowlands in the winter in immense numbers, and anywhere, except in the immediate vicinity of Victoria, can be easily shot. The ordinary price of venison in the Victoria market is from 2½d. to 5d. per lb., and elk (wapiti) a little dearer. Indeed, a great deal of the "beef" eaten in the Victoria restaurants is "elk's meat," the former costing 1s. per lb. However, about the country districts it can be bought much cheaper from the Indians; I do not speak of the wilds, where I have bought a deer for twelve leaves of tobacco; and at Fort Rupert the ordinary price of a deer for some years was a charge of powder and a ball, most of the deer being killed in pit-falls. About Alberni, where some hundreds of men were to be fed, deer were bought one winter for 6d. apiece, and in that vicinity one Indian killed seven before nine o'clock in the morning, but there being a demand, he got from 3s. to 4s. apiece for them. This deer is the black-tailed species (*Cervus Columbianus*), which is approached without much difficulty, and therefore easily killed with a fowling-piece loaded with buckshot. Some of the pot-hunters adopt the system of "fire hunting," but at best it is a murderous game, only excusable under dire necessity. A man stands in the wood waving a frying-pan of dry pine knots or gum wood; the deer is then attracted by the light. His companion takes aim between the eyes of the animal, guided by the reflection of the light in its eyes, and in this manner will kill several in a night. One has to be cautious, however, when in a settled district. I know a man who shot a neighbour's mare in this manner, and another who dropped an Indian, and got afterwards shot himself in return for his mishap! The wild fowl afford good sport, more especially on the southern coast, near San Francisco, where they are killed in punts, or stalked behind an ox. This method could be well practised near Seattle, on the flats near the mouth of the Somass River at Alberni, and at Boundary or "Mud" Bay, on the British Columbia coast, at all of which places the wild geese and ducks are very abundant. The ox shooter has a trained ox, which walks before him, and protects the "sportsman" until within shooting distance. The boat shooters go in parties of two and three each in a punt with three double-barrelled guns, and then subsequently return to their sloop, which is in waiting for them. The boat shooters average during the season thirty ducks a day, and a good ox shooter will sometimes kill seventy or eighty brace of geese in the same period of time.

To those in town, the little wooden theatre, where a Californian company performs, affords nightly amusement. Here a place is set apart for the Indians and squaws, who, though not understanding a single word of the play, yet laugh most vociferously. Negroes are not admitted to the best parts of the house, and though there have been numerous attempts, either by law, by trick, or even by force, to obtain what they consider their just rights, yet in every case have they been defeated, the manager siding with the majority of his patrons. Perhaps still more amusing than the theatre was the town crier—a half-crazed fellow, though remarkably sharp-witted and clever in some things—"calling out" the theatre, first in English, then the "laughing-house" in Chinook, for the benefit of a gaping crowd of Indians; and then in a flow of gibberish, which was supposed to pass muster for Chinese for the information of the sloping-eyed Mongol gentlemen, who, however, did not patronise it, much preferring to spend their spare dollars in opium or gambling instead.

Of the life of these settlers or gold-diggers, who have only too implicitly followed Mr. Tennyson's hero's design of taking to himself "some savage woman," I need not speak. To hint at this is almost more than sufficient, familiar though the practice is to all who have ever visited North-West America. Lillooet, in British Columbia, and Lytton, on the Thompson River, used to be favourite localities for men afflicted with the "squaw and buckskin mania," for which Captain Burton assures us there is *no* cure. The Hydahs and Tsimpsheans are the coast favourites, while the Lillooet and Lytton Sultans throw their handkerchiefs chiefly at damsels belonging to the Shouswap and allied tribes (p. 21). The amusements of the place are not high. "Melodeons," or music-halls, the "squaw-dance house," and cheap Jack auctions, with secret gambling-houses, though these are sharply looked after by the Government, supply the chief amusements of the miners. For those of higher tastes there was a good deal of informal hospitality. The police-court was a very amusing place to the student of human nature, as developed among the various nationalities found on the North Pacific Coast, and the local legislature, a burlesque on representative institutes, now merged into a somewhat more dignified assemblage, containing the collective wisdom of British Columbia and Vancouver Island, was a never-failing source of mirth.

These were free-and-easy times. You required little introduction in those days, except an honest face and a little common sense. Many pleasant evenings did we then pass in the hospitable houses of the old settlers, coming home by the moonlight over snow or through rain, careless of either, for we had a light conscience and a heavy coat. Many of those who shared in those now—seemingly—far-off days may wonder if those light-hearted times can ever come back, when we used to trudge, with the now dead and vanished men, ten miles on a dark night to pay a visit at some country house in the backwoods, where we used to keep an extra pair of boots. I have still an invitation from some friends in Washington Territory which I may copy for the amusement and instruction of the reader. It runs as follows:—"Come and stay as long as you've a mind to. We've lots of pork, lots of flour, apples, and cider, half a keg of whisky, and twelve sons and daughters all grown up, and all living at home. But if you think that's not enough company, bring over a couple of your friends—the more the merrier." No wonder that doleful letters now reach us from that once jolly region. The times are changed, and the people are changing with them. There is now no house where spare shoes are kept. The old settlers are dying off; the good old times are gone; and that the "country is going to dogs" is the conclusion of the North-Western Cassandras!

After the dreary winter an early spring around Victoria is very pleasant. The whole country not covered with fir forest or oak groves is bright with flowers, among which the blue gamass lily (*Gamassia esculenta*) and the yellow bedstraw (*Galium*) are particularly conspicuous. Frogs—those harbingers of spring—croak in every marsh, and

UTE INDIAN CHIEF.

the whole vicinity of the town looks very enticing. Every morning during the springs I spoke of long troops of miners were to be seen moving down Bastion and other streets leading to the Fraser River steamer, lying at the Hudson's Bay wharf, pack of blankets on back and tin kettle in hand, accompanied by two or three friends and their boarding-house keeper, and, with a general advice to "take care of themselves," moved off with cheers. I have seen the then attorney-general of the colony in a similar "rig" on his way to the gold mines. In another six weeks or two months the town was deserted, and people had settled down to their usual quiet jog-trot mode of existence—the picnicing and boating up "the Gorge" by the log

WARRIOR OF BLOOD INDIAN TRIBE.

bridge (p. 17), and making that show of doing business, which is the normal life of the Victorian in summer.

With these remarks, we shall conclude this sketch of men and manners on a region destined yet for great things, and the motto of which might well be the hopeful one of Washington Territory. "Alki, alki"—"By-and-by!"

CHAPTER II.

THE UNITED STATES: THE ROCKY MOUNTAIN STATES AND TERRITORIES: NEW MEXICO; ARIZONA.

THE Rocky Mountain States and Territories, properly speaking, are Montana, Wyoming, Colorado, and, in a less degree, Arizona and New Mexico. It is, however, convenient to also include in this group Utah and Nevada; and, accordingly, under this heading, we may have a few words to say regarding the latter, and also of Dakota, which, perhaps, more rightly belongs to the Missouri-Mississippi Valley. For the sake, however, of not interrupting too abruptly the sketch of the physical features of the country, we shall follow the arrangement indicated. We shall, therefore, first take up New Mexico, then Arizona, which was once part of it politically, and is yet geographically. Then will follow brief sketches of Utah and Nevada. Montana may suitably come next, then Colorado, and last of all Dakota and Wyoming, which will introduce us to a knowledge of the Yellowstone Park and the Prairies, and so gradually lead to a consideration of the Missouri-Mississippi Valley proper.

NEW MEXICO.

New Mexico extends on its eastern boundary 345 miles, and on its western 390, while its average breadth at the 32nd parallel of latitude is 335. Its entire area is 121,201 square miles. It is divided into thirteen counties, while the population in 1870 numbered 91,874 people, of which 172 were "coloured," and 1,309 were tribal Indians. Santa Fé, the capital, had 4,765 at the same date, while most of the other towns could not boast of over 1,000. New Mexico is thus next to the district of Columbia—in which Washington is situated—the most populous of all the United States Territories. Of the total population in 1870, 86,254 were native, 5,620 foreign born, while the males exceeded the females by less than 3,000. Of this number 18,668 were employed in agriculture, while mining and manufactories occupied 2,205. Most of the inhabitants are of Mexican descent, and speak the Spanish language.

The number of tribal Indians, according to the Report of the Commissioner of Indian Affairs, was, in 1875, 27,100, including 10,000 Pueblo people. The Apaches are probably the most troublesome. So much have these Ishmaelites harassed the borders that there used to be a disagreeable proverb prevalent in the territory when such-and-such a like youth was described as a smart boy. "Yes," would be the reply, "he may turn out a spry youth, *if the Apaches don't nail him to a cactus!*" The exception to the generalisation is especially characteristic of the uncertainty of life on the frontier. The Pueblo Indians, on the other hand, are among the most civilised of all the American aborigines. Several times they have been decided by the Territorial Courts to be citizens of the United States, but they have always preferred to retain their tribal existence, each of the nineteen pueblos, or villages, having its own government. In all they number 9,500.*

---
* See also "Races of Mankind," Vol. I., pp. 205—211.

Viewed from a physico-geographical point of view, the general surface of New Mexico consists of high level plateaux, traversed by mountain ranges, "between which are many broad fertile valleys, from which rise occasional peaks of great height." The Rocky Mountains, before entering the territory from Colorado, divides into two ranges, the western of which is called the Sierra Madre (p. 15), east of which lies nearly two-thirds of the territory. West of the Sierra are table-lands or *mesas*—often standing apart from each other, and resembling gigantic fortresses—detached ranges of mountains, many fertile valleys, and some extinct volcanoes. East of the Pecos River, and the Eastern range of the Rocky Mountains, the surface slopes gently towards the Mississippi River and the Gulf of Mexico, while the south-east part of the territory is occupied by part of the celebrated *Llano Estacado*, or Staked Plain, a steppe or plateau entirely destitute of wood, and except after the slight rainfalls which visit it of any other kind of vegetation. Through this treeless plain—the route across which is marked by stakes, hence the name—the Red River of the South branches into numerous forks all running into deep gorges. On the plain itself the eye rests on no land-mark or object of relief, while here, in common with similar plains when the ground gets heated by the vertical rays of the sun during the middle of the day, the mirage accompanies the incessant tremulous motion of the lower strata of the atmosphere. The very extraordinary refraction of the atmosphere upon these elevated plateaux causes objects in the distance to be distorted into the wildest and most fantastic forms, and often exaggerated many times their true size. "A raven, for instance," writes General Marcy, in his official report to the United States War Department of his exploration of this region, "would present the appearance of a man walking erect, and an antelope often is mistaken for a horse or buffalo. In passing along over this thirsty and extended plain on a warm day, the eye of the stranger is suddenly gladdened by the appearance of a beautiful lake, with green and shady groves, directly upon the opposite bank. His heart beats with joy at the prospect of speedily luxuriating in the cool and delicious element before him, and he urges his horse forward, thinking it very strange that he does not reach the oasis.

At one time he imagines that he has made a sensible diminution in the distance, and goes on with renewed vigour and cheerfulness; then, again, he fancies that the object recedes before him, and he becomes discouraged and disheartened. And thus he rides for miles and miles, and still he finds himself no nearer the goal than when he first saw it; when, perhaps, some sudden change in the atmosphere will dissipate the illusion, and disclose to him the fact that he had been following a mirage." The level of the "Llano Estacado" towers some 800 feet above the surrounding country—3,200 to 4,700 feet above the sea*—and is bordered by precipitous escarpments, "capped with a stratum of snow-white gypsum, which glistens in the sun's rays like burnished silver."

"Prairie Dog" towns often extend for miles through portions of this territory. This little marmot—for the reader need hardly be told that except in its popular name it has no relation to the dog—is known to zoologists as *Cynomys Ludovicianus*, to the

---

* "List of elevations principally in that portion of the United States west of the Mississippi River" ("Geological Survey of the Territories," Miscellaneous Publications, No. 1, 1875).

44 THE COUNTRIES OF THE WORLD.

Indians as the *Wishtonwish*, and to the French Canadians as the *petit chien*. Its towns are the most familiar features of the dog prairies of the West. It has been

A PRAIRIE DOG "TOWN."

said to be found in Oregon; but this is, I believe, a mistake, the large gopher squirrels (*Spermophilus Douglasii* and *S. Beecheyi*) having been mistaken for it, unless, indeed, as has been suggested, it was exterminated by some epidemic such as that which some years ago broke out among the prairie hares in the vicinity of Walla Walla, on the

SNOWY RANGE OF THE SIERRA MADRE, ROCKY MOUNTAINS.

Columbia River, resulting in nearly exterminating them. Often on the prairies the traveller will pass through these "prairie dog" villages or holes for miles. At first the inhabitants are all on the look out on the top of the mounds in the vicinity, but at the slightest alarm they disappear into their "houses" with the alacrity of rabbits. Sentinels generally keep watch, and give the signal of danger approaching by a peculiar yelp, and immediately, after securing the safety of their village, disappearing. The burrowing owl (*Athene cunicularia*) often inhabits the same holes, and, as will be seen in the illustration given on p. 44, the rattlesnake is occasionally also a tenant with the mammal and the bird. I apprehend, however, that the co-operative housekeeping is entirely as unwilling a one on the prairie dog's part as is that of the bailiff with the impecunious Briton!

Santa Fé—7,047 feet above the sea*—in spite of its romantic associations derived from the narratives of the early adventurers in the West, is anything but a picturesque town. Its streets are narrow, its churches poor, and its houses, for the most part, painfully prosaic in their utter newness. Yet the latter are very comfortable, being all one storeyed and often covering several acres of ground. The walls are of adobe, or sun-dried brick, either white-washed or bare, and are thus cool in summer, and sufficiently warm in winter. Protestantism does not burn brightly in Santa Fé, the one church for years being unable to obtain a roof through the contributions of the members of that faith in the city. Nearly all the inhabitants are Roman Catholics, and the town itself is the seat of a Romish bishop. Of course there is a plaza, or tree-shaded square, as there always must be in every town which owes its origin to the sleepy siesta-loving Spaniard. The names of the streets and other localities in New Mexico also bear the same impress, contrasting in their piously superstitious nomenclature—their "Santa Domingos," "Jesus Marias," "Spirito Sanctos," or "El Sangré De Christos"—with the "Buckskin Joe," "Strip-and-at-em-Mine," "Cash Creek," or "Hooked Man's Prairie" of the rougher, but more manly Anglo-Saxons who have taken the place of the indolent Castilian race. Many of the traders are of German Jewish extraction, as they frequently are in the West, and in Spanish America. "Their stores," writes Dr. Bell, "are well filled with everything required by the emigrant, and a good deal of rubbish to meet the demand of the Indian population. A large trade is done in paste and brass jewellery, and a still larger one was formerly done in firearms, some specimens of which I examined with great curiosity. The wholesale price of the single-barreled guns was two and a half dollars each, and they could not possibly go off without bursting. However, since the Navajos have been 'improved' off this country, the market for this kind of goods has ceased, and as Northern Mexico no longer produces the vast hoards of precious metals which formerly enriched its inhabitants, the Santa Fé trade has degenerated to local insignificance, and the great Santa Fé trader has now joined the other romantic characters of bygone days. No doubt he will again reappear on the scenes, but so changed that we shall scarcely recognise him. He will wear a frock coat, and a linen shirt; his goods will come by steam; and his stories will relate not to Indian fights, but to railway accidents."

Hot and mineral springs, salt lakes, or *salinas*, are common over this territory,

---

* Taos is 8,000 feet, El Moro 7,238, while Elizabeth Town is 8,200 feet above the sea.

supplying, indeed, a large proportion of the salt used in the neighbouring portions of Mexico and the territory itself. Anthracite is found, gold and silver are plentiful, and were even mined in the Spanish times, and copper is abundant; but the scarcity of mints and Indian raids have greatly retarded the progress of all mining enterprise. The statistical returns show that about half a million dollars' worth of gold is annually produced by such mines as are worked. Lead, iron, and other metals are also found, but the extent and value are as yet unknown. The climate varies much. North of Santa Fé the days are never sultry, and in the vicinity of that city the winters are even severe. In the south the climate is mild, but owing to the elevation of the country rarely sultry. The sky is clear, and the atmosphere so dry that meat may be preserved a long time without salt. As might be expected, pulmonary complaints are scarcely known, the number of deaths from chest diseases being smaller in New Mexico than in any other part of the Union, except Arizona. Indian corn, wheat, barley, oats, apples, peaches, melons, apricots, and grapes grow well, but agriculture is as yet in a rather primitive condition. The territory is, indeed, better adapted for grazing than culture, the valleys, foot-hills, and table-land supporting large flocks of sheep on the sweet buffalo grass (*Buchloe dactyloides*) and grama grass (*Bouteloua oligostachya*) with which they are covered. The total number of acres of "improved land" was, in 1870, 115,007, divided among 4,480 farmers; but of these 1,515 contained less than ten acres, only four more than 1,000 acres. As yet there are no railways in the territory, but several are projected. The country is ruled by a governor and secretary, appointed by the president for four years, and an auditor, treasurer, adjutant-general, and attorney-general chosen by the territorial legislature, which consists of a council of thirteen, and a house of representatives of twenty-six members, elected by the people for two years; in fact, with a few trifling differences, what commonly prevails in other territories, while the same may be said of the judicial powers. In 1870 the total value of the real and personal property of the territory was 31,349,793 dollars. New Mexico was one of the earliest parts of America visited by the Spaniards; indeed, it was entered by them almost a century before New England was settled by the Pilgrim Fathers. In 1848 it was ceded to the United States, and in 1863 Arizona was set apart from it as a territory.*

ARIZONA: THE COLORADO AND ITS CAÑONS.

Arizona † is a corruption of the old Aztec name Arizuma, which was the name the Spaniards gave to that portion of Sonora which it comprises. It probably means a rocky country, from *ari* (rock), and *zuma* (country). It contains 77,383,680 acres of land, and is divided into four counties. The first account given to the European world of this part of the American continent was in the romantic story of Friar Marco de Niça, who, as early as the year 1535, made an expedition among the Indians of this region.

* Brevoort: "New Mexico, her Natural Resources and Attractions" (Santa Fé, 1874); Hawes, &c., Article "New Mexico," &c.

† For some of the information which follows I am indebted to notes drawn up by the Hon. Charles D. Poston, formerly Superintendent of Indian Affairs, and Delegate to Congress for the Territory; see also Cozzens, "The Marvellous Country."

He reported a semi-civilised people living in stone houses, dressing in clothes of their own manufacture, tending flocks and herds, cultivating the soil, and practising the arts of peace—then a more singular occupation in the New World than nowadays. This account led to the celebrated expedition of Coronado, which was organised and conducted under the patronage of the Viceroy of New Spain in 1540. The Indians described by the early explorers lived in the north-eastern part of the territory, about 6,000 feet above the sea level, in seven towns, called by the Spaniards the "Seven Cities of Cibola" (or the Buffalo), but readers of the early Spanish narratives must be cautioned against accepting these pompous narratives too literally. The explorers were as fond of calling every stream a "river," and every village a "city," as are their successors in the possession of this portion of their once broad domain. The people described by Coronado were the Moquis Indians, whose villages situated in lat. 35° 55', long. 110° 42', W., now contain about 6,000 souls, and have excited more than ordinary interest, both on account of their civilised life, and from the theory that they were the descendants of that Welsh Prince Madoc, who is believed (by his countrymen) to have sailed to America in the year 1170.*

But in Arizona are the remains of a pre-historic race. Near the Pima villages are the *Casas Grandes*, or Great Houses of the Aztecs or Toltecs, or whoever inhabited this region thousands of years ago. They have left no history, but the relics of a civilisation which puzzles the antiquarian. To use Mr. Posten's words, "The *débris* and remains of broken pottery would indicate that this city covered an area of about ten miles; but of all the houses which formed the city but one remains. It seems to have been a citadel or granary, as it is situated near the centre of the city. It was built of mud pressed into moulds and dried in the sun (adobes), and was composed of many small apartments, none of them very high. Five rows of joists may yet be counted, indicating five storeys; but from the fact that they are all burnt off to the wall this house seems to have been destroyed by fire. About twelve years ago I extracted one of the joists from the wall, and placed it in the Smithsonian Institution at Washington. It bears evidence of having been cut with a stone axe. The city which formerly existed here was furnished with water by a canal from the Gila River, which also irrigated a valley now desolate. The remains of the canal indicate a width of ten yards and a depth of four. As to the former inhabitants and their history all is left to conjecture. We know nothing of their origin, their manner of life, their politics or religion, of their loves or hates, of their mortality, or their immortality.

The only monument of their existence left upon earth stands there, in the solitude of the desert, as mysterious, as silent, as unreadable as the Egyptian Sphynx. One hundred miles south of this monument of a perished race stands another monument of another civilisation. It is the mission church of San Xavier del Bac, erected by the Jesuits, A.D. 1668. In the dim mirage of the desert these architectural sentinels stand confronting each other. The first represents the shadowy past, the second the epoch of Christian civilisation. The latter would be an ornament to any city in Europe or America. The weary emigrant, who has made his perilous journey across the North

---

* See "Races of Mankind," Vol. I., p. 205 *et. seq.*, for description of the Moquis and other Indians.

THE GREAT CAÑON OF THE COLORADO: RAFT PRECIPITATED OVER A CATARACT.

American Continent in search of the 'Land of Gold,' is surprised as he emerges from the forest to behold a specimen of Saracenic architecture, with dome and tower, and fancifully decorated façade. It appeared to me a magical creation as it stood in bold relief against the western sun; and marvel of architecture as it is, in this remote place, the impression is heightened when you enter the sanctuary, and hear the same vespers chanted which follow the setting sun from Rome around the world. In the archives of the Society of Jesus may be found an interesting account of the wanderings of Father Kino in this mysterious country. The Jesuits followed up their explorations by establishing missions among the natives, many of which remain to the present date in a somewhat dilapidated condition. The avarice and tyranny of the Spaniards, who were engaged in mining in the vicinity of the mission, exasperated the Indians to revolt, and in 1680 the Apaches commenced a war of extermination. The wily Jesuits charged their neophytes to preserve the sacred buildings, and assured the converts—with the sublime faith of their order—that as sure as the sun shone, water ran, and grain grew, they would one day return and resume their sacred duties. It was one of the strange episodes of life that during my service as Superintendent of Indian Affairs for Arizona it was my fortune to re-install the Fathers of the Society of Jesus in their ancient Mission of San Xavier del Bac. The Indians received them with firing of rockets, ringing bells, strewing flowers, and every demonstration of joy." *

The Colorado of the West is one of the greatest rivers of the Pacific slope. It was first explored by Fernando Alarcon about the middle of the sixteenth century, and it is now navigated by steamers 500 miles from its mouth, while many thriving villages are situated upon its banks. Before describing the most remarkable features of the river, viz., its cañons, it may be well to sketch briefly the general features of its basin, drawing mainly for the particulars on Dr. Bell's admirable observations. Compared with the Columbia the basin of the Colorado is rather smaller. Its area comprises about 250,000 miles, while that of the Columbia is only 200,000, or within 10,000 miles of the whole area of France. On the other hand, the basin of Mississippi is 1,100,000 miles, and that of the Rio Grande del Norte 240,000. The Bear Mountains, a northern extension of the Wahsatch Mountains, and their connecting link with the Rocky Mountains, form the division between the Columbia and the Colorado Basins. In another direction—the east—the Colorado Basin is separated from that of the Rio Grande and Mississippi by the Rocky Mountains. It thus appears that the Colorado Basin forms a large triangle, limited "on the east by the continental divide of the Rocky Mountains; on the south by the highlands about the Mexican boundary-line; and on the north-west by the Wahsatch Mountains." In the northern angle of this area run the waters of Green River (p. 56), while the western slopes of the "summit plateau" of the Rocky Mountains collect the sources of Grand River. These unite to form the Rio Colorado of the West. Further down it receives the Rio San Juan, then the Flax River, or Colorado Chiquito (Little Colorado), and, lastly, the Rio Gila, which drains all the southern half of Arizona, enters the main stream at Fort Yuma—a very warm locality—and sixty miles

* Dr. Bell has also given a very full account of these and other ruins in his "New Tracts," Vol. I., p. 193.

above its mouth. "The entire Colorado Basin consists of a series of table-lands, piled up one above the other, and covering the whole country. In elevation they vary from 4,000 to 7,000 feet, and reach, in some places, a height of 8,000 feet above the sea. They succeed each other in a series of steps, which generally present abrupt and wall-like edges, the more recent stratum occupying the highest portion of the plateau. Complete barrenness is the rule, fertility the rare exception; scarcely any vegetation, save the artemesian scrub (wormwood or sage brush), is to be met with between the 36th and 42nd parallels; the earth for the most part is bare and naked, showing the wear and tear of ages, the erosion of the primeval ocean, and the cracks and fissures of the more recent water-courses."

This area seems once to have been the bed of the ocean, when the American continent was smaller than now, and the material from which the table-lands were formed derived from the wearing away of the shores and islands by the dashing of its waves; or, to use Dr. Newberry's language, by the erosion of the "broad and rapid rivers, which flowed from the mountains, and through the fertile valleys of a primeval Atlantic." These thousands of feet of sediment in strata seem to have been gradually converted into dry land by the slow upheaval of the Plutonic rocks in which they were embedded, which were then the San Francisco Mountains. Mount Taylor, and Bill Williams's Mountain, marks the scenes of volcanic eruptions which formed these extinct volcanoes, now standing more or less isolated, and all beautiful in the midst of the eroded "mesas," or table-lands, such as are figured in Vol. I., p. 288.

The Colorado and most of its tributaries flow through what are called cañons, a Spanish-American word, signifying a deep defile with high walls, through which many rivers in the dry region of this section of country run. The word is commonly used over the Pacific coast for any defile, but as we shall presently see, it is only applicable to cuttings such as that which the Colorado makes. They are formed by the action of running water on soil on which even periodical rains do not fall to any considerable extent. The streams wearing the soil must be constant, and the surface-strata worn must be of a character which will easily yield to the current. After the water has got once into a groove, the constant attrition of the water and the sand, and gravel carried down by it, will wear even through the hardest granite, as in this region it has to the depth of 1,000 feet. Wherever there are cañons, the country is sterile and utterly worthless, for the deep cuttings drain it to the utmost; the river lying buried, as it were, in the bowels of the earth, "lie far beyond the reach of animal or vegetable life on the surface." Five hundred miles above the mouth of the Colorado ocean is the "great cañon" of the river, which of late years has attained much celebrity. This cañon is 217 feet long; and the falls vary in height from 1,000 to 6,233 feet; or, in other words, the river winds like a white thread more than a mile below the surface of the surrounding country. At the greatest elevation, the width of the chasm is from five to ten miles. In 1867, this great cañon was, much against his own will, explored by James White, a native of Wisconsin. To escape from hostile Indians, he, with a companion, set himself afloat on a raft on this river, and after terrible dangers, during which the raft was many times all but broken up, and on one of which occasions—in shooting a cataract—his companion was thrown off and drowned (a scene sketched on p. 49), he reached Callville, at the Pacific end of the cañon. Unable to land, and even could he have done so he would have found it impossible to have obtained food, he nearly perished of hunger, being six days without any nutriment, save a few green leaves. He had been six days on the raft since starting, and still the uneven current bore on the rude float. "He saw occasionally breaks in the walls, with here and there a bush. Too weak to shove his raft ashore, he floated past, and felt no pain, for the overwrought nerves refused to convey sensation. On the afternoon of this, the sixth day, he was roused by hearing the sound of human voices, and raising himself on one arm, he looked towards the shore, and saw men beckoning to him. A momentary strength came to his arms, and grasping the paddle, he urged the raft to the bank. On reaching it he found himself surrounded by a band of Yampais Indians, who for many years have lived on a low strip of alluvial land along the bottom of the cañon, the trail to which from the upper world is only known to themselves. One of the Indians made fast the raft, while another seized White roughly and dragged him up the bank. He could not remonstrate. His tongue refused to give a sound, so that he pointed to his mouth and made signs for food. The fiend that pulled him up the bank tore from his blistered shoulders the shreds that had once been a shirt, and was proceeding to take off his trousers, when, to the credit of the savages be it said, one of the Indians interfered, and pushed back his companions. He gave White some meat and roasted mesquit-beans to eat, which the famished man devoured; and, after a little rest, he made signs that he wanted to go to the nearest dwelling of the white men. The Indians told him that he could reach them in 'two suns' by his raft; so he stayed with them all night, and, with a revolver that remained fastened to the logs, he purchased some mesquit-beans and half a dog. Early the next morning he tottered to the bank, and again pushed into the current. The first day out he gave way to the yearnings for food, and, despite his resolution to the contrary, he ate up his entire stock of provisions, which did not by any means satisfy his cravings. Three long days of hope and dread passed slowly by, and still no signs of friends. Reason tottered, and White stretched himself on the raft: all his energies exhausted, life and death were to him alike. Late in the evening of the third day after leaving the Indians, and fourteen days from the time of starting on his perilous voyage, White again heard voices, accompanied by the rapid dash of oars. He understood the words, but could make no reply. He felt a strong arm thrown around him, and he was lifted into a boat, to see manly bearded faces looking on him with pity. The great objective point was reached at last—the battle for life was won—but with the price of unparalleled suffering. The people of this Mormon settlement had warm, generous hearts, and, like good Samaritans, lavishly bestowed every care on the unfortunate man so miraculously thrown into their midst from the bowels of the unknown cañon. His constitution, naturally strong, soon recovered its terrible shock, and he told his new-found friends his wonderful story, the first recital of which led them to doubt his sanity." When he was carried ashore, he was found to be literally flayed from exposure to drenching from water and the scorching rays of the sun. "His reason was almost gone, his form stooped, and his eyes were so hollow and dreary, that he looked like an old and imbecile man."*

* Bell's "New Tracts in North America," Vol. II., p. 215; and "Illustrated Travels." (Vol. I.)

THE *CEREUS GIGANTEUS,* OR MONUMENTAL CACTUS.

White's feat has been since then accomplished by Major J. V. Powell, whose party passed through the cañon in 1874. The current in the great cañon equals in velocity a railway train running forty or fifty miles an hour. Mr. Poston rightly enough remarks that the ruins of London could be thrown into the chasm without filling it up. (Vol. I., p. 288, and Plate XIII.) The whole length of the Colorado, or "Red River—so called from its red sands—is about 2,000 miles; but it is at present only navigated, and perhaps only navigable, to Callville, 612 miles from the mouth. At the mouth, the "bore," caused by the spring tides, makes a rise and fall of some thirty feet. The principal Indian tribes inhabiting its banks are the Cocopas, the Yumas, the Mohaves, the Chemhueves, the Hualpis, and the Yavapis, or Yampais (already spoken of). The bottom lands of the Colorado—away from the cañon—are very rich, and well suited for the growth of sugar, cotton, rice, maize, melons, and vegetables; but owing to the small rainfall, irrigation is necessary before cultivation can be pursued to any great extent. In the Colorado desert, as the southern portion of California in the vicinity of the river is called, the Indians plant their maize several feet in the ground, knowing that at a less depth the soil would be soon parched, and the grain be unable to obtain sufficient moisture to germinate. In some parts of the Colorado desert the soil is composed of sand packed closely together, with a hard, smooth, shining surface, or piled into loose hills, which are constantly shifting. To sum up the physical appearance of Arizona in a few words—it is composed of vast plains, covered with grasses, and crossed from north to south by broken ranges of mountains, full of minerals. In the northern part of the territory, the San Francisco Mountains rise to the height of 12,052 feet, and with their white caps are visible for a radius of 100 miles. In the southern portion of the territory, the Santa Rita Mountains are the most conspicuous, being about 7,000 feet high.

Fort Bowie is situated six miles up the Apache Pass, and is a mere collection of adobe huts built on the summit of a hill, "which rises as a natural look-out station in the centre of the defile, and commands the roads both ways for two or three miles of its length." It is a pre-eminently lonely and dangerous place of residence, as the many wooden tombstones in the little graveyard, all marked "Killed by the Apaches," abundantly testifies.

Arizona is rich in silver mines. Indeed, the Spaniards found near the boundary-line the largest mines of pure silver which have been ever discovered in the New World. Gold, lead, copper, and iron have been discovered in many localities. Rock salt is found in the mountains, while the lagoons, bordering the Gulf of California, furnish an inexhaustible supply of this necessary. The indigenous trees of New Mexico and Arizona—and in respect of their vegetation and animal life the two territories may be considered as one—are all of a more tropical aspect than the usual American forest growth to the west of the Rocky Mountains. The mesquit shrub (*Algarobia glandulosa*) allied to the acacias, which are also represented, and other sub-tropical shrubs, are common. The beans of the mesquit are imported into this country for feeding purposes, under the name of algorobo. The Indians make bread out of them, and also distil an intoxicating drink from them: a prolific mesquit tree will yield ten bushels of beans annually. A cactus (the *Cereus giganteus*, p. 53), is the most singular tree of the forest growth. Many species of the order are found in this region, indeed, it is their favourite home; but this is the largest of them. It is peculiarly fond of volcanic soil, and seems to require scarcely any moisture. Its huge grooved columns can be seen thrusting themselves out from between the rocks to the height of forty feet, the secondary columns forming a circle of four or six upright arms around the parent trunk. Its peculiar appearance in the landscape has obtained for it the name of the monumental cactus. Its fruit ("pitahaya") is gathered by the Indians when ripe, and is expressed for the sake of juice, which is made into treacle, while the pulp is compressed into a cake, stored away for winter use. The river bottoms are lined with cotton-woods, while the dry watercourses are marked by *Cercidium floridum*, the "green-barked acacia;" the arborescent *Idria spinosa*, distinguished by its silvery, leafless branches, and the valuable iron wood (*Olneya Tesota*). The *Chilopsis linearis*, allied to the trumpet flower (*Catalpa*), is abundant, being known as the desert willow, on account of the Indians weaving its long, slender branches into baskets. In the country more particularly affected by the arborescent cacti, Dr. Parry, in his report on the botany of the region, notes tree Yuccas, or Adam's needles, as forming a conspicuous feature in the landscape, while the true desert flora, such as the neat evergreen Larrea, with its myrtle-shaped leaves, together with a sort of thorny *mimosa*, dull-coloured *Obione*, or grease-wood, and the prevalent artemesias, or sage brush, all serve to give a faded aspect to the vegetation. The mountains bear an abundant growth of live oak (*Quercus agrifolia* and *Q. Emoryi*), and firs and pines, such as *Pinus contorta* (the Piño real of the natives), *Abies Engelmanni*, and the Piñon, or nut pine (*Pinus edulis*), the seeds of which are eaten. Wild hemp, sunflowers, and poppies are everywhere common and marked flowers. The American aloe, or maguey plant, is abundant on the hill sides and mountains, and is an important natural product. To quote Mr. Posten's graphic description:—"The head is formed in leaves like the artichoke, and grows to the size of a cabbage, being protected by sharp, bayonet-like shoots, eight to ten feet high. These being cut away with long knives, the Indians and Mexicans gather the head and utilise it in various ways. If roasted, it makes excellent food, smelling like a roasted pumpkin, but more astringent. It may be boiled down to a syrup, and form a saccharine feast, but the highest delight of the natives is to manufacture this mountain luxury into an intoxicating drink called ' mescal.' This is done in a primitive way, by fermenting the mashed head of the maguey in a raw hide stretched on poles in the sun, and distilling the juice in a rude alembic. The extract has very much the appearance and flavour of Scotch whisky, and has consoled many a weary traveller in that region besides myself. The fibre is manufactured into ropes, mats, carpets, and saddle-blankets." Among other useful plants may be noted the indigenous potato, or "comote," found on the mountain sides in all its native coarseness. Even in the desert a food is produced from a parasite known as the sand food of Sonora, which resembles the sweet potato in shape and flavour, and the amole, or soap plant (*Chlorogalum pomeridianum*), is used by the natives as a substitute for soap, from California to Arizona. It is considered a great preservative of the hair, making it as glossy as if oiled. Moreover, it washes flannel without causing shrinkage; though, as this is a use the natives rarely put it to, the toilet application of the plant is more valuable. The ordinary American fruits grow wild in the territory; but, owing to the absence of extensive forests, wild game is not abundant; nor do the rivers furnish many fish, or great variety of what are got. Buffalo does not extend so far west; but bear, deer, antelope, hares, wild turkey, the top-knot quail (Vol. I., p. 289), and the peculiar Mexican "Paysano, or road-runner" (*Geococcyx Californianus*). It lives almost entirely on the ground, very rarely flies, and frequents the highways, along which it will run from any one approaching. Its speed is nearly equal to that of a common horse, and it often furnishes an exciting chase to the solitary rider. It is allied to the cuckoo, and like that bird lives among the bushes, though it disports itself in open places. It is

VIEW OF GREEN RIVER, A TRIBUTARY OF THE COLORADO, UTAH.

the subject of many anecdotes in reference to its power of killing snakes, which are about as apocryphal as "snake stories" usually are. It is said that, on perceiving the rattlesnake coiled up asleep basking in the sun, this bird will collect the cactus and hedge him around with a circle, out of which the reptile, unable to escape, and irritated by the continuous pricks he receives in attempting to do so, bites his own body, and generally dies from the effect of the venom. It is just possible that this, and a score of similar tales almost as extraordinary, may be true—though this is not probable. Among other animal annoyances of Arizona are various venomous snakes, the tarantula, or large spider, the bite of which is often dangerous, huge centipedes, scorpions, horned frogs, &c.

Arizona has progressed rapidly. In 1856 the first exploration of the territory was made by a company under the command of Mr. Poston. After a weary march of 1,500 miles through hostile Indian tribes, he established his head-quarters at Zubac,

VIEW OF THE VALLEY OF THE BUBBLING WATERS, UTAH.

on the Santa Cruz River, and in this far-away wilderness planted the seeds of civilisation. The nearest military post was 100 leagues, and a mail from Washington did not reach this remote outpost in less than sixty days. The territory is now in telegraphic communication with all the principal towns in America, and the world. Mail communication is everywhere quick and convenient, while daily and weekly newspapers are published in all the large towns. The capital is Tuscon—a town south of the Gila—containing a population of more than 3,000 people. North of the Gila the chief town is Prescott, named after the historian of Mexico, which contains about 1,500 souls. Altogether the population of the territory may be estimated at 50,000 Indians, and 25,000 whites; the latter chiefly engaged in farming, mining, and commercial pursuits. In its area of 120,000 square miles may be found almost any climate, and land capable of supporting millions of cattle on the natural grasses which cover the country. Taken as a whole, its inhabitants consider Arizona well entitled to the name they love to apply to it—"The Marvellous Country."

CHAPTER III.

THE UNITED STATES: THE ROCKY MOUNTAIN STATES AND TERRITORIES: UTAH; NEVADA; MONTANA; COLORADO; DAKOTA.

GLANCING at a map of North America the reader will notice a district lying between the Wahsatch Mountains and the Sierra Nevadas extending northward beyond the 42nd parallel, and southward into Lower California, marked as the Great Basin. None of its rivers reach the sea; but nevertheless the name, as Dr. Bell, who has written an excellent account of it, from which we shall take some facts, points out, it embodies a glaring topographical error. Like the Colorado Basin, it is triangular, the apex of the basin pointing to the south and south-west. From the level of tide-waters at the Gulf of California the ground rises to about 5,000 feet. This is also the general level of the whole basin in its broadest part, while north of the Humboldt River, where the drainage divides, this elevation is exceeded, and though there are many local depressions under 4,000 feet north of the 37th parallel, there are few places below this elevation. The general physical character of the basin is much the same, the country being traversed by low, volcanic mountain ranges, about 1,000 to 4,000 feet above the general level, extending generally north and south, but as a rule running parallel to whatever spur of the Rocky Mountains or Sierra Nevadas they happen to be nearest. They are usually perfectly bare of trees, or even shrubs, and the denuding action of rain is evidently washing them down, and filling up the valleys with the eroded *débris*. The whole country is extremely dry, and on this dryness most of its peculiar characteristics depend.

"Artemesian scrub (sage brush)* and grease wood (*Obione canescens*)," writes Dr. Bell, "alone sprung from the dry, parched earth, except when some stream of unusual persistence supports a row of cotton-wood trees (*Populus angustifolia*), and a few acres of grass, along its edges. From the decomposition of volcanic rocks, the soil in its ingredients is very rich, and when irrigation can be supplied, yields most abundant crops. There are broad, level districts, however, called by the settlers 'alkali flats,' which are covered with salts, usually nitrate of soda, and are thereby rendered perfectly barren. These white glistening sheets on the dry, unsteady atmosphere of the desert, form the most tantalising mirage to which a thirsty traveller could be exposed. At certain seasons they are covered for a short time with a thin coating of water—the local drainage of the surrounding district—which is soon dissipated by the scorching sun. The plateaux of the basin region were undoubtedly the last portion of the Western continent raised from the sea, the last from which the Gulf of California retired. Even now subterranean fires are active, and the process of gradual upheaval may still be going on. Earthquakes are frequent; mud volcanoes are still to be found in places; huge cracks in the earth's surface have occurred within the memory of living man; craters recently active dot the whole district; and hot springs are so numerous that I have counted fifty-two jets of steam issuing from the ground like pillars of smoke in one valley alone. When the Great Basin came into existence, or rather emerged from the water, there were dry lands and mountains east, west, and north of it, shutting out from it the moisture of the Pacific Ocean, as well as any that might travel thither from the far-off Gulf of Mexico. The climate may be considered to have been then not unlike that of the present time, so that the rainfall was far less, even in the new-born 'Basin region,' than it was over the Colorado Basin in its primeval state, which was then washed by a broad Pacific Ocean. The effect of these climatic peculiarities was that a sufficient quantity of rain never fell upon the 'Basin region' to form a complete system of drainage from the highest lands down to the sea."

There is nothing in the physical construction of the Great Basin to have prevented a great river emptying itself either as an independent stream or as a tributary of the Colorado into the Gulf of California. The Great Basin is not really a basin without an outlet, but only a collection of numerous small basins, each with its feeding stream; but owing to the little rainfall and great evaporation, the lakes are rarely so full that they require to be emptied by a stream powerful enough to break through the barrier which all streams must first encounter, and thus by the union of their waters to form a complete drainage system. The lakes are accordingly usually salt, if they have no outlet, and are becoming salter year by year. But when the lake has an outlet it is fresh, though the stream is generally soon lost in the desert by forming a shallow sheet of water or a "sink" which has no outlet. The Great Salt Lake is a type of the salt lakes without outlet; Utah Lake of those which are fresh, this sheet of water being emptied, as is usual with lakes in other localities, by a stream. Most of the other lakes are, however, not permanent, being only broad sheets of water after rain, and

---

* Dr. Bell calls it "*sedge* brush," but this is an error. It is a too familiar sight in America.

perfectly dry and barren during the greater part of the year. They also vary greatly in size and elevation. For example, Great Salt Lake is in Utah, 4,200 feet above the sea level; Servier, in the same territory, 4,690; Lake Tahoe, in California, 6,700 feet; Bear Lake, Utah, 5,951; Walker's Lake, Nevada, about 3,840; Monro Lake, 6,154;

A STREET IN SALT LAKE CITY, UTAH.

Pyramid Lake, 3,940; Humboldt River Basin (Lassen's Meadows), 1,200; Williamson's Lake, 2,388; Morongo Sink, 1,500; Mojave Lake, 1,000, and Perry Basin, 530 feet. On the other hand, Soda Lake Basin, a large saline flat a little north of the Mexican boundary-line, is about seventy feet below the sea level, and though nearly always dry, Hardy's Colorado, or the New River, flows through the desert towards it when the Great Colorado is flooded.

On these occasions it breaks over its banks about forty miles south of Fort Yuma, and sends the New River north-westward for a distance of 100 miles or more. It has, indeed, been proposed to cut a canal from the river to the low ground, so that the "desert"—

VIEW OF SALT LAKE CITY, UTAH (WESTERN SIDE, SHOWING THE TABERNACLE, ETC.)

which is said to be land of excellent quality—might be navigated and cultivated. But as yet nothing has been done to effect this desirable improvement.

Death Valley, the sink of the Amargoza, is, however, the most extraordinary depression in the Great Basin. It is 175 feet below the sea level; and though an area of 50,000 square miles drains into it, it is an arid desert, and appears at one time to have been a deep lake, overflowing in a river reaching the sea, when (as at one time must have been the case) the country was more humid than now.

The Great Salt Lake is the largest sheet of water in the Basin. It is about sixty miles long and ten broad, and there are not wanting evidences to show that at one time it must have covered twice or thrice its present area. Of late years it has been steadily rising; and it has been suggested that this is owing to the increased rainfall caused by the country in the Salt Lake Valley having been irrigated by the Morman settlers, and thus increasing the evaporating surface. West of it, on the borders of Nevada, lies the "American desert." The soil here consists of clay and sand impregnated with salt. When wet it has the consistence of mortar. During the driest season lightly-laden wagons may pass between Pilot Peak and Spring Valley, but whenever there has been any continuance of wet weather the desert is impassable. Moore's Lake, a small, but, perhaps, more than usually picturesque sheet of water in Utah, is sketched on Plate XII.

The Humboldt River, 300 miles long, is the largest river of the Basin. It empties itself into Humboldt Lake. Reese River flows through a narrow valley about 100 miles long, and the Truckee, Carson, Walker River, and other streams, all end in the desert lakes or sinks. In the vicinity of most of these streams are little settlements of agriculturists. But the mining capabilities of the Great Basin are its greatest attraction. The State of Nevada, and the Territory of Utah, are both comprised within it. The former is the area within which are situated the richest silver mines of the world, and this fact renders the region, of such exceeding little interest otherwise, of great importance. The best of mines are situated in the Comstock Lode, at Virginia City, but highly remunerative mines, both of gold and silver-bearing quartz, are found scattered all over the territory. We have already referred to their yields. But as a specimen of what they produced during the "flush" period, we may mention that in 1867 the Savage Company paid in dividends a larger sum than is derived from all the metallic mines of England and Wales put together.*

UTAH,

So called from the Utah, Yuta, or Ute Indians—for the spelling of an Indian name is very immaterial, as most usually under the best of circumstances it is only approximately correct— "the dwellers in the mountains" (p. 10)—is a territory containing 87,176 square miles, and a population numbering, in 1870, 86,786 (though now nearer 120,000), mostly members of the Church of the Latter Day Saints, or "Mormons." Its physical features we have already described. It possesses some good grazing land, but the vegetation is not luxurious, and timber, except pines and firs in the mountains, is scarce. The climate is, moreover, bleak and changeable—deep snows and intense cold in winter, scorching heat and no rain in summer,

* Bell's "New Tracks," Vol. I., p. lxv. (Introduction)

accompanied with thunder-storms and clouds of dust, which, for the time being, overwhelm everything. The minor earthquakes so common in California, and in all the Pacific region as far north as Alaska, however, rarely visit the Great Basin. The soil is barren; but when irrigated, produces good crops of wheat, oats, barley, maize, buckwheat, flax, hemp, and rye. Some spots, indeed, are amazingly fertile, producing as much as from 60 to 100 bushels to the acre. The chief manufactures are farming implements, furniture, wagons, woollen goods, leather, steam-engines, machinery, and cutlery, which industries have attracted many English mechanics to the country. There are now several newspapers, schools, and all the other appurtenances of Western civilisation. The Pacific Railroad runs through the territory, and a branch line connecting Ogden with Salt Lake City, the capital, is visited daily by hundreds of "Gentiles," many of whom are settled in the country. The Salt Lake country will speedily lose most of those characteristics which socially have so long distinguished it. Indeed, any account of "the City of the Saints" and the adjoining region, written prior to the epoch of the Pacific Railroad, would be manifestly misleading. Brigham Young is still the high priest of the hierarchy, but the "destroying angel," and other institutions of polygamy, are about as extinct as Bishop Lee, who in 1877 expiated his crimes as the representative of a state of affairs that can never again return. I do not consider it incumbent on me to inflict on the reader the thirty-times told tale of how Joe Smith, the founder of Mormonism, was murdered in Nauvoo, and his followers, led by the present "Prophet," fled to this then remote valley; how they suffered, how they fought, and how they prospered by dint of perseverance, self-denial, and rigid discipline. Neither—not being a German—will my literary conscience be disturbed if I omit all reference to the deep metaphysical principles which are supposed to underlie—but do not—the monstrous falsehood of which Latter Day Sainthood is the representative. The Mormons prospered because they lived in strict obedience to shrewd if unscrupulous leaders, owing to their fortune in being on the line of travel to the Pacific, and above all because they "hung well together." They had a hard task in contending against nature, but they conquered by dint of great industry, directed and stimulated by the knowledge that, as they had fled from civilisation in order to practice unmolested their peculiar tenets, if they did not make a home here they would either die, or, in seeking bread elsewhere, fall into the hands of the Philistines, from whose wrath they had for the time being escaped. They were more moral than similiar communities in the West; for, unlike most of these, their conduct is regulated by some moral rule of life, even though the leading one practically is a social sin. Still, a bigamous Mormon is not like a bigamous Englishman. The one breaks no law of the community he lives under, though the law of the land; the other invariably does so, and, accordingly, the rest of the latter's conduct is usually in keeping with this uxorious weakness. I have, however, repeatedly heard those who know Salt Lake City much better than I know London deny that it is a moral place. True, drunkenness is unknown among the faithful; but conjugal fidelity is not more universal than it is in other polygamous countries. The women live in a state of degradation. Most of them are Welsh, Scotch, English, German, and Scandinavian, and are usually grossly ignorant, and (poor creatures!) homely-faced and dowdy-figured in the extreme. A Mormon usually speaks of his wives as *his women*, though too much ought not to be made of this, because the same phrase is commonly applied to their own and to other men's lawful wives by the rougher of the Western settlers. The male Mormons are intensely greedy after money. Their creed is a purely utilitarian one. I never heard of an American who, in these latter times, at least, joined them from a conviction of the truth of their religious principles; and, considering that it must be a strange creed that will not find converts to it in the United States, the fact—and fact I believe it to be—speaks volumes against the Mormons. Their "converts" are usually Europeans; and even then, the material advantages held out to the land-loving Swede or Welshman has as much to do with the

AN HOTEL IN SALT LAKE CITY, UTAH.

matter as any idea about the truth of the faith compiled by Joe Smith and his successors. So far from considering the Mormon creed as the social and religious system of the best colonists in the world, I agree with those who think that it contains within it elements of decay. They have been successful, not on account of Mormonism, but owing to causes with which their faith had nothing to do. Already it is decaying. The sons of Joe Smith, the founder, have seceded from the main body of the church. Many of the adherents are grumbling at the heavy church-dues they have to pay, while others, finding that, now civilisation has overtaken them, they have nothing to gain, but much to lose socially by belonging to the Mormons, and that they will be protected by the Government, are falling away. Already the Law Courts have—as, indeed they could not do otherwise under the Federal laws—refused to recognise these polygamous "marriages." The wives so "married" are, therefore, no wives, and their children are illegitimate, and, unless under a direct provision made for them, cannot share in their father's property. This will

VIEW OF CLEAR CREEK CAÑON, COLORADO

be the end of it. The lawful children will assert their rights to the estate of their intestate father, the whole system will be broken down, and the wives, finding that they live in an almost womanless country—which, for long, the West must be—and can obtain husbands to themselves, will not be inclined to be subject to their present degradation and disadvantages. On pp. 57, 60, 61, and 64 we have engraved various views of Utah and Salt Lake City, which will serve better than many words to give an idea of the country.

EMIGRANT TRAIN IN COLORADO. (From an Original Sketch.)

NEVADA.

We have said so much about the physical geography of the Great Basin, about mines and miners, and about social life generally on the Pacific, that a very few words will suffice about Nevada, which, though politically separate from California, to which it was once united, is yet socially a part of it. It has an area of 112,090 square miles. Its population was, in 1870, 42,491, of which 38,959 were white, 357 "coloured," and 3,152 Chinese. Agriculturally, we have already sketched it; from this point of view it is of no moment. Mines constitute its only riches, silver and gold being abundant, but quicksilver, lead, and antimony are also found. The chief town is Virginia City (which in 1870 had over 8,000 inhabitants), but the capital is Carson, with a population of 3,042. Its gold and silver yield we have already spoken of.

Montana.

Montana is very correctly named. It is in reality the Rocky Mountain Territory, west of the 111th meridian; its borders, indeed, lie along the crest of that range and the Bitter Root Mountains. Its principal towns are Helena, the capital, 4,266 feet above the sea, and with a population of about 4,000; Virginia City (elevation 5,824 feet), Deer Lodge (at an elevation of 4,654 feet), Argenta (elevation 6,337 feet), Fort Shaw (6,000 feet), Boulder Town (5,000 feet), &c., none of which have 1,000 people. The census of 1870 showed that 20,595 people ought to be credited to the territory. Of these, 183 were "coloured," and 1,949 Chinese. The males exceeded the females by nearly 13,000. Only 2,111 were employed in agriculture, and 6,720 as miners. The tribal Indians are Flatheads, Pend d'Oreilles, Kootenays, Mountain Crows, River Crows, Blackfeet, Bloods (p. 41), Piegan, Santee and Sisseton Sioux, Yanktonais Sioux, Unepapa Sioux, Unepalena Sioux, Assinaboines, Tetons, Gros Ventres, and a few smaller tribes, the whole numbering, in 1875, less than 21,000, most of whom are tolerably peaceful. The eastern portion of the territory—about three-fifths—consists of rolling table-lands or plains; but the west is mountainous. There are various rivers flowing east and west, but the chief is the Missouri, which is navigable to Fort Benton—and indeed at certain seasons to the Great Falls 2,540 miles from its junction with the Mississippi—more than 500 miles from the boundary, and its tributaries also for greater or less distances. The territory is rich in minerals, being, indeed, in its gold yield, only second to California. Silver is also found, while copper-mining will in time, when communication is cheaper, become an important branch of industry. In the year 1874, the bullion product was estimated at 4,000,000 dollars, but this is a falling off from previous years. In fact, there has been a gradual decrease since gold was first mined in the territory in 1862. The climate is healthy. Little rain falls, and irrigation is in general required before crops can be raised. The average temperature is higher than in regions further south. Cattle winter out, so little snow falling that they can generally find food enough to subsist on. The climate is generally too cold for Indian corn, but most of the temperate crops grow well. Buffalo, bear, and antelope abound. In 1870, there were 851 farms, containing 84,674 acres of improved land, but quartz, flour, and saw-mills, breweries, and several manufactories of jewellery formed the chief industrial establishments of the territory. There are as yet no railroads, and the total value of all real and personal property was returned in the census of 1870 at 15,184,522 dollars. Part of the "Wonderland" of America is in Montana, but as we shall have occasion to describe it when speaking of Wyoming, in which the Yellowstone "National Park" is situated, it will be better to defer any account of this extraordinary region until we can describe it in its entirety.

Colorado.

Colorado is in shape almost that of a parallelogram, averaging in length 380 miles, and in breadth 280 miles. In 1870 there were 39,804 white people, and 3,000 tribal Indians (chiefly Utes) in the (then) territory. Rating, however, the voters as one to five of the white residents, the elections in the autumn of 1876 would indicate—unless "repeating" had been carried on in a more wholesale manner than usual—a population of 135,000. There can, however, be no doubt that for some time past Colorado has been looked on as a kind of emigrant's paradise. It is a favourite idea with young Americans, and with middle-aged people whose race in life has been rather tardy, that the proper thing for them to do is to remove to one of the newest States, and "grow up with it." To this practice may be attributed the rapid settlement of Nebraska and Kansas, and of the State under consideration. Its principal town is Denver, the capital (p. 69), which in 1870 contained 4,759 people. There are, however, now 1,000 children alone on the school register in this town, and its settled population—exclusive of visitors, averaging 2,000 to 3,000—may be set down at 25,000. To this may be added a floating population of miners of from 1,500 to 3,000, the gold-diggers of Colorado being no less addicted to wandering than those of other regions. In 1876 there was a hegira of 600 of them to the Black Hills, while hundreds who left a few years ago to try their luck in Utah, California, and other States are now returning. The other principal towns* are Central City, Boulder, Georgetown, Pueblo, and Trinidad, each with from 4,000 to 5,000 inhabitants; and Colorado Springs, Del Norte, Lake City, Greeley, Golden City, and Las Animas, with upwards of 2,000 each, while there are ten or twelve other towns which number about 1,000 each, with hundreds of smaller villages and miners' camps scattered over the entire State. A fair estimate of the present population of the entire State is 150,000. Of this number, about one-half are settled in the large towns and cities; some 25,000 in the mining camps along the mountains, from Sunshine District to San Juan. The other 50,000 are scattered about on the plains, principally engaged in farming, dairying, and cattle and wool growing.

The oldest and greatest industry of Colorado is mining. Since their first discovery in 1858, the gold and silver mines have produced upwards of 60,000,000 dollars, an average of over 3,000,000 dollars per year. Of late years there has been a large increase in the yield. That for 1874 was 5,362,000 dollars; for 1875, nearly 6,350,000 dollars; and for 1876, about 7,000,000 dollars; and it gives every promise of becoming larger still. Indeed, we think it safe to say that the time is not far off when California will be the only conspicuous rival of Colorado. Fortunes do not often suddenly thrust themselves upon people in this State, but there are some mines whose product runs as high as 350,000 dollars, a number that yield 100,000 dollars, and many from 50,000 dollars to 75,000 dollars per annum. It is chiefly quartz that is mined, placer diggings not being so widely spread or so generally profitable as in some of the States further west (Vol. I., p. 247). Next to the mines, the produce of the farms, or, as they are called, the "ranches," claims attention. At the outset, it may be well to say that the farmers manage to get through the season year after year without any scientific or laboured system of irrigation. The farms generally lie along the valleys, near streams, and a large canal is built to carry the water along the upper part of the farm, from which small lateral ditches bring it over all parts of the land. The water is shut off when not needed by means of gates. In

* For most of the statistics which follow I am indebted to the Denver correspondent of the *New York Times*, whose statements I have taken pains to satisfy myself were correct (1876).

many seasons there is rain enough for the crops to do without irrigation, but generally from the middle of June until harvest is the "dry spell," when they need attention. The cereals most abundantly produced are wheat, potatoes, barley, and all kinds of vegetables. Indian corn is not grown to any extent. All other kinds of grain and vegetables are produced in sufficient quantity to more than supply all the demands of consumption in Colorado. During the last few years, as in many other of the Western States, there has been great loss from the ravages of grasshoppers, notwithstanding which enough was

A PACK TRAIN READY FOR LOADING. (From an Original Sketch.)

raised to supply all the home markets. Wheat is the great crop. In 1876 over 1,000,000 bushels were produced. The total agricultural product for that year is estimated to have been worth 6,000,000 dollars, being double that for 1870, and had it not been for the spoliations of grasshoppers, the total amount for 1876 would have been nearer 9,000,000 dollars, thus exceeding the value of the mining product.

None of the industries of Colorado has developed faster or more profitably than cattle raising. There are at the present time about 625,000 head of cattle grazing in the State, the value of which at the assessed rate is very near 10,000,000 dollars. The natural increase of herds per year is about 50 per cent. That would make the number in 1877 about 750,000, and as 125,000 represents about the number marketed and driven east, the figures given, 625,000, as the number now in the State, are presumed to be about correct.

A STREET IN DENVER, COLORADO.

Wool growing comes next in importance. The assessment of 1876 showed 424,977 sheep, valued at 810,213 dollars. The increase in this industry is very marked, and estimates completed by the Rocky Mountain Wool Growers' Association places the number of sheep at present in Colorado at 800,000, value 1,500,000 dollars. The number marketed in 1876 was about 150,000 head. The wool clip reached 2,500,000 lbs., worth about 600,000 dollars.

Copper, iron, and other metals are found; and as coal has been discovered, they may be expected in time to be worked to a great extent. Already the Colorado coal is used by all the railways in the State. In 1876 about 300,000 tons, worth 1,500,000 dollars, were mined.

The climate is good and equable, the winters mild, and the summers cool and bracing. Sultry nights are unknown; while the beautiful scenery and suitable climate have made the territory a favourite resort of invalids. About one-third of the territory is suited for agriculture. In 1870 the number of improved acres was 85,594, a mere moiety in a State comprising within its bounds 106,500 square miles, or, 68,160,000 acres. Game birds — such as the wild turkey, mountain grouse, sage hen, prairie chicken, ducks, geese, swans, ptarmigan, &c., abound. The common mammals and other animals of the Western States are also frequent; but, perhaps, the most notorious of the Colorado fauna is the potato beetle (*Doryphora decemlineata*, p. 72), which feeds naturally on a wild species of *Solanum*, but since the country has been settled up it has attacked cultivated plants, and more particularly the stalks and leaves of the potato. Since 1859 it has travelled east 1,500 miles—in a straight line—carrying devastation wherever it goes. It has for some time past been plentiful on the Atlantic seaboard: it has also been seen in Germany; and in due time it may make its appearance in this country.

There are various other industries and manufacturing enterprises of large and increasing importance. The pine woods furnish material for about fifty lumber mills, though the once abundant forests are now greatly thinned by the reckless destruction to which they have been subjected. The wool product has suggested woollen mills, and two or three of large capacity are in operation. Tanneries have been started up to utilise the hides. Many classes of business growing out of the chief industries give employment to capital and labour. Stamp mills, smelting, reduction, and sampling works, ore dealers, assay offices, miners' supply stores, and the like, become the necessary outgrowth of our mines, while live stock markets and slaughter-houses for preparing dressed beef to ship to Eastern markets, are the natural attendants of the Colorado cattle interest.

Commercial facilities are greater than were enjoyed by any other new State at the time of admission. There were, in 1876, eight railways, working 919 miles of road, and there are eighty-two miles now under construction, which, when completed, will give the State over 1,000 miles of railroad. The emigrant's wagon (p. 65) is year after year getting a less familiar object, and the pack-train* (p. 68), now such a common sight all over the West, will in time be replaced by a more expeditious mode of transit for merchandise. There are

* That is—a train of mules or horses, fitted with pack-saddles, and employed in carrying goods to remote localities still unconnected with civilisation by means of wagon-roads.

five express companies (p. 16), reaching all the principal towns, stage lines to most of the remote hamlets and mining camps, and daily lines to Santa Fé, with connections to El Paso, Albuquerque, and Tucson, in Arizona. The Western Union and Atlantic and Pacific Telegraph lines have sixty-nine offices in Colorado. There are nearly 300 post-offices, twenty-one of which are money-order offices. There are eleven national, twenty-five private, and three savings banks, having a paid-in capital of over 2,000,000 dollars.

Educational and religious interests are well cared for. There is an excellent public school system; the number of school districts being 375; the number of school-houses, 225; and value of school property, 550,000 dollars. There are besides a State university, agricultural, mining, scientific, and theological schools, seminaries, and academies. For the dissemination of news, there are forty-six newspapers and periodicals. All religious denominations are represented. The Methodist Conference have thirty-five churches and societies supplied with regular preaching; the Methodist Episcopal Church, South, ten; while the Presbyterian, Congregational, Episcopal, Catholic, and other societies are represented in all the larger towns.

The value of property, real and personal, in the State is about 75,000,000 dollars. State, county, and municipal taxes of all kinds in the larger towns vary from six to twenty mills on the dollar, and in the country and mining districts much less.

"With the mines turning out a constantly increased product under a steadily reduced expense of operation; the farms yielding abundant returns; cattle and sheep multiplying rapidly, and raised at a much less cost than in any other Western State; new business enterprises starting up day by day; trade in all branches in good condition, and capital coming in freely for investment in mines, stock growing, farming, and merchandise, and an excellent climate—Colorado seems destined, in population and industries, to multiply and increase."

All this it may be remembered is in a State comparatively new, even dating from the commencement of its territorial probation, and where the first gold was discovered not twenty years ago.

The State may be naturally divided into three divisions—the mountain region, including the "park" system, the foot-hills, and the plains. The park system is the most remarkable feature of Colorado, and we shall accordingly confine ourselves to describing this point in its orography. These parks are in reality basins or depressions surrounded with high mountains, their elevation varying from 7,000 to 8,000 feet above the sea, and being well watered and wooded. The climate is throughout most of the year delightful, and always healthy. The only drawbacks are the violent storms of wind, and, in some parts of the country, heavy hail showers. The humidity seems on the increase since the settlement of the State, a fact which we have also noted about Utah. Streams which formerly dried up in the summer now flow all the year round, while the volume of others has for late years doubled. All the vegetable products of the North grow here; and, owing to the rich luxuriance of the grass, they will in time become the great grazing localities of the Western United States. San Luis Park, drained by the Rio Grande del Norte, is the most southern of them. It is also the lowest and largest of them all—7,000 to 8,000 feet—and having been long settled by a population of 8,000 to 10,000, chiefly Mexicans, is the most cultivated of all of these upland valleys. It is, in fact, a great amphitheatre of 9,400 square miles, with a surface as flat as a lake, showing every sign of it having been once the bed of an inland sea. South Park—the Valla Salada of the Spaniards—is the next in order going north. It gives rise to the South or Main Platte, which flows to the north-east, and then eastward to the Missouri. The park is twenty to forty miles wide, and sixty to seventy long—in fine, a vast meadow at the height of 8,000 to 10,000 feet above the sea, which supports thousands of cattle. The mountains surrounding abound in gold and silver, and rich deposits of gold are worked in some places. Middle Park (averaging 7,500 feet above the sea) is drained by the Grand River to the west, and thence by the Colorado to the Gulf of California. Its outlet by the

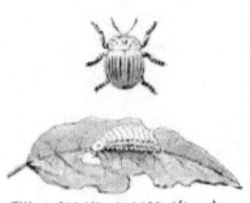

THE COLORADO BEETLE (*Doryphora decemlineata*).

Grand River is through a magnificent defile or cañon. It is as yet unsettled, but in time will become a favourite summer resort for those who do not dread temporary isolation. The North Park is near the northern boundary of the territory. It gives rise to the North Platte, which, after flowing north and east, joins the Missouri. It is heavily timbered, and of an average elevation of 8,000 feet. It is circular in shape, and twenty or thirty miles in diameter. As yet it is little resorted to, and has no settlers, though its scenery is fine. Animas Park, a later discovery, is about 7,000 feet above the sea. These parks are separated one from the other by narrow but lofty ranges of mountains. They all abound in fish and game, and present very varied and romantic landscapes.* "The colour of the landscape" is, to use the words of Sir Charles Dilke, "in summer green and flowers, in fall time yellow and flowers, but flowers ever."

DAKOTA.

Dakota comprises an area of 150,932 square miles, but Yankton, the capital, 1,000 feet above the sea, and containing, in 1870, 737 people, is the only town of any consequence. The whole population of the territory at that date was 14,181, which included 94 "coloured" persons, 1,200 non-tribal Indians, and 4,815 people of foreign birth, chiefly Norwegians, Canadians, Irish, Germans, and Swedes. There are in addition about 29,000 tribal Indians belonging to the Sioux and other septs found in the neighbouring territory, gathered on reservations. Most of the white population are settled in the south-eastern portion of the country along the Missouri River, and are chiefly engaged in agriculture. The country east and north of the Missouri is principally undulating prairie, without "swamp, marsh, or slough," but traversed by many streams, and dotted with endless lakes. Along the eastern border extends for 200 miles from the south, fifteen to twenty miles breadth, an elevated plateau (1,800 to 2,000 feet above the sea), called the Plateau du Coteau du Missouri. Open grassy

* See Hayden's "Geographical and Geological Survey of the Territories" (1874); Hollister: "Silver Mines of Colorado" (1867); Bowles: "Parks and Mountains of Colorado" (1868); Blackmore: "Colorado" (1869); Greeley: "Summer Etchings of Colorado" (1874); Porter & Coulter: "Synopsis of the Flora of Colorado" (1874), &c.

plains form the basin of the Red River, while in the south-west, extending into Wyoming, are the Black Hills and Mauvaises Terres, or Bad Lands, full of the remains of extinct animals, the discovery of which in late years has added so much to our knowledge of the ancient fauna of North America. The Black Hills contain gold, silver, iron, coal, lead, salt, and petroleum; and, most probably, will in time take a high rank among the mining regions of the United States. There is, as in all the prairie region, comparatively little rain; and accordingly the climate is dry and pure, and consumption an almost unknown disease. Grazing is extensively followed, while the vast herds of buffalo, elk, deer, and antelope ranging over its western portion gives "the land of the Dacotahs" a reputation among hunters, as its black bear, wolverine, musk-rat, otter, mink, marten, and wolf do among the North-Western trappers. In 1870, the number of improved acres was 42,645, and the entire value of the real and personal property in the territory 5,590,752 dollars. Dakota has, moreover, the distinction of having no public debt. Much of the territory west and south of the Missouri was "unorganised," but in 1877 a new territory called Huron was projected out of this wild land.*

A DAKOTA OR SIOUX INDIAN.

* In reference to the nomenclature of the territory, a writer in the *New York Nation* remarks "that there is not much to be said against the name of 'Huron' for a new territory, though its selection is but a questionable 'tribute to the aborigines;' but I trust that the derivation of Pembina from *pain bni* will not make its way from the Senate Chamber to school-books and gazetteers. The name is perfectly harmless. It comes from the Chippeway and Cree name for the high-bush cranberry, *nipiminu* (and *neheminh*, literally 'water-berries.' The French of Canada corrupted this to Pemine, and the voyageurs and trappers made it *Pmbina*, and gave it to two or three streams and lakes where they found the berries plentiful.

A great deal of ingenuity has been wasted in inventing French etymologies for Indian names. The northern Indians use the dry leaves of the bear-berry (*Arctostaphylos Uva-ursi*), which the Chippeways call *sagaccomi—*i.e. 'ground-berry'—to mix with their tobacco for smoking. Sir John Richardson was made to believe that 'on account of the' Hudson's Bay officers carrying it in bags for a like use, the voyageurs gave it the appellation of *sacca-commis*. A curious etymology of this sort is the one which has been found for the River Quinechien, a point at which it joins the Ottawa River, near the Long Saut. The name is Alonkin, and properly belongs to the 'long rapids'— *Quinechouan*, Champlain wrote it; but the Abbé Ferland informs us that it is said to have been given in memory of the first company of settlers at this point, fifteen Frenchmen, who, because they were always quarrelling, were called by their countrymen and neighbours, '*les quinze chiens*.'

The French have, on the whole, dealt more mercifully than the English with Indian names. They have mutilated, but without mangling. A Frenchman would not have resolved Cupping—which means 'a harbour'— into 'Cape Poge,' or Potopaco into 'Port Tobacco.' 'A tale of love in Acadie, home of the happy,' flows more

CHAPTER IV.

THE UNITED STATES: THE PRAIRIES WEST AND EAST OF THE ROCKY MOUNTAINS.

DAKOTA is essentially a Prairie State. The great grassy plains called prairies—from the French word for a meadow—are in America especially the treeless fertile regions lying between Ohio and Michigan on the east and the arid desert on the west (Vol. I., p. 264). The western part of Ohio, nearly the whole of Indiana, Illinois, and Iowa, the southern part of Michigan, the northern part of Missouri, and portions of Wisconsin, Kansas, Nebraska, and Dakota are prairie regions, though to a less extent they are found in all the States immediately to the east of the Rocky Mountains beyond the reach of the streams flowing from their melting snow, and to a smaller extent to the west of that range. The altitude of the prairies above the sea differs. At Prairie du Chien in Wisconsin the elevation is 100 feet above the Mississippi; about the centre of the State of Illinois the average height is 650 to 750 feet, while near the northern border of the States some of the most elevated portions of the prairies are about 1,000 feet above tide water. In Iowa the *Plateau du coteau des prairies*, which divides the waters of the Mississippi from those of the Missouri, is about 1,400 or 1,500 feet above the sea. Usually the prairies are more or less undulating, though occasionally they are level and smooth. The first kind are called rolling prairies, the second flat prairies. Among the plants of these grassy plains compositæ, or the daisy and dandelion order, and especially sunflowers and their allies, are the most frequent and numerous in species and individuals.* On the borders of streams, or in other exceptionally moist localities, they are found, but west of the Mississippi they become rare, and near the longitude of 98° all but disappear. The prairies are among the richest soils of America. They are usually free from stones, and though in the rich "swales" and "bottom lands" the vegetable mould is very deep, yet the usual thickness on the Upper Prairies is from one to two feet. The sub-soil is almost invariably a clayey loam, in its lower parts mixed with sand, and occasionally pebbles. Between the 99th and 104th meridians are treeless plains, differing from the prairies in being dry and partly desert,

---

smoothly than would be possible for any tale of love in 'Quoddy.' The French softened one name, and the English roughened the other out of the same two syllables of the Abnaki original.

As for *Huron*, there is no evidence that the name is Indian, even by derivation. Father Lalemant, in the "Relation" for 1639, and Charlevoix, assure us that it is French. The Ouendats, as the Hurons called themselves—the name is not written Wyandot—had strange fashions of wearing their hair; some cut it short on one side and let it hang in long locks on the other, others trimmed it to a bristling crest, from forehead to crown. When the French saw them for the first time, some soldier or sailor, as the story goes, cried out, *'Quelles hures!'*— 'what beastly heads!'—and the savages soon got the nick name of Hurons. I am not aware that the truth of this story has been questioned, but I confess that I do not believe a word of it. It is too much like the *quim chicu* and *pain bou*. *Ron* or *ronon*, as a termination means 'people' or 'nation;' and *Houeron* or *Houronon* may be only another form of *Ouendat*, or [as Segard gives it] *Houendate*—in which *endate* probably means 'towns' or 'villages.' The name, however, is of questionable origin, at best. The genuine one would be unexceptionable. Why not Wyandot or Wyandots?"

* Asa Gray: "American Journal of Science," Second Series, Vol. XXIII., p. 397 (1857).

although usually capable of supporting crops if irrigated. Some Western men affect to apply the name of "plains" to these tracts, covered with the short curly buffalo grass, reserving the word prairie for the undulating regions covered with tall luxuriant grasses we have already described, though the distinction is not strictly kept up, or, indeed, possible in every case. The buffalo grasses change colour early in the summer. Hence the "plains" look yellowish-green long before the autumn comes, while the rolling hills and valleys of the "prairies" are still fresh and green.

THE PRAIRIES WEST OF THE ROCKY MOUNTAINS.

The "mound" prairies are found near Puget Sound and in other portions of the North Pacific Coast, particularly on the Nisqually Plains—grassy openings in appearance not unlike English parks—(Vol. I., p. 309). These plains are ranged among the dry prairies—in contradistinction to the wet ones—of the region to the west of the mountains. Among these may be classed the pleasant little prairies on Whidby's Island, off the coast of Washington Territory, the Willamette Prairie, Oregon, the Squak Prairie near Seattle, the small Cowichan and Comox Prairies on Vancouver Island, &c. They are generally found in the vicinity of rivers, and their soil seems to have been deposited from the waters of the neighbouring streams; though they are now elevated more than 100 feet above the bed of the river. They are composed of a thick black loam one to three feet deep, almost entirely vegetable in composition. The soil is sometimes too rich for grain, though that of other prairies near the shore is poorer and sandier. Often these prairies are scattered with oaks, chiefly *Quercus Garryana*, and a few pines, which give a peculiarly pleasant home-like aspect to these openings in the gloomy forest around. They are often—as in the case of the Nisqually Plains—dotted here and there with little lakes swarming with the fresh water turtle,* surrounded with pleasant groves of maple,† pine,‡ oak, and occasionally the Oregon ash. Scattered over these prairies are also little hillocks covered with trees—looking, at a distance, like wooded islands in a grassy sea. These mounds, I am inclined to believe, are not dissimilar in formation to the much more marked "mound prairies" between Olympia and Monticello, on the Cowlitz portage in Washington Territory. The form of these mounds is very distinctly circular, and average, from a mere swell in the prairie, to a height of six to eight feet. They are composed of gravel, apparently brought together by the action of water; but the mounds never coalesce, though often in close proximity to each other. Amid the endless theories which are advanced by local *virtuosi* and other more scientific visitors, I can find none which will at all reasonably explain these remarkable natural mounds, though they appear not to be peculiar to this prairie, but are also found in Texas. I am inclined to hazard an opinion that their origin is due to the same causes—whatever these may have been—which went to form the kames, escars, or gravel mounds generally associated with the glacial epoch, and found scattered over many portions

* *Actinemys marmorata* (Agassiz).
† *Acer macrophyllum* (Pursh).
‡ *Pinus ponderosa* (Douglas).

of the British Islands and the north of Europe. Professor Joseph Le Conte considers them due to surface erosion under peculiar circumstances.*

Very frequently these dry prairies are covered with a thick crop of fern (*Pteris aquilina*), which is a great nuisance to the farmers, who resort to liming and other means to rid the soil of it. I am told by some experienced Western farmers that if this fern be continually cut down it will in a very short time disappear, the sap being drained off by this constant bleeding—the result of which is that the bracken very soon dies of exhaustion. In considering the origin of these prairies, we are struck

STEAMER ON THE LOWER MISSOURI.

by the abrupt termination of forest surrounding them. They seem like oases in a forest desert. At one time these prairies must have been much wider than now, and, indeed, there are evident signs that they are only remains of much more extensive grassy plains, which once extended over the valleys, until the forest encroached on them. The climate must have been at that period very different from what it is now; but everything tells us that there has been some wonderful revolution in the climate of North-West America since first it took its present contour. The prairies by the side of rivers, such as the Comox, and even the Willamette Prairies, were probably at one time of the nature of meadows, overflowed by the rivers when the volume of water in them was much more extensive than now; and it seems not unreasonable to suppose that the absence of trees on them is due to the sapping by water to which they were then subject, if, indeed, there were not continuous swamps; for

---

* For a more particular account of these extraordinary mounds on the "mound prairies," I may refer the reader to Gibbs: "Pacific Railroad Report," Vol. I., pp. 469 and 486; Cooper: "Natural History of Washington Territory," p. 18, &c.

swamps will, in due course, by the formation of soil, and by the decay of their own vegetation, become dry prairies—a fact I have more than once observed—or by some of the convulsions so common in a volcanic region like North-West America, they may have been suddenly drained of their superfluous moisture and become dry land. At all events, we know that the forest is steadily encroaching on the present prairies. Old settlers have pointed out to

A FLGID INDIAN.

me, on the Willamette Prairies, places on which thirty years ago they grazed their cattle, though now they are covered with a dense thicket of firs. Again, on the Nisqually Plains, is a place where some officers of Her Majesty's ships were buried about twenty-five years ago. It was then open prairie. When I visited the spot it was with difficulty that we could find our way through the bush and trees which covered the ground. The Indians tell the same tale—that the forest in Washington Territory and the neighbouring country is encroaching on the prairies. These prairies support a luxuriant herbaceous vegetation; probably one-half of the flowering plants found to the west of the Cascade range grow on them. Among these may be classed various species of buttercup, *Potentilla*, *Collomia*, *Collinsia*, *Dodecatheon*, *Fritillaria*, *Cypripedium*, *Aquilegia*, &c., most of which have been introduced into England by Douglas, Jeffrey, Burke, and the author, and are getting to be common garden flowers. A very characteristic plant of some parts of these prairies is the tall yellow *Oenothera biennis*, now introduced into our English gardens. Other common plants are *Delphinium azureum*, *Gaillardia*, *Lupinus*, *Sanicula Menziesii*, *Erigeron*, *Linaria Canadensis*, wild strawberries, &c. Finally everywhere, from May to June, the prairies are covered with the beautiful blue flowers of the gamass lily (*Camassia esculenta*), which are among the most charming and characteristic flowers of North-West America. Several species of mammals are closely confined to the dry prairies or their borders. Among these are the gopher, meadow mice, Oregon mole, &c.; and in a lesser degree the Sewellel (*Aplodontia leporina*), one of the most extraordinary of Western mammals,* the Western vole, the prairie mouse, &c. Very few birds are peculiar to prairies, though some of the smaller species, shunning dense forests, frequent their borders. The shore lark and the Savannah sparrows are, perhaps, the only land birds never seen in the Far Western woods, while some water-birds frequent the marshy portions, along with the brown crane and the Canada goose, which are rarely or never seen along the sea shore.

Wet prairies are on the Pacific Coast generally formed around the mouths of various tidal rivers, such as the Fraser. They are only overflowed by the tide at its highest periods two or three times annually. The soil is rich, and produces good vegetables, but in the summer season they are so infested by mosquitoes as to be almost uninhabitable. They produce a coarse grass, which is cut when in the vicinity of a market, and sold under the name of swamp hay. The extensive flats on either side of the Fraser mouth are accordingly unused, except for this purpose, though there can be but little doubt that eventually they will be regarded as valuable agricultural lands, and proper means taken to embank them. In Nova Scotia, similar tracts form the richest agricultural lands of the province.†

Sometimes the tide prairies get covered with bush, as at the mouth of the Columbia and Chehalis. At other times they are dotted here and there with clumps of trees, like islands in a grassy sea. Near Fort Vancouver, on the Columbia, there are meadows annually overflowed in some period from June to August, by the rising of the river, but it is found that if the seed is put in after the flood has subsided it prospers very well.‡ The absence of trees is no doubt due to inundation, or to the icy flood from the mountains which saps the ground. There are often found small prairies about the head waters of streams, particularly in hollows in the mountains. They are generally marshy, from the continual welling up of springs, and are covered with a coarse grass, or, where drier, with bushes. These little oases among the wooded mountains are particularly pleasant to the traveller, and even more so to his horse, as they generally form the only grassy stops on his route—that is, if it be one by which horses can be taken, a rarity in this region. Of this nature are the cranberry marshes along the coast, which curiously produce a group

* For a list see Lyall: "Journal of the Linnean Society," Vol. VII., p. 135 (1863).
† Dawson: "Acadian Geology," p. 16.
‡ Cooper: "Natural History of Washington Territory" Botany', p. 18.

of plants, many of which are identical with the Alpine Flora, at 5,000 feet elevation on the mountains, and are composed of species found in the northern portion of the American and other continents, forming exceptions, brought about by peculiar physical circumstances, to the ordinary geographical distribution of plants in North-West America. In the smaller rivers and lakes are also found similar cosmopolites, but more truly aquatic, such as *Scirpus lacustris*, *Typha latifolia*, *Polygonum amphibium*, &c., almost universally in company with the characteristic "skunk cabbage" of the Western swamps *Symplocarpus kamtschaticus*).

The wet prairies also support a group of plants almost peculiar to them. On the Sumass River Prairies, off Fraser River, the North-West Boundary Commission pastured their horses, but so terribly were they tortured by mosquitoes, that the animals, maddened with pain, would sometimes spring into the river and get drowned. Even the tough skin of the Indians do not escape with impunity. It is almost impossible to engage them to paddle up the sloughs leading into these swamp prairies in July, August, or September, except at very high wages. Some friends of ours, engaged in surveying these places, had to give three dollars, or 12s. 6d. per diem, with food, to their Indian canoemen, and even then they bargained to be supplied with mosquito bars. Such is the torture inflicted by these pests, that I do not wonder at the Indian mythology ascribing mosquitoes to have originated from the ashes of the *Slalacum culcultith*, or wicked witches, who haunted those parts—

> "In days no more remembered,
> When the heavens were closer to us,
> And the gods were more familiar."

and whose evil deeds I have recorded elsewhere.*

I doubt not, however, that as settlements and cultivation extend, the mosquitoes will disappear, smoke causing their destruction. When first I visited New Westminster, a little town on Fraser River (great on *paper*,† however, and the then capital of British Columbia), sleep was scarcely possible for mosquitoes, and the faces of the people would have led the stranger to suppose that an epidemic of small-pox was desolating the town. When I last visited it, the smoke had quite driven them off.

THE PRAIRIES EAST OF THE ROCKY MOUNTAINS.

The reader must pardon this long digression regarding a part of the country we have already travelled over, and which is many hundreds of miles from Dakota, the proper subject of this section. It is, however, better to consider the interesting and important question of prairies as a whole, more especially as these most remarkable features in North American geography are very imperfectly understood, and are sometimes classed as due to the same causes, while in reality those of the West and those of the East have really nothing in common, except that all of them are more or less treeless. On the

---

* "Races of Mankind." Vol. I., p. 139.

† At that period the Municipal Council was advertising for tenders to "cut down the *standing timber* upon Argyle Crescent, Lytton Square," and so on, perfectly unconscious of any joke.

Pacific coast, circumstances no longer in force have been shown to have formed the prairies—now in the changed character of climate getting rapidly encroached on by the forests. The true prairies—which we have already noticed—are on the eastern side of the Rocky Mountains. The position and extent of these great plains we have already indicated (p. 74), by stating that the prairie country is comprised, speaking roundly, within the meridian line on the western side of Louisiana, the boundaries of Arkansas, Missouri, and Iowa, for their eastern limit; the Rocky Mountain crest for the western, with Texas at the south, and the "Barren Grounds" on the north, embracing a longitudinal parallelogram of somewhat less than 1,000 miles in width. They have a gentle slope from the west to the east, intersected by rivers running into the Missouri, Mississippi, and St. Lawrence, and into the sea on the Texan coast. They are of homogeneous formation, slightly undulating and continuous, without timbered spaces or lakes. The soil, though compact, is of a fine calcareous mould, producing an abundance of herbage peculiarly adapted to the climate. During a temporary prevalence of moist atmosphere in the spring, the delicate "gramma"* and "buffalo grasses" flourish, and are converted into hay upon the ground, by the gradually returning drought. It is upon this longitudinal belt of perennial pasture that the buffalo finds his winter food; and here, also, are found vast herds of wild horses, the elk, the antelope, and numerous other animals.† These plains are not the same throughout. Towards the north, chiefly within the British possessions, and in the influence of the Saskatchewan, is found the celebrated belt of fertile land.‡ Further to the south, the absence of rain has caused the country to be very barren, and covered with sage brush (*Artemesia*). The country in the vicinity of Salt Lake City, and much of that now gone over by the Pacific Railroad, is of this nature. However, this soil is not irretrievably barren. Indeed, that quite the contrary is the case, is evinced by the magnificent crops raised by irrigation in Utah. Luscious peaches and other fruit are there growing in abundance; but look over the fence, and outside is the dreary sage brush. When first the Mormons came into this valley, they found an old trapper—Jim Bridger—then trading among the Indians. He laughed to scorn the notion of ever raising wheat there, and even went so far as to say that he would give a hundred dollars for the first ear he saw grown in the dismal-looking desert around. Irrigation, has, however, accomplished wonders, and Mormondom, in addition to feeding a population of over 120,000, supplies Idaho and the neighbouring gold mining districts with flour from wheat grown in the valley. Further to the south, the breezes from the Mexican Gulf bring sufficient moisture to render the Texan prairies fruitful. This leads us to the inevitable conclusion that the origin of these prairies and plains is due to the absence of the moisture necessary for the growth of trees. The central region is comparatively dry, and consequently treeless, except near the mountains, which act as condensers, and precipitate an amount of rain sufficient to sustain a forest growth. The rainfall is greater in that part of the continent east of the Mississippi, and hence we

* This name is applied to several species of grass. In Arizona, for instance, to *Aristida purpurea* (Nutt.); *Muhlenbergia pungens* (Thunb.); *Pleuraphis Jamesii* (Torr.); as well as to *Bouteloua oligostachya* (p. 47).

† Gilpin and Brown: "U. S. Patent Office Report."—Agriculture, 1857, p. 294 (with map. Plate vi.)

‡ Mullans: "Miners' and Travellers' Guide to Oregon," &c. (1865) p. 71; as well as to Palliser: "Red River Explorations" (Blue Book).

VIEW IN THE VALLEY OF THE UPPER MISSISSIPPI

51

have an almost unbroken forest along the line where treeless and forest districts meet. Local causes determine the presence or absence of trees. Belts of timber border the streams and cover the more porous and absorbent soils, while level surfaces, with a firm and unporous soil, sometimes very wet and sometimes very dry, sustain only a growth of grass which can endure the alternations fatal to trees. Annual fires have their influence in extending the area of grassy surface; and over much of this middle ground the causes limiting the growth of trees could be removed, and their forest area extended.* The forces of nature are here so nicely balanced, that slight causes would make one or other preponderate. I think it is now almost universally agreed among those who have thought over the subject, that the many theories which attribute prairies to any other cause than the want of water are erroneous, or only of local value. On the great prairies west of the Mississippi, every variety of soil and surface fails to sustain trees, and there only a change of climatic conditions will convert the grass-covered surface to forest. We must, however, separate the operations which originally made these prairies from those which keep them in the condition of treeless tracts. With the first we have nothing to do: it might have been a geological revolution, or it might not. We only know that the absence of trees now is due to the causes we have explained. Heat alone has little influence in the growth of trees. Take, for instance, the rainy tropics, and the bare plains or deserts of an equally tropical country. The one is covered with dense forest, the other is treeless.

On the other hand, the country in the vicinity of the great lakes is, though cold, densely wooded, almost to the limits of perpetual frost; while the Steppes of Russia and Asia, though under similar climatic conditions, have little or no wood, owing to their being supplied with little moisture. The equal distribution of rains is also important. It should also be remembered that the absence of rains, necessary for forest growth, does not interfere with ordinary crops. In fact, most crops will succeed better with less rain than is necessary for most trees to thrive, and in some years there is even a greater supply of rain in the Texas and Illinois regions than eastward. A consideration of the source of rains will explain why prairies have their present limits. Coming north from the Gulf of Mexico, the moisture-bearing clouds are carried more and more eastward by the westerly winds, and as the greater part of the moisture is precipitated before reaching the Ohio River, the Illinois region is deprived for many years of its due share of rains. The Texan region, lying considerably west of the line of tract of these gulf storms, has to depend on less abundant sources of rain. Now, as we go westward, the supply rapidly diminishes, until in the prairie country proper it is entirely inadequate to the growth of trees, as well as of many other cultivated products, and in some cases even grain and other herbage entirely disappear over vast tracts. From the great bend of the Missouri northward, however, there seems to be an improvement in the country. On the banks of that river, above Fort Union, there is no long interval without trees, as there is further south on nearly all the streams; and on the Saskatchewan† there is even less.

* In Iowa, for instance, it is calculated that every three years not less than 5,000,000 trees are planted, and that within considerably less than half a dozen years the forest area has been increased by 25,000,000 trees.

† For which see, inter alia (Vol. I., p. 259). Lord Southesk's "Saskatchewan" (1875), and Hind's "Canadian Red River Exploring Expedition" (1858).

The nature of the soil and of the underlying rocks assists much in the aridity of the country; and we, therefore, find that the line marking the junction of the carboniferous rocks of the Illinois region with the cretaceous and tertiary is a distinct limitation of many trees, as, indeed, rocks are elsewhere. Much more could be said on the same subject, but for our purposes it is sufficient to say that though different trees may require, individually, different degrees of moisture, yet, at least, sixteen inches of annual rainfall *during the growing season* are essential to the prosperity of a forest region. These facts are exemplified in the regions between the Rocky Mountains and the Cascades. The eastern slope of the Rocky Mountains, moistened by the rains condensed from the moist breezes impinging on the lofty peaks, and also by the melting of their snows, is covered with trees. The same is true of the western slope, and here a vegetation not unlike that of the Western Cascade Mountains makes its appearance, similar physical circumstances leading to similar organic products. As we go further west, out of the influence of the Rocky Mountains, the country gets more arid, until, at length, a region is reached out of the influence of either the moisture of the Rocky or the Cascade Mountains. This causes the great desert between these two ranges. When we cross the Cascades we come into a region where the moist breezes of the Pacific reach. Accordingly, we find a dense forest stretching down to the shores of the great ocean, open lands being the exception, and prairies only occurring of very limited extent, and under local conditions, with which the absence or presence of rain has little or nothing to do. But this is bringing us round to the point from whence we set out; and having thus arrived at the starting point we may dismiss the vexed question of the origin of prairies, the most reasonable explanation of which I have adopted, though so unfortunate as to have a geologist of Professor Whitney's eminence in the opposite camp.* Above all, I consider that the theory of Professor Hall, which attributes the treeless character to the finely comminuted condition of the underlying strata, especially the shales and limestones, is untenable. It is too sweeping to supply even a plausible explanation applicable in every case. In time, I believe, some parts of the Western prairies may again be forested. Settlers are beginning to plant trees around their houses, and along the banks of the streams. Trees—it is a very familiar bit of knowledge—greatly affect rainfall. In Asia Minor, Greece, Spain, Portugal, Malta, India, parts of California, Australia, Switzerland, Austria, South Africa—indeed, in nearly every country—the same facts have been noticed. When forests are cut down the rain speedily runs over the surface, causing floods for a time, and aridity for long afterwards. Hence it is looked upon as excellent State policy to at once replant such deforested tracts, in order to increase and economise the showers which fall. This has been done in various countries with the best effect.† In California it is always noticed that

---

* See his papers—now reprinted—in the *American Naturalist*, 1876. The more general view is also and illustrated by Dr. Cooper in "Smithsonian Report," 1858, pp. 246–279, with corrections in Report for 1860, p. 438; Newberry: "Transaction of the American Association for the Promotion of Science" (Buffalo Meeting), 1864; Forster: "Mississippi Valley" (1869), &c.

† Markham: "Journal of the Royal Geographical Society," Vol. XXXVI, p. 180; Clegharn and Eilet: "Proceedings of the Royal Geographical Society," Vol. XIII, pp. 75–82. Bolander: "Proceedings of the

the fogs driving inland from the Pacific invariably turn to rain when they come in contact with the great red wood forests on the coast. Hence, springs in or near the red woods are never in want of a good supply of water, and owing to the same cause crops on the Coast range are not liable to fail. The destruction of forests in North-West America is controlled by no law. In Washington Territory, and some other places, the State exacts a small royalty in the shape of "stumpage," but beyond this there are no regulations affecting the destruction of timber. Indeed, a tree is looked upon as a natural enemy—the sooner destroyed the better. A man in the woods, in order to provide a "backlog" for his camp fire, will fell—without the slightest compunction—a noble pine, which would be the pride of an English forest, and the money value of which would eventually be great.

Prairie and Forest: Their Sanitary Aspects.

The Western man, with all his hard-headed shrewdness in most affairs of life, is but a "pig-headed" sort of fellow in some other things. In the matter of a dwelling, and the selection of a site for it, he has no medium. He either goes into one extreme or another, as circumstances lead him or gain tempts him. If he fixes his residence in a wooded country, his whole object in life is to slash and hew at the timber, until he has effected a clearing. Be the climate ever so warm, he cares nothing for that. No shady trees cast their cooling influences over his picturesque log cabin. The very sight of a tree seems an eyesore to him, and that handy little American axe of his—so different from the cumbrous, absurd, Flemish tool which we so persistently use in this country—soon helps him to remove the object of his dislike. Fire, the augur, and blasts of gunpowder, assist, and soon he has what he delights to style "a right smart chance uv a clearin'." The maxim of the Laird of Dumbiedykes * has no place in his philosophy —quite the reverse. From sire to son the injunction descends, "Bring 'em down, Seth, why cumber they the ground?" And at the same time, the practical effect of it is pointed out in the "powerful heap" of pumpkins, corn, and sweet potatoes he can grow in their place. The result is, that in a few years his log cabin stands shelterless from sunshine and storm; and the tasteless, gaudily-painted mansion, which in course of years of prosperity replaces the pioneer "shanty," is equally bare, the sunshine beating down upon it in summer, and the wind howling through the seams between the contracted boards in winter. On the other hand, let us suppose our friend "locates" near prairie land. Then he goes into an opposite extreme. He commonly selects for his residence a position in a dense grove of timber, where, by the exclusion of sunbeams, and almost of the atmosphere, a continual dampness prevails. This is frequently in a low bottom, or on the banks of a stream. Though he has no intention of passing his life there, yet the fancied wealth of the timber region, its superior value, as he thinks, over the prairie, decides

California Academy of Science," Vol. III. (1866); Wilson: "Journal of the Royal Geographical Society," Vol. XXXV., p. 106, and in numerous other less accessible works; while in Croumbie Brown's "Reboisement in France" (1875), the whole question is discussed.

* "Aye be stickin' in a tree, Jock, it grows while you are sleepin'."

his choice in favour of that position; and, neglecting the high, open, healthy prairie that spreads before him—sea-like in extent—a virgin soil, unsurpassed and all but inexhaustible, where in two years he might be the possessor of a rich farm, he seeks the immersion of a dense and damp forest, where, with his poor cabin and his habits of life, his exposure and hardships, combined with the atmosphere and the decaying vegetation, intermittent fever, or ague, is soon added to his other discomforts, and sets its pallid mark of emaciation on

A HERD OF BUFFALOES (*Bison Americanus*) ON THE PRAIRIE.

the countenances of the family. Incarceration may be a term less applicable to the condition of a roving backwoodsman than any other, and is especially unsuitable to his habits. Possibly the cabin you see him in is the third or fourth he has built within the twelve months, and a very slender motive would place him in a fourth before the ensuing winter. Labouring under no restraint, his activity is only bounded by his own physical powers. He has no veneration—no associations which bind him to a place. He is, or was, always "calculating to move West." Still he is incarcerated—shut from the common air—buried in the depths of a boundless forest. The breeze of health never reaches these poor wanderers. They are all tall and pale, like vegetables that grow in a vault pining for light. They are all of one pale yellow, without the slightest tint of healthful bloom. In peering through a vast expanse of the backwoods, I have been so much struck with the effect that I fancy I could determine the colour of the inhabitants if I were apprised of the depths of their immersion! and *vice versâ*, I could judge of the extent of the clearing if I saw the people. The blood is not supplied with a proper dose of oxygen, from their gloomy atmosphere, crowded with vegetation, growing almost in the dark, or decomposing, and in either case extracting from the air its vital principle.* Nearly all of these new countries are subject to ague when the virgin soil is turned up. It is the chronic malady of some districts, and is really thought little of, though I cannot say I ever met any of the citizens of the numerous "Edens" of the West who, like Colonel Chollop, was "fever-proof, and likewise agur." On the contrary, my experience was that most of the citizens of a malarious neighbourhood found it convenient to arrange their engagements in accordance with their "shaking days." Most of these primitive people were rather surprised when told that in England their familiar trouble was almost unknown. "No fever—an' aguey!" would be the exclamation; "then I cal'late the old country's pretty well cleared. That's it, cap'n. Mos' uv the land's taken up in your section, I reckon?" If the country is swampy, then the trouble is much magnified. Those who have seen the miserable sickly wood-cutters on the banks of the Mississippi, where, according to Mark Tapley, the toads are "on the wisitin' list," will understand what I mean. When the river rises, the farms of these cachectic wretches are generally flooded. A humorous writer, who in this case scarcely exaggerates, describes at this season "crazy rail-fences sticking a foot or two above the water, with one or two jean-clad, chilly-racked, yellow-faced male miserables roosting on the top rails, elbows on knees, jaws in hand, grinding tobacco, and discharging the result at floating chips through crevices left by lost milk teeth; while the rest of the family and the few farm animals were huddled together in an empty wood flat, riding at her moorings close at hand. In this flat boat the family would have to cook and eat and sleep for a lesser or greater number of days (or probably weeks), until the river should fall two or three feet, and let them get back to their log cabin and their chills again—chills being a merciful provision of an all-wise Providence to enable them to take exercise without exertion. And this sort of watery camping out was a thing which these people were liable to be treated to a couple of times a year—by the December rise out of the Ohio, and the June rise out of the Mississippi. And yet these were kindly dispensations, for they at least enabled the poor things to rise from the dead now and then, and look upon life when a steamboat went by. They appreciated the blessing, too, for they spread their mouths and eyes wide open, and made the most of these occasions. Now what *could* these benighted creatures find to do to keep from dying of the 'blues' during the low-water season?" This refers especially to the Mississippi, and the reader can put the sketch to its discredit when, in a later chapter, we have occasion to say something about that mighty river, which, in its latter end, is worse than at its beginning. The prairie has, however, none of these drawbacks. It is a gladsome, healthy home for man. In these breezy expanses the traveller feels free. He experiences a sense of escape from the trammels of his past life, whatever they may have

---

* Birbeck: "Notes on Illinois," &c., p. 138 *et sq.* Bradford: "Notes on the North-West or Upper Valley of the Mississippi," p. 148.

been, and a sense of rejuvenescence rarely experienced under any other circumstances. A new world seems to have opened itself before him, in which he has, at least, an equal chance with the rest of his kith and kin. And just as he who has lived on the bosom or by the shores of the ocean cannot be satisfied with green fields and babbling brooks, so a child of the prairie, or one who has learned to love its vastness and wandered over its corn-covered fields, can never leave it without casting a lingering glance behind, and hoping in his heart that his lot will lead him there once more. It is no wonder that men sell out and go West. It is no wonder that those whom fortune is pushing to the wall in older communities escape to find solace for their woes in the solitude of the plains. They feel that man is in closer communion with nature there than elsewhere, and that his neighbour has no right and will have no inclination to bring with him the restraints and woes of towns and capitals. This is sentiment, of course, but that is the nature of the sentiment begotten of the plains.* After travelling on the prairies for days at a stretch, one feels that the world is not so narrow as we think it is after toiling through one little valley after another, never seeing more than half a mile ahead. The impression is the same that strikes one after sailing day after day on sea, out of sight of land. On the seas of grass and of water we appreciate as we can nowhere else Göthe's words, when he asks—

"To give wide space for wandering, is it
That the world is made so wide?"

At one time the only inhabitants of the Far Western prairies were the buffalo and its hunters. Now the buffalo is disappearing, or is getting year by year more and more circumscribed,† and the Indian is following it. In 1718 the herds of bison and other animals was so great along the Ohio River, that travellers were often obliged to discharge their guns to clear a passage. Boone and his associates found buffalo more abundant in Kentucky and Tennessee than they "ever had seen cattle in the settlements." This was as late as 1780. The indiscriminate, reckless slaughter has been going on, until now the buffalo (p. 85) has retreated far beyond the Mississippi. There the massacre is still in progress. From the time when the white settlers first invaded its haunts to the present day, thousands have been killed annually in miscalled sport, while of many thousands more only the tongue, or other choice morsels, has been saved, the carcase being left entire as food for the wolves, and other wild beasts, or to poison the air by putrefaction. At other times, they have been slaughtered by hundreds, and even by hundreds of thousands, in a single year, for their hides alone. Such has not only been the case in late years on the plains of Kansas, but during the last century was practised east of the Mississippi, from Kentucky to Western Pennsylvania, so many being sometimes killed in single localities that their murderers would be driven away from the immediate scene of slaughter by the effluvia arising from the remains of their defenceless victims. The same is true of the various species of deer, and of almost every other animal, the slaughter of which could afford profit, amusement, or gratification of any sort to man. The extermination of the Indians has been almost as rapid. The whole route over which the Pacific Railroad now runs is dotted with the

* St. John: "The Sea of Mountains," Vol. I., p. 50 (1877).
† Allen: "The Buffalo" (Memoirs of the Museum of Comparative Anatomy: Cambridge, U.S.) 1875.

graves of the early pioneers and their enemies. The Platte is to the American almost as classic ground as the Tiber is to the European. The one is redolent of the early history of the New World—the other of that of the Old. The wild men—white and red—are giving place to hordes of Eastern emigrants, who are covering the land; and where only a few years ago were Indian wigwams, towns are springing up along the line of the great Trans-Continental Railroad.

STATION OF THE PACIFIC RAILWAY AT OMAHA.

CHAPTER V.

THE UNITED STATES: WYOMING; THE WONDERLAND OF AMERICA.

OF *Wyoming*, after the account we have already given of the prairies, little need be said. It is 350 miles long from east to west, and 280 miles from north to south. Its area is 93,107 square miles, or 59,588,480 acres, of which very few are improved. In 1870, indeed, only 338 were. The country is in general mountainous, and comprises high level plains. The Pacific Railroad runs through it. It enters the Laramie Plains by a pass 8,262 feet in height, and crossing these prairies, which are on an average 7,000 feet high, to Bitter Creek, attains, at the watershed at Benton, near Bridger's Pass, a height of 7,534 feet. Two hundred miles west, the Wahsatch Mountains are reached, and passed at an elevation of 8,271 feet, when the line descends the Weber and Echo Cañons to Salt Lake Valley. The Laramie Plains was long a battling-ground of the Indians. Here they frequently made a stand, and disputed the right of way to the railway-makers. The point at which the railway passes the mountains is the small village of Sherman, but the traveller is unconscious of it. He is even disappointed. He sees nothing but a number of mud and sand-hills, and can scarcely realise that this is the summit of the Rocky Mountains. Yet, if he attempts exertion, he speedily discovers that the atmosphere is

VIEW OF YELLOWSTONE LAKE, WYOMING.

entirely rarified. He feels lighter and brighter, and everything seems in clearer relief than in the valleys. The line enters Echo Cañon soon after passing over Dale's Bridge, a fine work spanning over a stupendous height. This point is nearly 1,000 miles from Omaha (p. 88). "Before reaching that city," writes Mr. St. John, "he has travelled 402 miles from Chicago, through a country of cultivated prairie, and that has been succeeded by days and nights of rolling prairie, and level plains, the greater part of which has been wild as when the waters first receded and gave birth to the land. Fifteen hundred miles of prairie in its different forms and stages have accustomed the eye and mind to a land of pasturage and corn-fields. Then, with only a little preparation, one darts into a valley from which the mountains rise in diverse precipitancy on either side. On the south side, their face, though serrated by numerous little gulleys cut by the melting snow, and rocky throughout, is comparatively smooth when placed within a field of sight that takes in the opposite side of the pass, and beyond receiving an occasional glance they are forgotten and unnoticed by the many.

It is north of the cañon that all eyes are directed as the cars move slowly through. It is not the heights of the precipitous cliffs that give them grandeur, for though they are high they are below the snow level, and not higher than, if so high, as the more sloping mountains opposite. But they have thrust themselves forward in bold, bright red bluffs and promontories without vestige of plant or soil, their faces varied by caves and weather-worked indentations and sinuosities; pitted into extraordinary irregularities by the storms of ten million seasons, and capped by rocks of fantastic shape, that in one instance look like a sentinel on duty, in another like the bastion at the angle of a fortress, in another like the prow of a ship, and so, throughout the length of the valley, ever presenting on the summit of the mountain, or on some ledge or peak of its bold red face, an exaggerated lineage of some familiar form." The train runs through this picturesque pass in a few hours, and enters Weber Cañon, which is as beautiful if less wondrous strange than that which lies in such close companionship to it. It is practically the commencement of that Mormon land which we have left. It is the long funnel through which we enter the beautiful valley of the same name, entirely surrounded by high snow-mottled peaks of "The Rockies." Wyoming is drained on the south-east by the north branch of the Platte, while the Yellowstone, rising in the Yellowstone Lake (p. 89) and its branches, flows through the eastern and northern parts. In the south-west is the head stream of the Green River (p. 56), which rises in the angle made by the Wind River and Wahsatch Mountains, near the source of the Lewis River and the headwaters of the greater Missouri. The climate of Wyoming is, owing to its elevation, cold. No grain will ripen; but, on the other hand, the grass is always fresh, and accordingly stock is largely reared, wool to the extent of 30,000 lbs. being produced, while the Sweet Water Diggings have yielded on an average 50,000 dollars *per annum*. There are extensive mines of tertiary coal, and no doubt in time useful metals will be discovered, so that Wyoming, though it is hardly likely ever to be a competitor of Colorado, or even of Montana or Dakota, may attain a greater population than 10,000. The Indians in the territory are the Shoshones and Bannacks. They are mostly peaceable, though in 1876 there was a good deal of ill blood stirred up amongst them by the Mormons. These polygamous "saints" started a propaganda among the Indians, holding out, among other inducements to the aborigines, that, as they were the "chosen people," they should inherit the "Gentiles'" houses when they were driven out of the country. Instantly an unwonted desire to be baptised, or "washed," as the Indians called it, seized on the natives. A Mormon bishop stood up to his waist in water all day long "washing" his unsavoury converts. His zeal resulted in nothing, for beyond the unwonted ablutions the Indians had not the most remote notion of what it was all about, and having received their presents they returned to the agency. Still for a long time afterwards they looked upon the rite as a something which absolved them from the consequences of their acts. When caught in such familiar peccadilloes as horse-stealing or petty theft generally, they would assume an air of great dignity, and indignantly reply to the soft impeachment, "Me good Mormon; me heap wash!"

The "Wonderland" of America.

We have lingered long in the West. We must linger yet a while longer. The Eastern States with their teeming industries are still before us, and the South, yet smarting under the recollection of the lost cause, is close at hand. Still a steam-engine or a plough is only a steam-engine or a plough all the world over. One industry is much the same as another industry, whether it be in Europe or America, and after a country has been settled up for a century it ceases to have many characteristics beyond those with which nature has already supplied it. Accordingly the Eastern States, so often described and so accessible, will have what may be thought unduly small space devoted to them compared with that which the wild lands west of the Mississippi have had. The better-known and more civilised portions of the Western Continent are uninteresting so far as their scenery goes. Excepting the lovely Hudson, the beautiful Lake George, the mountainous districts of New England, of Virginia, and other spurs and offshoots of the Alleghanies, I quite understand Lord Dunraven thinking the country between the Alleghanies and "the Rockes" flat, dreary, and uninviting. "Exceedingly graceful is the maize plant when its silken tassels droop in the hot sunshine amid the dark green covering leaves. Yet the eye wearies of interminable corn-fields, bounded by untidy and tortuous snake-fences. Nothing is more vulgar-looking and unkempt than recently-cleared land. The face of Nature, shorn of the beauty of its natural covering, looks mean and dirty; and as, compared with its appearance when clothed in the forest, resembles the contrast between a man's countenance when disfigured by a coarse stubby growth of a week old, and the same when adorned with the soft flowings of a patriarchal beard. Blackened stumps stand thickly in the foreground amid rocks and weeds, and the forest seems to huddle itself back out of reach of the fatal axe. The beauty of nature is destroyed, and is not succeeded by the artificial beauty of civilisation." The plains are charming from their vastness, their glorious sunsets, and exquisitely delicate covering; but though fascinating at first, and always pleasing, they cease after a time to charm. I cannot, however, coincide in the opinion of the pleasant writer whom I have quoted, that they are "inexpressibly sad and mournful." The great attribute of the wilder parts of the continent is the stupendous scale on which everything is found. Rivers are great, plains are great. The mountain peaks, if not beautiful like the much smaller ones of the Alps, are yet high; the lakes are seas; the forest jungles stretches for hundreds and hundreds of miles. Even the so-called Rocky Mountains extend the whole length of the continent, and in some places are 500 or 600 miles wide, and comprise important ranges dividing great and fertile plains, and containing important valleys. As for the defiles which the currents have sawn out for themselves, they are simply unparalleled in any other part of the world. The cañons of the American rivers stand *per se*.

Lord Dunraven, to whom, as well as to Professor Hayden, I have much pleasure in acknowledging myself indebted for much of what follows, gives so excellent a *coup d'œil* of the region in the centre of America which affords the finest scenery, that though it has already been briefly sketched in connection with the geography of the country, I cannot omit quoting it:—"Comprised," writes this lively traveller, "in the territories of Montana and Wyoming there is a region which contains all the peculiarities of the continent in a remarkable degree, and which, moreover, is exceedingly interesting, on account of its scenery, its geography, its mineralogy, and its sport. Although the altitudes are not so high as in other parts of the continent, it may be truthfully called the summit or apex of America. There the waters flow in all directions, north, south, east, and west. There it is that great rivers rise, running through every clime, from perpetual snow to tropical heat. On the one side glance the currents destined to mingle with the tepid waves of the Gulf of Mexico; on the other, up the rapids, leap the salmon ascending from the distant waters of the Pacific Ocean. It is the geographical centre of North America. It is essentially 'the Great Divide.' Roaming at will through the trackless wastes that compose and surround this region are to be found the best representatives of the fast-vanishing aboriginal race. By the great bend of the Yellowstone are grouped the tall lodges of the gigantic Crows, men of six feet four or five inches in height, with long hair reaching in heavy plaits to their knees. From Idaho come parties of Bannacks, great raisers of stock, and traders in horses. Pend d'Oreilles, Gros Ventres, Flatheads, Bloods (p. 77), and Piegans, warlike Blackfeet, Assinaboines, and Sioux, wander through the hunting-grounds, seeking their meat from God, stealing horses, hunting, and warring upon one another in something like their natural freedom. Some of these are very hostile to the pale-faces, and much to be dreaded, like the well-armed and numerous Sioux, or, as they style themselves, Dakotas (p. 73), others are friendly to the whites, like the Crows, Snakes, Bannacks, and their kindred tribes. A few, such as Nez Perces and Bannacks, are semi-civilised, cultivating a little land, and rearing horses and cattle, possessing farming implements, and using in war or for the chase the newest fashion in repeating rifles, and the latest thing out in revolving pistols. Others, such as Blackfeet and Assinaboines, are primitive and unsophisticated, depending in a great measure upon their ancient weapons, the bow, the lance, and the club, and subsisting entirely by the chase—wanderers who have their homes far to the north in British territory. In this same region are still to be found great herds of wapiti—noblest representatives of the deer tribe, and soon to be numbered among things of the past. In the swampy flats, among old beaver dams, where willows and alders grow, or among the thickest groves of young firs, still linger the largest of existing elks, the moose. Poor *Cervus alces!* your ungainly form has an old time look about it. Your very appearance seems out of keeping with the present day. The smoke of the chimney, the sound of the axe, are surely, though slowly, encroaching on your wild domains. The atmosphere of civilisation is death to you; and, in spite of your exquisitely keen senses of smell and hearing, you, too, will soon have to be placed in the category of things that have been. In the valleys are both the white-tail and black-tail deer. On the little prairies, open glades, and sparsely-wooded slopes, grazes the small mountain bison or buffalo, whose race has also nearly vanished from the scene; and not far distant are to be found herds of his congener, the great bison of the plains;* for down in the Judith basin lie the hunting-grounds where the Crows go every summer and winter in search of the prairie buffalo.

* This one is only a race of the other. Both are the *Bison Americanus*.

In summer they kill them for their flesh, in winter they utilise their skins to make robes and houses for themselves. Prong-horn antelopes, the only specimen of the species

VIEW OF THE CLIFFS IN THE GRAND CAÑON OF THE YELLOWSTONE, WYOMING

on the continent, and the only known variety in the world that carries a branching horn (Plate VI., Vol. I.), are very numerous on the plains and foot-hills. Clear against the sky-line, standing on some jutting crag, may not unfrequently be seen the massive, stately outline of a bighorn, or mountain sheep,* a near relation to the mouflon or argali; and far up in the wildest fastnesses of the range, among untrodden peaks, wild goats,† distantly allied to the Eastern ibex, have their inaccessible abode. If this last be not sufficient, and it is to be considered that an element of danger is necessary, the sportsman will be glad to learn that nowhere, save perhaps in Southern California, will he be more likely to encounter *Ursus horribilis* (the grizzly bear). If he has ever pursued or been pursued by that unpleasant beast, he will be gratified to learn that, as a rule, pine-trees are numerous and not difficult to climb."

Every year the wilds of America get more circumscribed, and though there will be always localities impossible to cultivate, and without attraction to settlers, these spots will not be the fertile valleys and prairies of the West. One year nothing is heard in one of these solitary places save the barking of foxes, the dismal howls of coyotes, or the "coughing" of deer. Next summer, when the hunter, oblivious of events—the Rip Van Winkle of twelve months—arrives he finds that wheat-fields have taken the place of the buffalo grass, and the silky tassels of maize wave where only a short time previously rank weeds or thickets of reeds grew; and if he comes back still later he will hear the whirr of the threshing machine, or witness the citizens of some mushroom town, with a high-sounding name, assembled to celebrate the Fourth of July, and listen patiently while some local orator drawls out that famous declaration in which such hard things are said of "George, King of Great Britain," and his subjects who thought with him. One of these few thoroughly wild regions is that which has been called the "Wonderland of America," from the fact of its containing so many natural objects of interest. It is situated in the region we have sketched in Lord Dunraven's words, but is principally in the territory of Wyoming. Here the United States Government have set aside an extensive tract as a "national park," and, looking forward to the time when the country would be more densely populated than now, have reserved it from settlement. The region is usually called the "Upper and Lower Fire Hole Basin," or the "Geyser Basin," or more comprehensively the "Upper Yellowstone Country." The area of the reservation is 3,578 square miles; and, roughly speaking, it is contained within the meridians of 110° and 111° west longitude, and 44° and 45° north latitude. Its general elevation is about 6,000 feet above the sea level, while the wall of mountains which surround it on every side attain a height of from 10,000 to 12,000 feet. Frosts occur every night in almost every month in the year, so that agriculture is out of the question. Indeed, in much of Montana this is also the case. At Deer Lodge (4,654 feet above the sea), two years ago the squirrels went into winter quarters on the 12th of August, and snow fell on the 18th of the same month.

Most of the rocks being volcanic, mines are not likely to be discovered, so that in setting it aside as a national park the Government were not depriving the people of anything which could be of much value to them commercially. But though in the eyes of the farmers or miners a worthless section, the "Wonderland" is, perhaps, the most interesting district on the American continent. Its scenery is not so magnificent as that of the Yosemite Valley (Vol.

* *Ovis montana* (Cuvier).
† *Aplocerus montanus* Richardson).

I., p. 319, and "Races of Mankind," Vol. I., p. 29), but it does not depend for its attractions on this alone. The remains of volcanic activity in this basin are among the most remarkable in the whole world. "The Mammoth Hot Springs of Gardiner's River, and both geyser basins, are situated in it. Entombed in its forests, at a height above the sea of 7,788 feet, lies a large and lovely lake, which is, with four exceptions,* the highest body of water of any considerable size in the world; and in the snow that falls upon its summits are born four of the largest rivers of the continent. On the north are the sources of the Yellowstone; on the west, those of the three forks of the Missouri; to the south-west and south rise the springs of the Snake and Green Rivers, the former gaining the North Pacific [as a tributary of the Columbia], the latter finding its way to the Gulf of California [as a tributary of the Colorado]; and lastly, in the south, head the numerous branches of Wind River. Thus it is, as auctioneers would say, a most desirable park-like property; and if Government had not promptly stepped in, it would have been pounced on by speculators, and the beauties of nature, disposed of to the highest bidder, would have been retailed at so much a look to generations of future travellers."† We have figured some of the localities and other features of the "Wonderland." Though these sketches (pp. 89, 95, 96, 97) must stand in the place of any extended description, we may append a few words explanatory of some of the more interesting of the places mentioned.

The "Mammoth Hot Springs" on Gardiner's River (p. 96) are the remains of the gigantic volcanic fires that shook the whole of this region at one period in its history. The springs are calcareous, and vast quantities of lime are deposited from them. Many invalids flock to the springs, the waters from which are conveyed into rough wooden troughs, which in a short time become coated with a fine smooth marble-like enamel derived from the springs. An area of about two miles is occupied by the calcareous deposit from these springs, and in some cases even the trunks of dead trees are standing in six or eight feet of this material, the accumulation of which gives a strange weird aspect to the whole of the district. The temperature of the springs varies from 94° Fah. to 200.9° Fah., and their elevation above the sea level is 6,779 feet. Lime, silica, ferric oxide, alumina, and a trace of soda and magnesia, are the materials shown by analysis to enter into the composition of these hot springs. The only other similar spectacle equal to these is said to be found in the hot springs of Te Tarata and Otukapuarange of New Zealand, where similar terraced fountains are found. At these springs there are, however, fine geysers, which are not found at the Mammoth Hot Springs. "The central spring merely bubbles constantly, and the overflow of water from it is moderate, and does not vary much in the quantity at different times discharged." Enormous geysers have, however, to all appear-

---

* These exceptions are Lakes Titicaca in Peru, and Uros in Bolivia, which are respectively 12,871 12,359 feet above sea level; and Lakes Manasarowak and Rakas Tal in Thibet, both of which lie at an elevation of 15,000 feet. In the United States, however, are small lakes at a higher elevation than Yellow Lake. For instance, Twin Lakes, Colorado, are situated at the height of 9,357 feet; Osborn's in the same State at 8,824 feet; and Lewis, in Wyoming itself, at 7,882 feet.

† Dunraven: "The Great Divide," p. 15. In this work will be found the best account as yet of this country, and the routes to it.

ance, at one time existed on Gardiner's River, but they are now silent. The Yellowstone Falls is another of the sights of "Wonderland." Where the Yellowstone River leaves the lake of the same name, it flows in a certain steady current until it reaches the mountains, which oppose its passage to the north. It then, to use Lord Dunraven's words, "performs a series of gymnastics over rapids, cascades, and waterfalls, as if exercising its muscles and sinews, preparing itself and getting strength for the mighty effort by which it tears a passage through the granite flanks of the range. A mighty effort, truly, or rather a vast expenditure of force, has been employed in clearing the

VIEW OF THE HOT SPRINGS ON GARDINER'S RIVER, WYOMING.

Grand Cañon, a rent in the mountains over twenty miles long and of vast depth (p. 93). Where the river enters the cañon the sides are from 1,200 to 2,000 feet high, and further down they rise to a greater altitude, an altitude which has never been determined, for the depth of that chasm has not yet been explored or trodden by human foot."

Both the Falls of the Yellowstone are caused by basaltic dykes or walls, which traverse the rocks at right angles to the course of the stream. The falls are neither very high nor very romantic, but still they have a certain strange beauty of their own—"a wild loveliness peculiar to them—and what they lack in volume, power, and general grandeur is amply atoned for in the pre-eminently distinctive character of the scenery about them, and by the lavish display of colour and strange forms of stratification which distinguish their surroundings. The scene is so solitary, so utterly desolate, the colouring is so startling and novel, the fantastic shape of the rock so strange and weird, that a glamour of

VIEW OF THE FALLS OF ST ANTHONY, MINNESOTA.

enchantment pervades the place, which, though indelibly impressed upon my mind, is yet quite impossible to describe," and the falls are in reality a series of falls, the highest cascade being 397 feet. At "Crater Hills" are hot sulphur springs, and near by this two "buttes," of 150 to 170 feet respectively, which give the name to the locality. These mounds are composed of calcareous deposit, impregnated with iron and sulphur. The water in the springs also contains a considerable amount of sulphuric acid. At Mud Springs are some geysers, which cast up mud and water about twenty or thirty feet high, or sometimes as high as forty feet; but beyond this fact do not differ greatly from any other geysers of the kind. The river close by is full of trout of a large size and fine quality to look at. But they are full of insects, not only in the intestines but in the

VIEW OF TOWER FALLS AND COLUMN MOUNTAIN, WYOMING.

flesh also. From the scars on the outside of the fish, it would seem as if the insects ate their way completely through the muscles; and when these get the upper hand of the trout he usually becomes "a lanky, dull-coloured, ugly-looking brute." The prevalence of these parasites seem due to the warmth of the water, or to the presence of the mineral substances in solution. For it is remarkable, as the author whom I have so frequently quoted remarks, that whereas a trout entirely free from them is almost unknown above the falls, he never heard of a wormy fish being taken below them, or even between the Upper and Lower Cascades. The Lower Fire Hole Basin is another remarkable place in the National Park, though not equal to the upper one we have already noted, and from which it is separated by a low divide. Still there are in it a great number of geysers exhibiting an infinite variety of fine structure, appearance, and size. "Some small, some large, meriting almost to be called little lakes, and containing vast volumes of boiling water; others mere cracks or fissures in the surface, occasionally ejecting air or liquid, like the diminutive puffing-holes one meets with on the sea-shore." Sometimes it is silica or flint that is deposited; at other times iron, or silica and iron together, mixed in some cases with sulphur also. Old springs are here constantly dying out and new ones springing into life, and pine-trees standing near springs, coated with deposit, and yet in full life, show that the outbreak of these springs must be something of very modern date. Numerous bare patches in the forest indicate where springs and geysers once have been, and in many little lakes or ponds are buried the remains of extinct geysers. There are also various mud volcanoes in this basin. The Fire Hole River flows through the basin. Its waters are warm, and much appreciated in cold weather by flocks of geese and ducks. The whole of the river sides are "honeycombed and pitted with springs, ponds, and mud-pots furrowed with boiling streams, gashed with fissures, and gaping with chasms, from which issue hollow rumblings, as if great stones were rolling round and round, or fierce, angry snarls and roars. The ground sounds hollow underfoot. The track winds in and out among holes that puff sulphur flames or squirt water at you; by great caverns that reverberate hideously, and yawn to swallow you up, horse and all; crosses boiling streams which flow over beds composed of a hard crust, coloured yellow, green, and red, and skirted by great cisterns of boiling, bubbling, seething water. The crust feels as if it might break through at any moment and drop you into fire and flames beneath, and animals tread gingerly upon it. You pass a translucent lovely pool, and are nearly pitched into its hot azure depths by your mule, which violently shies at a white puff of steam maliciously spitten in its face through a minute fissure in the path. You must needs examine into that ragged-mouthed cavern, and start back with more agility than grace to escape from a sudden flood of hot water, which spitefully, and without warning, gurgles out and wets you through. The air is full of subdued, strange noises—distant grumblings as of dissatisfied ghosts, faint shrieks, satirical groans, and subterranean laughter—as if of imprisoned devils, though exceedingly uncomfortable, were not beyond being amused at seeing a fresh victim approach. You fancy you can hear the rattle of the loom, the whirl of wheels, the clang and clatter of machinery; and the impression is borne upon the mind that you are in the manufacturing department of Inferno, where the skilled hands and artizans, doomed to hard labour, are employed. I can compare it only to one's feelings in an iron foundry, where one expects every moment to step on a piece of hot iron, to be run through the stomach by a bar of white glowing metal, to be mistaken for a pig and cast headlong into a furnace, or be in some other way burned, scalded, or damaged."

In this valley is also the Castle Geyser, which spouts to an elevation of at least 250 feet, with a roar like the sound of a storm driving the wild waves against the cliffs added to this, which can only be compared to the shrieking of the steam-pipes of many steamers blowing off. There is another geyser known as "Old Faithful," because it plays regularly every three-quarters of an hour. It throws a column to a height of from 100 to 150 feet, but though a fine geyser, it is not to be compared to the "Castle."

The "Giant" will play for an hour and twenty minutes, all the time throwing up a column of water to the height of 140 feet. There are numerous others, but as they are all similar in kind, though not in degree, we may omit noticing them by name. The borders of the Fire Hole River, and its confluent Iron Spring Creek, and a great part of the plain enclosed by these two streams, "are dotted in all directions by mud pools, solfataras, fumaroles, warm pools, boiling springs, and the remains of many extinct geysers of considerable size." The still, deep, quiet wells are very lovely. They are circular basins fifteen or twenty feet across, and fifteen to sixty feet in depth, filled with exquisitely transparent water. The ground which surrounds these basins slopes very "gently back from the brink by an inch to three or four inches in height at a time. The edges of these steps are carved into a series of semi-arches, and adorned with mouldings of pearly beads, ranging in colour from a dull white to a coral pink. The rim of the basin is convoluted and gathered in, into a system of irregular curves, scalloped and beaded. The interior is of a most delicately rich cream colour, intensified in places to rose; and over portions of it is spread a fine network of lace-like fabric. Deeper down the ornamentation becomes larger, and the sides are composed of rounded sponge-like masses. The basin is filled to the brim with water, more transparent than anything you can imagine, and deeply blue. As the sun, rising or sinking, strikes at a greater or smaller angle the surface of the water, its rays, refracted more or less obliquely by the revolving element, give a constantly varying, but ever lustrous, appearance to the interior ornamentation and colouring of the pool, that baffles all attempts at description. One never tires of looking at these fairy lakes, for though language fails to convey the impression of variety, and the character of sameness would appear to be inseparable from them, yet it is not so at all; on the contrary, a constant and beautiful change is going on at every succeeding moment of the day."

With this notice of one of the many wonders of "Wonderland," we leave at once Wyoming and the Rocky Mountains—that familiar range, and which conveys so little meaning to those who live under its shadow—the Rocky Mountains being in reality a vague term for the many separate ranges, which together extend through the whole length of North America. Five or six years ago this interesting region was unknown, except by the dubious tales brought to the white settlements by Indians and trappers. Even yet we know little about it. A few parties visit it from Virginia City and come out at Bozeman; but they go in each other's footsteps, examine the same objects, and halt at the same places. Yet doubtless to those who have enterprise or time to stray from the usual route, there are in "Wonderland," to use the language of the only Englishman who has yet published descriptions of it, "hundreds of valleys into which no human foot has ever burst, thousands of square miles of forest whose depth has never yet been penetrated by the eye of man."

CHAPTER VI.

THE MISSISSIPPI BASIN: NEBRASKA; MINNESOTA; IOWA; KANSAS; WISCONSIN; ILLINOIS.

IN a work of this nature it would be scarcely possible to fix a strictly natural arrangement of the regions to be described, and it would be still more difficult to keep accurately to that arrangement if once made. Accordingly, we have described as Rocky Mountain States some which should come, strictly speaking, under the head of the Mississippi Basin. But other reasons, already explained, have decided us to depart now and then from what physical geography might strictly demand. However, we have now altogether bidden farewell to the Divide of Waters. We enter the region drained by the Mississippi River and its tributaries, some of which are as large, if not larger, than the river into which they flow. But it has always been a moot question where a tributary ends, or a river of which it is a feeder begins; and so we shall see it is in the Mississippi Basin. The Gallatin, the Madison, and the Jefferson, after each short separate course, unite and become the Missouri, which, according to the song, "rolls down to the sea." Undoubtedly it does; but there are the great falls, which at a distance of 2,540 miles from the Mississippi interrupt all navigation. Even there it carries a volume of water three or four times greater than the Thames at Richmond. It is sometimes only 300 yards wide, at other times it spreads from bank to bank over a distance of a quarter of a mile. One hundred or one hundred and twenty miles below Fort Benton—after flowing through a wild savage country—it is joined by the Marias, and then, suddenly turning to the east, forms the "Great Bend." It now passes through the Judith Basin, a country full of buffalo and other game; then, after flowing easterly for about 250 miles, it is swelled by the Milk River from the north-west; a little further down it is joined from the Yellowstone, which flows out of the beautiful lake of that name (p. 89). Yet, strange to say, though we have been journeying southward, the climate has not been improving. Indeed, in winter the Indians, when travelling from the neighbourhood of Fort Buford and the Lower Yellowstone, are frequently compelled to use their snow-shoes till they get near the spurs of the hills, where not a vestige of snow is to be seen. They are then, owing to warm radiating masses of the mountains, and probably the soft breezes from the Pacific also, able to cache their raquets, and pursue the journey on unencumbered feet.* From the rolling prairies, the home of the bison, and the vast solitudes of the Dakotas or Sioux, we have been entering a dismal, most peculiar region of brown flats, only allowing of the growth of a few cacti and sage brush shrubs. The river banks are no longer picturesque. The waters wash against brown clay, and the Missouri, literally the "Mud River," gets muddier and yellower every hour. "Nothing breaks the meaningless stupidity of the round, plain, flat, sad-coloured grey or olive-green, bounded by blue, watery sky. Not a single bit of bright colour, no object of beauty, not a shade even of pleasant verdure, refresh the tired eye. Everywhere is brown mud, grey clay, or white alkali; everything is

* "The Great Divide," p. 303.

VIEW ON THE UPPER MISSISSIPPI, MINNESOTA.

graceless, hideous, depressing." A startled Indian drinking at the river tells us that we are in the country of the Blackfeet and Minneconjou Sioux—their near relatives—Ishmaels in whose presence no man's cattle are safe, and no man's scalp sacred. We are passing through the *mauvais terre*, the Bad Land of the Dakotas, "a desert of clay, alkali, and sage brush, uninhabited and uninhabitable." Still we follow the river through the Coteau du Prairie, a plateau 1,800 to 2,000 feet in height, on through the Indian's land. But now pine-trees appear floating on its yellow waters (pp. 76, 104), and as the current swells with the contributions which the Big Cheyenne, the Niobrara, and a score of lesser tributaries pour into it, we emerge from utter savagedom into semi-barbarism. Indians are still the chief people we see on the banks, but they are Indians glossed with civilisation—they are red men in white men's clothes. We are passing through the Tetons and Yankton Sioux's country. East of this region is the Red Pipe Stone Quarry, famous in Indian song and story as the place where from time immemorial the Indians obtained the material out of which their calumets were fashioned, and where, accordingly, they met for the time being in mutual peace. Among the sand-hills that fringe the western banks marauding parties of Sioux, Brulés, Ogallalas, Pawnees, Otoes, Winnebagoes (p. 109), and Omahas may be found warring amongst each other, but ready at any time to rob the whites. They are Indians who have lost whatever virtues they once possessed, and by contact with civilisation have imbibed everything that is bad, but little that is good, in the white men. Steamers are now numerous, and just before the shallow broad waters of the Platte pours into its sandy yellow tribute, the river sweeps between the rival "cities" of Omaha and Council Bluffs, the latter once famous in the annals of the West as the rendezvous of trappers and the "jumping-off place" of explorers bound for the then mysterious region. Then comes barbarism again. We are now going through the Indian territory, that tract set apart for the children of the soil who, amid torrents of blood and wrong unspeakable, have been removed from their ancient homes to this region. Soon after, according to geographers, we should be entering the Missouri. But long below its junction with the clear Mississippi (pp. 81, 101) the river has been changing its appearance. It now imparts its muddy character to the Mississippi. It is, moreover, the largest stream, and drains the greatest extent of country, and should, therefore, still be called the Missouri. But fate has decided otherwise; and soon after leaving the once important, and still quaint French settlement of St. Charles, we glide into the shadow of the huge arches of the St. Louis Bridge, and step ashore in the queen city of the West (p. 117). "What a change," writes Lord Dunraven, to whose graphic pictures we must for much of the preceding sketch be in debt:—"What a change has taken place! Can this turbid, sullen flood, reeking with the filth of cities, and rushing sulkily through the arches, frothing on its slimy banks, torn and beaten by the paddles of countless steamers, be the same stream that leaped into life in the Northern Sierras, and, sweet with 'the odour of the forest, with the dew and damp of meadows, with the curling smoke of wigwams,' rushed through its 'palisade of pine-trees?' How utterly incongruous and out of place do we appear, and our little birch-bark canoe, in this busy hive of men, this great city of 300,000 [now nearer 500,000] inhabitants! What do the men who jostle and stare at us know of the far-away life of the prairie and the woods, though in their warehouses are stored thousands and thousands of buffalo robes and skins? The best thing we can do is to get out of our moccasins, buckskins, and flannels, and with the help of the barber and the dry-goods store, transform ourselves into civilised beings in white shirts, black store clothes, and plug hat. How horribly uncomfortable we shall feel! How red and weather-beaten our faces will appear! And as to our hands—well, the less said about them the better."

Taking, then, the Missouri as a part of it, the Mississippi is the longest river in the world, excepting, perhaps, the Nile. It is literally the *Miche Sepe*—the "Great Water." It drains a region nearly half the size of Europe from its source in Lake Itasca—the Lac la Biche of the French, 1,575 feet above the sea—until it falls by many mouths into the Gulf of Mexico. From the mouth of the Missouri to the gulf it is 1,286 miles long; from its head waters to its mouth 2,616 miles, or from the Gulf of Mexico to the sources of the Missouri about 4,400 miles. Its numerous tributaries, many of them large rivers, drain all the states and territories between the Rocky and Alleghany Mountains, and constitute a natural system of waterways, having an aggregate extent of 1,500 miles. The following table shows the names of the chief of these tributaries, the area of their basins, the rainfall, and annual drainage*:—

Basin.	Area in Square Miles.	Annual Downfall in Cubic Feet.	Annual Drainage in Cubic Feet.	Ratio.
Ohio River	214,000	20,700,000,000,000	5,000,000,000,000	0·24
Missouri River	518,000	25,200,000,000,000	3,780,000,000,000	0·15
Upper Mississippi	169,000	13,800,000,000,000	3,300,000,000,000	0·24
Small tributaries	32,400	2,600,000,000,000	5,240,000,000,000	0·20
Arkansas and White River	189,000	13,000,000,000,000	2,000,000,000,000	0·15
Red River	97,000	8,400,000,000,000	1,800,000,000,000	0·20
Yazoo River	13,850	1,500,000,000,000	1,250,000,000,000	0·30
St. Francis River	10,500	1,100,000,000,000	9,900,000,000,000	0·90
Entire Mississippi, exclusive of Red River	1,147,000	78,900,000,000,000	18,500,000,000,000	0·23

"Below the mouth of Red River," writes Mr. Hawes, "the Mississippi is divided into numerous arms and passes, each of which pursues an independent course to the gulf. The highest of these is the Atchafalaya on the west side of the river. Below its point of separation from the Mississippi, the region of swampy lands of bayous and creeks, is known as the delta. Above this the alluvial plain of the river extends to the Chains, thirty miles above the mouth of the Ohio, and to Cape Girardeau in Missouri, where precipitous rocky banks are first met with. These are the lower secondary limestone strata lying in a nearly horizontal position. The total length of the plain from the mouth of the Ohio to the gulf is estimated at 500 miles. Its breadth at the upper extremity varies from thirty to fifty miles; at Memphis it is about thirty miles, and at the mouth of the White River eighty miles. The extreme width of the delta is rated at 150 miles, its average breadth is probably ninety miles, and its area 12,300 square miles. The

---

* Humphrey's and Abbot's "Report on the Hydraulics of the Mississippi River" (1861).

elevation of the bottom lands at Cairo above the sea level is about 310 feet, while the slope of the high water surface from that place to the gulf is from 322 feet, to 0. These bottom lands are subject to inundations, and consequently annual enrichment. Under the system of slave labour, large plantations were cut out of the dense forests which cover them, but vast tracts of unsurpassed fertility are yet covered with canebrakes and cypresses. The alluvial plain extending from above Cairo to the gulf is terminated in the east and west by a line of bluffs of irregular height and direction, composed of strata of the eocene and later tertiary formations. Down this plain the river flows in a

RAFTS OF TREES ON THE MISSOURI.

serpentine course, frequently washing the base of the hills, on the east side, as at Columbus, Randolph, Memphis, Vicksburg, Grand Gulf, Natchez, and Baton Rouge, and once passing to the opposite side at Helena. The actual length of the river from the mouth of the Ohio to the gulf is 1,097 miles, increasing the distance in a straight line by about 600 miles, and by its flexures also reducing the rate of its descent to less than half the inclination of the plain down which it flows. The range between high and low water at Cairo, near the head of the plain, is 51 feet, and at New Orleans it is 11.4 feet. The river flows in a trough about 4,170 feet wide at the head, and 3,000 feet at the foot. The immense curves of the stream in its course through the alluvial plain sweep around in half circles, and the river, sometimes, after traversing twenty-five or thirty miles, is brought within a mile or less of the place it had before passed. In heavy floods the water occasionally bursts through the tongue of land, and forms what is called a 'cut off,' which may become a new and permanent channel. The height of the banks and the great depth of the river bed check the frequent formation of these cut offs, and attempts to produce them artificially have often failed, especially when the soil is a tough blue clay, which is not readily worn away by the flowing water. This was the case at Bayou Sara, when in 1845 an excavation, intended to turn the river, was made, by which a circuit of twenty-five miles would have been reduced to a cut off of one mile; and also at Vicksburg in 1862-1863, when the Union army endeavoured to make a cut off out of range of the Confederate guns. Semicircular lakes, which are deserted river bends, are scattered over the alluvial tract. These are inhabited by alli-

VIEW OF THE SOURCE OF THE ARKANSAS, ONE OF THE TRIBUTARIES OF THE MISSISSIPPI.

gators, wild fowl, and garfish, which the steamboats have nearly driven away from the main river. At high water the river overflows into these lakes. The low country around is then entirely submerged, and extensive seas spread out on either side, the river itself being marked by the clear, broad band of water in the midst of the forests that appear above it." One social result of this overflow has been referred to already (p. 80).

During the spring freshets, the river often undermines the banks, and the current is then often so strong that steamboats are sometimes swept into the bayous in the forest beyond. The lower portion of the delta is often not over a few inches above the level of the surrounding waters, and in no place more than ten feet. The delta extends far into the Gulf of Mexico, and year by year is increasing, owing to the immense quantity of fine mud brought down by the river. Humphrey and Abbot calculated that more than 812,500,000,000 lbs. of sedimentary matter, constituting one square mile of deposit 241 feet in depth, are every year carried down in the Gulf of Mexico by this great river. This, however, only refers to the sediment in suspension. In addition to this the current pushes on before it large quantities of earthy matter—calculated at about 750,000,000 cubic feet, which would every year cover a square mile about twenty-seven feet deep. The uniform width of the Mississippi is one of the curious features in connection with it. At New Orleans it is 3,000 feet wide, and it maintains pretty much the same breadth for 2,000 miles, except in the bends, where it swells out to one or even one mile and a half. The depth is, however, very variable—the maximum being usually about 120 or 130 feet. The great impediment to its navigation, as to the navigation of all other American rivers, is often the shoals, the rafts, and trees brought down by the freshets. Sometimes these will block up the course or get fixed in the mud, or swing backwards and forwards. In the one case the tree is known as the "Snag," and in the other a "Sawyer." Both are dangerous obstacles to navigation, and frequently damage the fragile steamers. These are, however, usually built in air-tight compartments, so that if a hole is knocked in the bows the water only fills a portion of the vessel, and does not therefore sink it. The greatest drift in the Mississippi was that which began to accumulate about the year 1778 in the Atchafalaya, until in 1816 it had extended to more than ten miles in length, over 600 feet in width, and about eight feet in depth. This great accumulation rose and fell with the water, but, nevertheless, it afforded a soil for the growth of bushes and trees, some of which attained the height of sixty feet. In 1835 the work of removing it commenced, and the costly task was accomplished in four years. The Red River raft is even more famous, not only for its size—it was thirty miles long—but for the sums of money which have been spent by the Federal Government in attempting to remove it. Up to a late date this amounts to 7,789,810 dollars, and the raft has not yet entirely disappeared. The "Lower Mississippi" properly begins where the Missouri, so-called, joins; but the name is not usually applied until the river has crossed the "rocky chain" (p. 103), which extends across the channel between St. Louis and Cairo. All below this "chain" in the Mississippi Valley is alluvian, through which the river winds along from bluff to bluff. "Touching these bluffs at Commerce, Missouri, on the west bank," writes Mr. King, one of the students of this changeable flood, "it courses across the valley, passing the vast prairies of Lower Illinois, known as 'Egypt,' on the east, meets the Ohio at Cairo, then strikes the bluffs again at Columbus, on the eastern or Kentucky shore. It skirts these bluffs as far as Memphis, having on its west the broad earthquake lands of Missouri and Arkansas. It then once more crosses its valley to meet the waters of the White and Arkansas Rivers, and skirts the bluffs at Helena in Arkansas, flanking and hemming in the St. Francis, with her swamps and sunk lands. Reinforced by the White and Arkansas, it again crosses its valley to meet the Yazoo, near Vicksburg, creating the immense Yazoo Reservoir on the east bank, extending from the vicinity of Memphis to Vicksburg, and the valleys and swamps of the Macon and Tensas on the west side. These latter have no terminus, save at the Gulf of Mexico, as the river does not approach the Western Bluffs after leaving Helena. From Vicksburg to Baton Rouge the river hugs the Eastern Bluffs, and from Baton Rouge to the mouth, the pure 'delta country,' for a distance of more than 200 miles."*

The Mississippi and the people, and their cities and villages on its banks, would afford ample material for many volumes. But with these introductory remarks we must now say a few words about the political divisions of the country on either side of it; in other words, about the states and territories of the Mississippi Valley proper.

NEBRASKA.

The State of Nebraska has an area of 72,995 square miles, and is divided into sixty-five counties—the north-west portion being still "unorganised," extending in width from north to south about 210 miles, while the length in the central portion is about 420 miles. Lincoln, the capital, had, in 1870, 2,441 inhabitants, and in 1874 about 6,500. Omaha (p. 88) is, however, the chief city, having 20,000 inhabitants in 1874; then comes Nebraska City, with 6,050 (in 1870). The total population of the State, exclusive of Indians not taxed, was, in 1870, 122,993, including 122,147 white, 789 coloured, and 87 Indians. There were 70,125 males, and 52,668 females, 92,245 of native, and 30,748 of foreign birth, one-third of these being Germans. In 1870 the percentage of deaths to the population was only 0·84 per cent. In 1873 there were 6,579 Indians in Nebraska on reservations comprising 892,800 acres, and by the Commissioners' Report of 1875, 6,446. These comprise Santee Sioux, Pawnees, Winnebagoes (p. 109), Omahas, Iowas, Sacs and Foxes, Otoes, and Missouris. The general surface of the country is nearly a great plain, rising by a gradual slope from the Missouri to the Rocky Mountains. The prairies, of which the State chiefly consists, are either undulating or broken into low hills and ridges. There are, however, no hills of any magnitude, except at the west and north-west, where the Black Hills and Rocky Mountains make their appearance. There are no large lakes, but many streams, none of which are navigable except the Missouri, which forms the N.E. and E. boundary of the State. The Platte is a wide but shallow river, and passes through a very fertile valley. The country is not rich in minerals. Iron and coal have been found, but neither in great abundance. Saline deposits and alum are, however, more plentiful. The eastern portion of the country is a good agricultural region, but the west is better adapted for grazing purposes. Large numbers of cattle fatten on the nutritious natural grasses of Nebraska, while numbers of sheep are kept on the luxuriant pastures of the bottom and table-lands. Wood of a natural growth only flourishes on the bluffs and river banks, but planted trees grow rapidly on the prairie. The climate is "dry and exhilarating," the mean temperature in winter being about 22°, and in summer 70°. Of the total area of the State (46,716,800 acres) 647,031 acres were in 1870 in "improved farms." These numbered 12,301, of which 787 embraced less than ten acres each, and only eleven from 500 to 1,000 acres. The greatest number of farmers had from twenty to fifty acres of land. The total valuation of the State was, in 1874, $1,218,843 dollars, which is a rise of more than 25,000,000 dollars in two years. Nebraska is noted for the

* "The Southern States" (1875), p. 262.

liberality of its laws, and more especially for the full exercise of "women's rights" to the fullest extent. Any property which a woman may possess at the time of her marriage, or which she may acquire by her own exertions afterwards, is free from the control of her husband, and not subject to his debts. A married woman may convey her estate, and in every contract or business transaction is treated in the eye of the law exactly as an unmarried woman. She can plead for a divorce on almost any reasonable, and a few perfectly unreasonable, ground. In addition to the usual pleas allowable

VIEW IN NEBRASKA.

in most civilised countries, the Nebraska wife can claim to have the marriage knot untied if her husband has been sentenced to imprisonment for three years or more, if he is an habitual drunkard, if he has deserted her for two years, if he is "extremely cruel"—in fact, if he does anything inimical to the wife's good pleasure.

MINNESOTA.

Minnesota was the nineteenth State admitted into the American Union, and is the twenty-eighth in point of population. Its extreme length, from north to south, is 380 miles; its breadth, from 183 miles in the middle to 262 miles on the south line, and 337 miles near the northern boundary; while its area is 83,531 square miles. St. Paul's, the capital, near the eastern border of the State, had, in 1870, 20,030 inhabitants. At that date the other chief towns were Duluth, Hastings, Owatonna, St. Cloud, Rochester, and Mankato, none of which had 4,000 people; Minneapolis, which had nearly 14,000; Winona, with over 7,000; Red Wing, with more than 4,000; and St. Anthony, with over 5,000. In

WINNEBAGO INDIAN.

1873 the whole population of the State was calculated at 552,150, of whom 36,000 were born in Norway, 21,000 in Sweden, and 12,000 in Germany, while about 30,000 are natives of Great Britain. The density of the population was 5.26 to a mile. Over 75,000 of the 133,000 working people are engaged in agriculture, including 21,000 labourers and 55,000 farmers and planters. The death average is about 1.035 of the whole population. In 1875 5,973 Indians were reported in Minnesota: these, consisting of Chippewas, who were settled on reservations in the central and northern parts of the State, were for the most part engaged in agriculture. Minnesota lies nearly in the centre of the Continent, and occupies the most elevated plateau between the Gulf of Mexico and Hudson's Bay. It thus forms the water-shed of the three great river systems of Eastern North America—that of the Mississippi, that of the St. Lawrence, and that of the Red River of the north. Generally the surface is an undulating plain, with an average elevation of 1,000 feet above the sea, and "presents a succession of small rolling prairies or table-lands, studded with lakes and groves, and alternating with belts of timber." There is no coal in the State, but lead exists, and copper abounds in the mineral belt stretching along the north shore of Lake Superior, chiefly in the form of the virgin metal. Iron of good quality is also found in some places, while, as if to compensate for the want of fuel of other kind, great deposits of peat exist in all parts of the State. Salt springs are found; slate, limestone, &c., are mined; while gold and silver have been discovered, though in small quantities. The soil is fertile, two-thirds of it being well adapted for the growth of all kinds of cereals and roots of the temperate zone. The climate is pleasant: cold, but clear and dry, during the winter, with no great amount of snow; while the summers are warm, with breezy nights, during which time most of the rain falls above lat. 46°. The country is well timbered, pine forests extending to the north, and birch, aspen, ash, and maple abound. Over the central portion of the State a great forest, known as "The Bigwoods," composed for the most part of hard-wood trees, extends. These Bigwoods—the *Bois Franc* of the early French settlers—cover an area of about 1,000 square miles. "Many objects of natural interest," writes Mr. Hawes, "are found in the State. The Mississippi, studded with islands and bordered by high bluffs, presents a succession of picturesque scenes (pp. 81, 101). Mountain Island, with an elevation of 428 feet; Maiden's Rock, celebrated in Indian tradition, on an expansion of the river called Lake Pepin, about 400 feet high; and La Grange Mountain, on the same lake, are all notable. St. Anthony's Falls are celebrated as much for their surrounding scenery as for the descent of the waters, which have a perpendicular fall of only eighteen feet, are further up the river. A few miles beyond, between Minneapolis and Fort Snelling, are the Minnehaha Falls, a romantic and beautiful cascade, with a perpendicular pitch of forty-five feet, flowing over a projecting rock, which permits a passage underneath; Brown's Falls, which have a perpendicular descent of fifty feet, and including the rapids of 100 feet, are west of the Mississippi, on a narrow stream, which is the outlet of several small lakes. There are also falls or rapids on the St. Croix, about half a mile below which is a noted pass, through which the river has forced its way, called the Dalles of St. Croix, and others of less note on various streams. About two miles from St. Paul's is Fountain Cave, an excavation in the white sandstone, with an entrance about fifteen feet in diameter, opening into a chamber 150 feet long and twenty feet wide." It is as an agricultural State that Minnesota must be considered; as a manufacturing region it ranks low. However, the abundant supply of water-power, which is found everywhere throughout the State, will eventually lead to its industrial prosperity. St. Anthony's Falls are estimated as being equal to 100,000 horse power, capable of being utilised through nearly the entire year; and as sentiment does not usually form a marked feature in the character of the Western settler, there can be little doubt that in time this beautiful feature in the scenery of the State will be utilised for such a prosaic purpose.

On Plate XIV we have engraved the Falls of St. Anthony as perhaps the most remarkable natural feature of Minnesota, and the one which it is to be hoped will be most unchanging, though, as already mentioned, the Philistinish eyes of the searchers after "water-power" have their eye upon these beautiful cascades. They are situated about an hour from St. Paul's, and are now a great place of resort for visitors to the West, and deservedly so. They are by no means the finest of the cataracts of the American rivers, but still they have a wild picturesque grandeur almost their own. One can understand the enthusiasm of Père Hennepin, who was so carried away with admiration when he first saw them, that he named them after his patron saint. Even the stolid Indian has perceived the beauty of them. In the Dakota language they are *Rara*—from "irara," to laugh—a name which may be partially translated as the "laughing waters." Long before coming in sight of them the ear hears the deep solemn roar, "the sound of many waters." When they burst into view the appropriateness of the aboriginal name at once strikes us. They are indeed the "laughing waters." They have not the overwhelming grandeur of the Niagara, nor the same height of fall, deafening roar, or lofty character of the scenery. St. Anthony is more within the grasp of human comprehension, and, therefore, as one of the earliest visitors to them remarked, looked upon with more real pleasure. Niagara appears to wear a kind of threatening frown, while St. Anthony greets the pilgrim to its base with a more winning and complacent smile. Yet, on account of the vast body of water continually rushing over the rocky mass in the river's bed, the scene is one of greater sublimity, as well as of beauty and loveliness.* Lumbering is also another great occupation of the State, one-third of the State being estimated as timber land. The total taxable property in the State was, in 1874, valued at 217,427,211 dollars. Educational and other advantages are abundant.

IOWA.

This State presents an area of 55,045 square miles. North and south it extends 208 miles, and east and west about 300. The largest cities in 1873 were Burlington, with (in round numbers) 27,000 inhabitants; Davenport, with 21,000; Dubuque, with 23,000; and Des Moines, the capital, with 16,000. The whole population of the State was, in 1873, 1,251,338, including a few hundred tribal Indians. Diseases of the lungs and chest are very fatal throughout this State, whooping-cough showing a higher mortality than in any other State, except Nebraska and Arkansas. Iowa has usually an undulating surface. There are no mountains, or even high hills, but on the banks of the rivers there are frequent bluffs or calcareous strata, intersected by ravines. These bluffs are the breastworks of table-lands that stretch away from behind them in gentle undulations. The southern portion of the State excels in picturesqueness, abounding with grassy lawns and verdant plains, interspersed with groves and meandering rivulets, and intersected by the larger rivers which flow to the Mississippi, or by the numerous affluents of the Missouri. In

* Bond: "Minnesota and its Resources," p. 150.

the north-eastern part the surface is more elevated; hills and mounds are not uncommon, their tops covered with towering oaks, and the rivers tumble over precipitous ledges of craggy rocks. The north-east section abounds in lead ore, and various other metals, but, nevertheless, contains much excellent land. The unique and admirably-diversified prairies of Iowa are, however, its most distinguishing feature. These natural meadows are covered with a rich coating of coarse grass, forming excellent pasturage, and are not unfrequently interspersed with hazel thickets and fragrant shrubs, and in the season of flowers are decorated with a brilliant garniture of honeysuckles, jessamines, wild roses, and violets.* Iowa is essentially an agricultural State. According to the census of 1870 it produced more wheat and Indian corn than any of the States, except Illinois, and ranked fifth in the production of oats. The State at that date contained 9,396,467 acres of improved land, 2,524,796 of woodland, and 3,620,533 "of other unimproved land." About one-third of the farms contained from twenty to fifty acres, over 41,000 from 50 to 100, 30,000 from 100 to 500, 321 from 500 to 1,000, and only thirty over 1,000. In 1872 the number of acres of "improved land" was 9,987,788. In 1873 the total taxable value of the State property was 364,336,580 dollars; the State taxes yielded 909,464 dollars, at the rate of 2½ mills; while the rest, including the county hospital for the insane, county school, district school, bridge, road, special, judgment and bond, corporation, and railroad, brought the entire revenue up to 10,711,925 dollars.

KANSAS.

This State owes its name to the Kansas Indians, a sept of the Dakota family. It is in the form of a rectangle, extending east and west 410 miles, and north and south about 210 miles, the whole area containing 81,318 square miles. Its chief cities are Atchison, 7,054 inhabitants; Leavenworth, 17,873; Lawrence, 8,320; and Topeka, the capital, which in 1870 had 5,790 people within its borders. In 1873 it was calculated that in the entire State there were 610,863 inhabitants, showing a density of population represented by forty-eight persons to the square mile. The general surface of the State is an undulating plateau, sloping gently from the western to the eastern border, which is not elevated more than 750 feet above the sea at the mouth of the Kansas River. The river bottoms are generally from one-fourth of a mile to three miles wide; but towards the western part of this State, on the Arkansas and Republican rivers, they are from two to ten miles wide. Back from the bottom lands bluffs run to the height of from 50 to 300 feet, with a slope of 20° to 30°. From the summits of these bluffs may be seen a succession of rolling or upland prairies, whose tops are from a quarter of a mile to a mile apart, and from twenty to eighty feet above the intervening valley. The gentle inclination of the ridge is north and south. There is no portion of the State which is flat or monotonous. The surface of Eastern Kansas is chiefly undulating, and presents a succession of rich prairies, grass-covered hills, and fertile valleys, with an abundance

* See also Hall and Whitney: "Geological Survey of Iowa." (1858–59); White: "Geological Survey Reports" (1872).

of timber on the streams. The western half is not so diversified in its scenery, but it has a rolling and varied surface, with every requisite for a fine grazing country.

VIEW OF "THE PICTURED ROCKS," SOUTHERN SHORE OF LAKE SUPERIOR.

The soil is favourable to agriculture, Indian corn, potatoes, and barley being raised in abundance, while the natural grass of the western portion of the State affords excellent fodder for cattle. Coal and stone are plentiful; while oak, elm, black walnut, cottonwood, box-elder, honey-locust, willow, hickory, sycamore, white ash, and hackberry abound. The buffalo, elk, deer, antelope, squirrel, horned frog, prairie dog (p. 44), grouse, wild turkey, wild geese, and other animals, form the most prominent features of its fauna. According to the census of 1870, 5,656,879 acres were in farms, while in 1873 2,982,559 acres were cultivated. The people are chiefly devoted to agriculture, and stock-rearing, and fruit-growing. There is, accordingly, little manufacturing industry, though in time the water-power of the State will force attention to this. In 1873 the total taxable property was returned as worth 125,684,176 dollars, and the tax actually levied was, at 6 mills, 754,105 dollars. The laws of Kansas, like those of most Western States, are extremely favourable to woman. Her pre-marital property is protected, subject entirely to her own control. Neither husband nor wife can bequeath more than one-half of his or her property without the written consent of the other. Divorces may be granted by the district courts for, among other causes, abandonment for a year, drunkenness, cruelty, gross neglect, and imprisonment in the penitentiary subsequent to marriage. Kansas, being one of the border slave States, was long one of the battle-fields of the *pro* and *contra* slavery parties. Finally it adopted the free state principle; and though it suffered greatly during the Civil War, remained loyal to the Federal Government.*

WISCONSIN.

Of Wisconsin, or Ouisconsin, a few words may suffice to give an idea. In length it is 285 miles long from north to south, 255 broad, and its area 53,924 square miles, or 34,511,360 acres, of which 5,899,343 acres are improved land. In 1871 its population was 1,054,670. The surface of the State consists of rolling table-lands; the streams or the watershed between the Mississippi and Lake Superior having falls and rapids. Our friend, the late lamented Dr. Bryce, to whom we are indebted for these notes, remarks that two other ridges cross the State—one in the south-west near the middle, and the other in the south-east, dividing the waters which enter Green Bay from those flowing direct to Lake Michigan. Near the base of the second ridge there is a depression across the State in the line of the Fox River and Lake Winnebago and lower course of the Wisconsin, so that the headwaters of the former come near the Great Bend of the latter. In a very wet season, or on the sudden melting of the snow, the intervening plain near Portage City becomes flooded, and a natural connection is established between the Mississippi and Lake Michigan. Advantage has been taken of this singular feature to maintain a constant connection by means of a canal. Thus, a vessel can pass up the Mississippi into the Lakes, and thence by the St. Lawrence into the Atlantic. Wisconsin is a poor agricultural State. The climate is severe and the soil bad, though towards the south it improves, and supports good crops and excellent pasture. Iron, zinc, lead—the latter in great abundance—are mined; plumbago, saltpetre, gypsum, and fine marl also occur, while the Illinois fields enter the State on the south. The neighbouring lakes, the Mississippi, the Chippewa, and Black Rivers, afford great natural facilities for carriage. Hence the trade of the State is very prosperous; and, in addition to the above, gives employment to

* Hutchinson: "Resources of Kansas" (Topeka, 1871).

about 2,000 miles of railway. The amount of wheat—grown chiefly in the park-like southern prairies—was, in 1870, estimated at 25,606,344 bushels, some of which was sent to Milwaukee from England without transhipment. Among the more interesting features of Wisconsin may be mentioned the earthworks in the form of men and animals formed by the same prehistoric race who mined the copper of Lake Superior; the Devil's Lake, 600 acres in extent, on the summit of a mound 300 feet high; the precipitous shores of Lake Pepin, rising to the height of 500 feet; the high bluffs of the Mississippi and Wisconsin Rivers; the falls of St. Louis (320 feet in sixteen miles); and of the Menomonee (151 feet in one and a half miles); the wild cliffs of Lake Superior, the beauty of which culminates at "The Pictured Rocks" (p. 113), on the shores of Michigan, &c.

ILLINOIS.

This State takes its name from the Illinois tribe of Indians, members of the Abenquin family. It was the eighth admitted into the Union, and is now the fourth in population. It has an area of 55,410 square miles, and had a population, in 1870, of 2,511,096 free whites, and 28,762 free negroes. Springfield, which possessed at that date a population of 17,364, is the capital, but Bloomington, Aurora, Cairo (said to be the "Eden" of "Martin Chuzzlewit"), Galena, Jacksonville, Ottawa, Pekin, and Rockford, are all considerable towns, while Peoria, Quincy, and, above all, Chicago (which had, in 1872, more than 365,000 people), are larger than the capital. Next to Louisiana and Delaware, Illinois is the flattest State in the Union. Its greatest elevation is 1,150 feet, and its mean elevation about 550 feet above tide-water. Accordingly, though lead, coal, and other mines are worked in the State, agriculture and manufactures are still its greatest industries. "In every part of the State the plough may pass over thousands of acres without meeting even so much as a pebble to impede its course." Timber is, however, abundant, but is unequally distributed — whole districts being bare. The climate, though in most places healthy—except in the low-lands near the river—is excessively cold in the winter, and unusually hot in the summer. Pleasant breezes, however, modify the heat, so that the climate is not so unfavourable for out-of-doors labour. Illinois, in 1870, showed a return of more acres under cultivation than any other State, and its yield of wheat, Indian corn, and oats was also the highest. The total number of farms was 202,803, 53,240 of which were from twenty to fifty acres, 68,450 from 50 to 100, 35,940 from 100 to 500, and 502 containing 1,000 acres and over. According to the same census, Illinois ranked sixth in the amount of capital invested in manufactures and in the value of the products. Its products of butchering, distilling liquors, planing timber, and packing pork, brought it ahead of any other State of the Union. In 1871-2, 1,631,025 pigs met death at the hands of the Illinoisians, the aggregate cost of the hogs being 22,691,394 dollars, and the total weight of the lard alone 69,808,463 lbs. The total valuation of the State property was, in 1873, 1,341,361,842 dollars, and the receipts into the public treasury for the two years ending in 1872 were 9,899,603 dollars. The State debt diminished from 12,280,000 dollars in 1863 to 1,706,750 dollars in 1874, a circumstance rather phenomenal in the history of these public luxuries.

CHAPTER VII.

THE UNITED STATES: MISSOURI; ARKANSAS; THE INDIAN TERRITORY; LOUISIANA.

In this chapter we enter upon what used to be called the Slave States, and which, now that slavery is abolished in the great Republic, are usually comprised, with others yet to be spoken of, under the name of the Southern States. They have much in common, owing to the great preponderance of negroes in most of them, and the fact that they were for four years leagued together in a struggle—still fondly, or, it may be, bitterly, remembered as the "lost cause"—and by the fact that in many of them the population is of a different origin from that in the New England or the northern portions of the Union. We can most conveniently say what we can afford space for under the heads of the different States.

MISSOURI,

Speaking of it simply as a geographical entity, is 277 miles long from north to south, and from 200 to 312 miles from east to west, with an area of 67,380 square miles, or 43,123,200 acres. Its chief towns are Jefferson City (the capital), St. Louis (p. 117), Boonville, Hannibal, Independence, and Lexington. Its great rivers are the Mississippi, which forms its boundary for 470 miles, and the Missouri, which borders the State in the west for a certain distance, and, with its affluents, the Osage, Gasconade, &c., passes through it from west to east. South of the Missouri the State is undulating, rising into mountains towards the borders of Arkansas, while the north portion is level prairie-land, with rich bottoms, and lofty picturesque bluffs on the rivers. Coal, iron, lead, and limestone are abundant; but the severe winters and hot summers do not prevent the rich land from producing maize, wheat, hemp, tobacco, peaches, nectarines, grapes, &c. Cotton is grown in the southern portions; and the Germans, who have settled in the State in great numbers, have introduced wine-making. The manufactories of Missouri are chiefly ironworks, distilleries, and breweries. St. Louis has a large trade, and now comprises upwards of 300,000 inhabitants, though in 1775 it was a French trading post, with only 800 people settled in the vicinity. The whole population of the State was, in 1870, 1,715,000, an increase in fifty years of 1,618,514. The population is now more than 2,000,000. Politically, Missouri is the child of a compromise, whose epitaph was written in letters of blood. During her early days the soil was disputed by Gaul and Indian, and afterwards by Frenchman and Spaniard. St. Louis, which was the centre whence the early explorers set out, experienced many vicissitudes. During the early days of the American Revolution it suffered from Indian raids, and the terrorism of the banditti who then haunted the Mississippi, until it began gradually to get acquainted with the gaunt American pioneers who floated into it on their way to the trapping-grounds of the mysterious West, and knew it as *Pain Court*—short bread—owing to their painful experience that in St. Louis provisions were not always to be had. The Osages were in those days ever hanging about the outskirts of the town, and many an unfortunate pioneer, in the years closing last century and beginning this, was burned at the stake, impaled, or otherwise slowly tortured to death by the ruthless savages, whose name and appearance are now in the region which once knew them mere matters of tradition. In 1804 Missouri was formally surrendered to the United States, and with the advent of the Stars and Stripes the new territory and St. Louis took giant strides on the road to prosperity. "Looking down," writes Mr. King, "upon the St. Louis of to-day

VIEW OF ST. LOUIS, MISSOURI.

from the high roof of the superb temple the Missourians have built to the mercurial god of insurance, one can hardly believe that the vast metropolis spread out before him represents the growth of only three-quarters of a century. The town seems as old as London. The smoke from the Illinois coal has tinged the walls a venerable brown, and the grouping of buildings is as picturesque and varied as that of a Continental city. From the water's edge, on ridge after ridge, rise acres of solidly-built houses, vast manufactories, magazines of commerce, long avenues bordered with splendid residences; a labyrinth of railways bewilder the eye, and the clang of machinery, and the whirl of a myriad wagon-wheels, rise to the ear. The *levée* is thronged with busy and uncouth labourers, dozens of white steamers are shrieking their notes of arrival and departure, the ferries are choked with traffic; a gigantic and grotesque scramble for the almost limitless West beyond is spread out before one's vision. The town has leaped into a new life since the war, has doubled its population, its manufactures, and its ambition, and stands so fully abreast of its wonderful neighbour, Chicago, that the traditional acerbity of the reciprocal criticism for which both cities have so long been famous is latterly much enhanced. The city, which now stretches twelve miles along the ridges branching from the water-shed between the Missouri, the Merrimac, and the Mississippi Rivers, flanked by rolling prairies, thickly studded with groves and vineyards; which has thirty railway lines pointed to its central depôts, and a mile and a half of steamboat at its *levée*, 1,000 miles from the sea; whose population has increased from 8,000 in 1835 to 450,000 in 1874; which has a banking capital of 19,000,000 dollars; which receives thousands of tons of iron ore monthly; has bridged the father of waters, and talks of controlling the cotton trade of Arkansas and Texas; is a giant in comparison with the infant settlement wherein, in a rude cottage, Colonel Stoddard had his head-quarters when the United States assumed territorial jurisdiction. In those days the houses were nearly all built of hewn logs, set up on end, and covered with coarse shingled roof. The town then extended along the line of what are known as Main or Second Streets. A little south of the square called the *Place d'Armes*, Fort St. Charles was held by a small garrison; and in the old stone tower, which the Spaniards had built, debtors and criminals were confined together. French customs and French gaiety prevailed. There were two diminutive taverns, whose rafters nightly rang to the tales of hairbreadth escapes told by the boatmen of the Mississippi. The Chouteaus, the Lesas, and the Lablodies were the principal merchants. French and English schools flourished. Pelts, lead, and whiskey, were used as currency, and negroes were to be purchased for them. The semi-Indian garb of the trapper was seen at every street corner; and thousands of furs, stripped from the buffalo and the beaver, were exported to New Orleans. The mineral wealth lying within a hundred miles of St. Louis had hardly been dreamed of; the colonists were too busy in killing Indians and keeping order in the town to think of iron, lead, coal, and zinc. The compromise which gave the domain of Missouri to slavery checked the growth of the State until after it had passed through the ordeal of the war. How then it sprang up like a young lion, confident of the plenitude of its strength, all the world knows. St. Louis under free institutions has won more prosperity in ten years than under the old *régime* it would have attained in fifty."

There is now no more cosmopolitan city than St. Louis. "Conservatism" is a reproach, and "go-aheadishness" the order of the day. After the war, the difficulties of "reconstruction" troubled the Missourians little. The negro element was never great in the country. Accordingly, the 100,000 freed-men never constituted, as they have in other of the ex-slave States, an element of trouble; while the 1,100,000 dollars' worth of taxable property is quite well able to bear, and in time will clear itself of, the burdens imposed by the war.

The ferry across the Mississippi was at one time a great feature in St. Louis. Its approaches were crowded with a multitude of the most cosmopolitan and motley character. "There may be seen the German emigrant, flat-capped, and dressed in coarse black, with his quaintly-attired wife and rosy children clinging to him; the tall and angular Texan drover, with his defiant glance at the primly-dressed cockneys around him; the 'poor white' from some far Southern State, with his rifle grasped in his lean hand, and his astonished stare at the extent of brick and stone walls beyond the river; the excursion party from the East, with its maps and guide-books, and its mountains of baggage; the little groups of English tourists, with their mysterious hampers and packages, bound towards Denver or Omaha; the tired and ill-uniformed company of troops 'on transfer' to some remote frontier fortress; the smart merchant in his carriage, with his elegantly-dressed negro driver standing by the restive horses; the horses of over-clothed commercial men from the northern and western cities, with their mouths distended by Havana cigars, and filled with the slang of half a dozen capitals; and the hundreds of negroes who throng the *levées* in summer, departing in winter, like the swallows, at the slightest tinge of snow, or of the fog which from time to time heightens the resemblance of the Missouri capital to London. Before the bridge was built the *levée* on each side of the river was a kind of pandemonium. An unending procession of wagons, loaded with coal, was always forcing its way from the ferry-boats up the bank to the streets of St. Louis, the tatterdemalion drivers urging on the plunging and kicking mules with frantic shouts of 'Look at ye!' 'You dar!' These wagons on busy days were constantly surrounded by the incoming droves of wild Texan cattle, that, with great leaps and flourish of horns, objected to entering the gangways of the ferry, and now and then tossed their tormentor high in the air; and by troops of swine bespattered with mud, and dabbled with blood drawn from them by the thrusts of the enraged horsemen pursuing them. Added to this indescribable tumult were the lumbering wagon trains laden with iron or copper, wearily making their way to the boats; the loungers about the curbstones singing rude plantation songs or scuffling boisterously; the nameless ebb-tide of immigration scattered through a host of low and villainous bar-rooms and saloons, whose very entrance seemed suspicious; and the gangs of 'roustabouts' rolling boxes, barrels, hogsheads, and bales from wagon to wharf, and from wharf to wagon, from morning to night. Below the bridge, the river, gradually broadening out, was covered with coal-barges and steam-tugs, and above it, along the banks, one saw, as one still sees, dark masses of homely buildings, elevations, iron foundries, and various manufactories; while along the shore are moored thousands of logs, fastened together in rafts." The building of the bridge has changed much of this, though the crossing of the Mississippi is to this day a point at which a varied mass of men and manners run into a narrow space. Except in the names of the streets, little of the old French character in the town remains. On the Illinois side of the river still lingers the old village of Cahokia—a moss-grown relic of the past—its venerable Church of Notre Dame des Kahokias, the oldest building in the West, standing a strange monument of a decayed civilisation among the magnificent residences of the merchant-princes, whose hives are on the opposite shore.

The street-life of St. Louis is varied and interesting, and the endless auction-houses which here, as in most Western and Southern cities abound, is a curious feature of commercial life. From morning to midnight the doors of the establishments are open, and the jargon of the seller may be heard ringing out above the rattle of wheels, as the tramway runs past his *locale*. "The genius who resides behind the counter," writes a visitor, " is usually some graduate of the commerce of the South. Accustomed to dealing with the ignorant and unsuspecting, his eloquence is a curious compound of insolence and pleading. He has a quaint stock of phrases, made up of the slang of the river and the slums of cities, and he begins by placing an extravagant price upon the article which he wishes to sell, and then decreasing its value until he brings it down to the range of his customers."

The Irish element in St. Louis keeps it lively in one sense, and the Germans impart animation to it from a pleasanter point of view. In even the fashionable sections of the city, ladies receive their friends seated in the porches and on the front doorsteps of their houses. On a summer evening hundreds of groups of ladies and gentlemen may be seen thus seated, behaving, however, towards each other with all the etiquette of the drawing-room. Society is, as may be imagined, exceedingly free and agreeable. The ladies especially, having in many cases been educated both in the East and West, have the culture of the former region, with the frank cordiality of the latter. Added to this, German manners prevail. In the summer evenings, the more fashionable beer-gardens may be seen crowded, not only with Teutonic, but with American families drinking läger beer and listening to the music, without—as they would in the Eastern and, indeed, in many of the Western States—regarding the custom as a dangerous "Dutch" innovation. The German migration began in 1830, but it was not until 1848, when the revolution set multitudes seeking a home across the Atlantic, that the best foreign element in the United States came in any great number. Since that year, however, multitudes of these have year after year arrived in America, and Missouri has ever been a favourite place of settlement for them. At the present moment it is calculated that, in St. Louis alone, there are more than fifty thousand of them, but, including children born in the city of German parents, the number must reach nearer 150,000. The Teutonic element, though an excellent drag on the more excitable American, and one which has kept the State from rushing into many excesses in the troublesome times preceding and during the war, is now fast merging into the American, though imparting to native society many of the heartier features of European life wanting among the austere New Englanders. In another generation or so, the fusion between the two races will be almost complete—thus giving "the Western people" features even more distinctive than they possess at present. In commerce, agriculture, literature, journalism—indeed, in every enterprise—the Germans are among the foremost; while in politics—a *rôle* for which they do not much care in "the States"—Germans, like Carl Schurz, have played not an undistinguished part. There are several newspapers in their own language, but any other attempt to preserve a separate national feeling has been abandoned, as being neither practicable nor, if it were possible, anything but mischievous. Still there are whole districts in the city which might be in Dresden or Hamburg, and there naturally, from the number of compatriots gathered in one place, the manners of the Old World prevail. They have their dancing, singing, and gymnastic clubs; but when it was attempted to establish the "Germania," into which no American was to be admitted, the scheme utterly broke down. German influence, however, has put its impress on the public schools of the city, and on the more thoughtful class. In no city in America is Kant's

"Critique of Pure Reason" better known, or Hegel more laboriously tried to be understood, as witness in the discussions of the "Philosophical Society" and in the "St. Louis Journal of Speculative Philosophy."

The sweltering heat of a Missouri July day, a day in which the stranger incontinently melts, has little influence on the keen St. Louis merchant. He clothes himself in the

thinnest of linen, and the broadest-brimmed of straw hats, and then, armed with a fan, seats himself in his office, or plies his trade in the Exchange, determined that in cotton he shall bring his city ahead of Houston, Galveston, or New Orleans, and as a grain distributor "whip" Chicago and the lake cities. St. Louis is already the railroad centre of the Mississippi Valley, and is fast becoming the metropolis of the steamboat traffic of the mighty river which flows past its wharves. Low water and railroad competition have for some years past made the steamboat men draw long faces. Still, the day when the railways will absorb the traffic of the Mississippi Valley is far off, and the river yet remains picturesque and vivacious, by dint of the quantity of all kinds of craft that grate their keels against the St. Louis *levée*—boats, barges, rafts, the floating palaces, the strong, flat-bottomed Red River packets, the cruisers of the Upper Mississippi and of the turbid Missouri, the coal, copper, lead, and iron-laden barges, the huge grain-laden arks, each capable of receiving 100,000 bushels of grain; while rafts of every size and shape are scattered along the giant stream, to use Mr. King's simile, "like chips and straws on a mountain brook." Nearly 3,000 steamboat arrivals are annually registered at the port of St. Louis. "Drifting down on the logs come a rude and hardy class of men, who chafe under city restraint, requiring now and then stern management. Sometimes one of these figures, suddenly arriving from the ancient forests on the river above, creates a sensation by striding through a fashionable street, his long hair falling about his wrinkled and weather-beaten face, and his trusty rifle slung at his shoulder." Steamboating on the Mississippi is a proverbially dangerous business. In 1872 there were over 550 disasters on the river and its tributaries, not many attended with loss of life, but all resulting in great loss of property. The record of these disasters is not without its grim humour. One can hardly suppress a smile at the announcement in the terse, expressive language of the river, that "*Phil. Sheridan* broke loose at St. Louis," or that "*Hyena* broke her engine," "*Lake Erie* ran through herself," "*Mud Hen* blew up at Bellevue," "*Enterprise* broke a wrist at Cairo," "*Lady Johnson* blew out a joint near Alton," "*Wild Cat* sunk a barge at Rising Sun," "*Humming Bird* smashed a shaft," "*St. Francis* broke her doctor," "*Daniel Boone* was crowded on shore by ice," or "*John Kilgore*, trying to land at Evansville, broke nine arms." The river men have not been satisfied to confer upon their beloved crafts the names of heroes and saints. "They rake up all fantastic cognomens which the romance of the century or the slang of the period can afford, bestowing them upon clumsy and beautiful crafts alike, while they pay but little regard to congruities of gender or class. The *Naid* may be a coal barge, or *Dry Docks* a palace steamer." The cotton steamers have often to load at a plantation somewhat in the manner figured on p. 121. The valuation of the property in St. Louis was, in 1874, 475,000,000 dollars.

Kansas City, built on the southern bank of the Missouri, just below the mouth of the Kansas, was, at one time, a station for the wild "bull whackers," or bullock-wagon drivers, who came there to load their "prairie schooners" from the Missouri river boats. At that period it was touchingly characterised by one of the frontier men, who gave it its unenviable notoriety, "There's no railroad west of Junction City, no law west of Kansas City, and no God west of Hays' City." Railways gave it life, as they gave life to a hundred other similar "cities." More than 50,000 people have now settled here. They have bridged the Missouri, and control the markets from the Missouri to the Rocky Mountains, have a valuation of 42,000,000 dollars, and a jobbing trade alone of 17,000,000 dollars or more. "Packing" pork and beef is one of the great trades of Kansas City. In 1872 more than 20,000 cattle and 120,000 swine passed through the hands of its butchers. In twenty years Kansas City will become one of the great centres of the West. The influence and mark of Southern manners, Mr. King tells us, have vanished from the western sections of Missouri, just as have the hard drinking, hard riding, blustering, bowie-knife-wearing "border ruffian," to whom the old state of matters was so congenial. A new type has arisen, and swept out of sight those who prevailed "befo' the waw." New England people are arriving, so that manners in such sections settled by the Northern people have a curious mixture of Colorado and Maine. In the wilder parts of the State ruffianism still prevails, though it finds it rather a difficult task to hold its own. The "Pike County" man, once a proverb and still a familiar name, has a healthy idea of rough, average justice, and of the necessity of keeping down future rascality. He never troubles a court of law with trying a horse thief. Such a miscreant always obtains short shrift and a long lariat. Missouri still requires many more people before it can attain the maximum of that prosperity of which it has the factors within it. The river bottoms are as rich as the Valley of the Nile. "In journeying beside them on the Missouri Pacific Railroad one sees immense spaces but recently cleared of forest, dotted with log cabins and barns, and their omnipresent appendages, the hog-yards, filled with dozens of swine; yellow cornfields, acres on acres, extending as far as the eye can reach among the gnarled trees; men and women cantering to market on bare-backed horses, and young children staring from the zigzag fences." The life is like the product of the soil, dirty and coarse, but it is full of vigour. Towns spring up in Missouri, as they do in all the West, as Mark Tapley remarked, "spontaneous like, owing to the fertility of the soil," and the spread of railways. The capital (Jefferson City) is a prosperous town of 7,000 inhabitants, and has been the seat of government since 1828, previous to which it was rather peripatetic. But it is a principle in America generally to have the capital near the centre of the country, and, if possible, never to submit it to the influences of a large metropolis, which are often for evil, and rarely for good. Though Missouri suffered grievously by the war, yet she withheld herself from the cause of Secession, and found her profit in so doing.

ARKANSAS

Is, in length, 242 miles; in breadth, 170 to 229; in area, 52,198 square miles; and in 1870 its population was 484,471. In other words, the State is the size of England, and its population that of Manchester. The surface is, in the eastern part, to a great extent, low and marshy, though the centre and west are hilly, or covered with undulating prairies; while, beyond the Ozark Mountains, it consists of an elevated plain stretching on towards the Rocky Mountains; and the climate is more salubrious than that in the low lands bordering on the Mississippi. The Washita, White River, Red River of the South, St. Francis, &c., drain it. The soil varies greatly, though in nearly every part cotton, maize, wheat, and oats grow well; while among the mineral products are iron, coal, zinc, lead, manganese, gypsum, and salt."

Little Rock, the capital (population 20,000), originally a French settlement, is situated on the Arkansas River. The State was as early as 1674 a French colony; and during the period of John Law's Mississippi scheme, Louis XV. made a grant of land on the Arkansas to that well-known adventurer. In 1763 it was handed over to Spain, and again in 1800 returned to France. In 1803 it was purchased by the United States; in 1836 it was received into the Union as a slave State; and during the Civil War threw in its lot with the Confederacy. The Arkansas River, one source of which we have engraved

* Owen: "Report on the Geological Survey of the State" (1858 and 1860).

(p. 105), rises in the Colorado Rocky Mountains, at an elevation of 10,000 feet above the sea, and after describing a course of 2,170 miles, and draining an area of 178,000 square miles, falls into the Mississippi at Napoleon. It is navigable by steamboats of slight draught to forty miles above Little Rock, and during high water to Fort Gibson, 150 miles higher up. Napoleon, it may be remarked in passing, was once a flourishing town, but is gradually slipping away into the stream. It did not bear a good reputation in past days. Illustra-

VIEW OF "THE TOWER ROCK," ROCK ISLAND, ON THE UPPER MISSISSIPPI.

tive of the manners and customs of the Napoleonites, various grim anecdotes are related. It was here that a man showed a casual passer-by on a steamboat a pocket full of coin, with the explanatory remark, that "he had bin among the boys last night, when they were having a frolic." Brawls were frequent; and brawls at Napoleon—as is the custom of "Arkansaw"—"always ended in burials." In those days an Arkansas neck-blister was a familiar name for a bowie-knife, from the fact that the desperado of these parts usually concealed his "wepen" down his back, when he did not carry it inside his knee-boot. It is a common phrase to say that "Arkansas is all swamp and backwoods," but this is erroneous. Many tracts along the Mississippi are unmitigated swamps, but the lands which extend from Napoleon to Memphis (on the Arkansas side) are fine, rich lowlands, which may in time be drained, settled, and if carefully cultivated, become a home where the negro may grow prosperous.* Out of seventy-three counties, fifty-one are watered by navigable streams. "The climate varies with the location, but none could be healthier than that of the romantic mountain region, more invigorating than that of the thick

VIEW OF A LEVÉE ON THE LOWER MISSISSIPPI.

pine forests in the lower counties, or more malarial than are the undrained and uncleared bottom lands." For long Arkansas was unpermeated by railways; while the evil reputation, more or less exaggerated, of the citizens, kept out of the country quiet-loving travellers who had no desire to eat at table with a revolver alongside their plates, or to whom the pistol-shots, heard almost daily along the banks of the river, were not musical sounds. Hence the settlers who were rough remained so, while the quiet people, of whom there were always numbers in the territory, were in a hopeless minority. Matters are,

* Henry: "Resources of Arkansas."

however, greatly altering for the better, and Little Rock, since "the iron horse snorted in its streets," has wonderfully changed. It is now a pretty well laid-out town, and many of the streets are beautifully shaded with azaleas, japonicas, China, and peach trees, the lovely magnolia, box elders, and elms affording a striking contrast to some of the rude lowland towns near the river, or the "log-built unkempt settlements in the interior, where morals are bad, manners worse, and there are no comforts or graces." Still, in the more remote parts of the State, society is very rude and very bad. Whiskey and the universal habit of carrying arms are the chief causes of this, while the slight regard for human life which prevails throughout the country operates badly against the preservation of order, which has not yet recovered from the shock of the Civil War.

Perhaps, on the whole, the "coloured" people of Arkansas exhibit the most cheering prospects. Many of them are men of education and refinement, and having obtained a prominent place in the Government, exhibit more fitness for it than the liberated slaves have usually done in the reconstructed States. The whites are, to a great extent, ambitionless, and even the most enthusiastic seem dubious about the future prospects of the country since the "surrendah," as they style that final scene in the drama of the Confederacy, enacted at Appomattox Court House. Yet the resources of the State are immense. It contains nearly 12,000 square miles of coal, while the lead belt extends diagonally across the State, and the lead and silver mines promise well. Kaolin (China clay), gypsum, copper, and zinc are found in profusion; manganese, ochre, and earth-paints exist in many places, while there are great quarries of slate, whetstone, limestone, and marble. Finally, iron has been discovered at various points, and, combined with the stores of coal, ought to enrich the State. The forests of yellow pine, cypress, cedar (juniper), cottonwood, mulberry, oak, hickory, and pecan (*Carya oliræformis*) are also valuable, while there are still 8,000,000 acres of State lands ready for settlers to occupy. All that is required is a decently good government, and some education. Surely it is not hoping against hope to expect these to come in time to unhappy "Rackensack?"

THE INDIAN TERRITORY.

The "Indian Territory" is neither a territory nor a state. It is simply a something devised to avoid a difficulty. From time to time the aborigines who once inhabited the country east of the Mississippi, and some of those also living west of that river, have been removed thither, and little by little the government is encroaching on the domain assured to them, as so many other domains have been ensured from time to time, and as often, to the disgrace of the Government, broken. Its area is 52,780,000 acres, and the population scarcely enough to make a town of moderate size. The number of Indians scattered over its vast plains and among its mountains has been estimated as follows:—Cherokees, 17,500; Choctaws, 17,000; Creeks, 13,500; Chickasaws, 5,500; Seminoles, 2,500; Osages, 3,500; Sacs and Foxes, 468; Shawnees, 670; Cheyennes (p. 128) and Arapahoes, 3,390; Confederate Peorias, 170; Eastern Shawnees, 80; Wyandottes, 150; Quawpaws, 236; Senecas, 188—in all, about 65,000, who inhabit a territory, the great distances in which are only partially unabridged by railways, and are thus separated from each other by barriers of language and custom, that there is hardly any intercourse among them. "The land lies waste because there are not hands to hold the plough, and the country remains a desert because the Indian jealousy refuses to allow the white man to make it blossom like a rose;" and the Indians are quite justified in so doing. Blossoming like a rose may be a very pretty natural operation to look at, but it may be a sight purchased too dear when it is performed at the expense of the Indian for the sake of a white man, to whom he is indebted for nothing but much wrong. No white man is allowed to settle here, unless by permission of the Indians, and even then leave is never given unless the applicant be of fair character, has married a dusky bride, and is willing to relinquish his allegiance to Uncle Sam. The Indian holds firmly by the treaty of 1837, by which he was to be allowed to live undisturbed here so long as he settled in it. Previous to that date most of the country belonged to the Osages, who have now almost entirely disappeared. Yet before the tribes could be removed into it the country was deluged with blood. From the unwillingness of the Seminoles to remove from their fair land originated the Florida War, and time after time smaller difficulties have arisen. The Indians in the territory receive annuities, derived from the sale of lands in other parts. These amount to about $95,000 dollars *per annum*. A railway spans their territory, but the Indians keep away from it, and travellers by it rarely see one of the aborigines. The Cherokees have made the greatest advance in civilisation, and have a ruling voice in all that concerns the polity and resources of the aborigines. Their general status, we are assured by one of the latest visitors to them, is not below that of the white backwoodsman. They are good agriculturists, though living remote on farms which they hold in common, yet to which each individual has a perpetual right of occupancy. All the Cherokee land is vested in the nation. "A man may sell his improvements and buildings, but not his land." Yet the Indian will never be a farmer in the true sense of the term. He has no idea of making gain; he only desires to "make a living." Throughout the various nations settled in the Indian territory there is an utter neglect of internal improvements. "An Indian highway is as difficult as the Vesuvian ascent, and none of the magnificent rivers were bridged before the advent of the Missouri, Kansas, and Texas Railway." The Indian agent, who serves the double purpose of being Indian trustee and diplomatic agent from the Government of Washington, or, to give a closer simile, bears to the chiefs of the tribes much the same relation as the Residents at the native courts of Hindostan do to the Princes to whom they are accredited, has, however, here and there, suggested a few improvements, which have been followed up. The government of the Cherokees I have sketched in another work.* It is, to a great extent, modelled on that of the American State governments. The chiefs—a "principal" and a second chief—are elected every four years. They have an upper and a lower house of legislature, the former continuing in power four, and the latter two years. There is a supreme court, with three judges, and the usual staff of district judges and sheriffs. Tahlequah is the capital. It looks like an ordinary South-West town, with nothing particularly Indian about it. The Choctaws and Creeks have the same general form of government, while the Seminoles have vested their executive authority in twenty-four band-chiefs, all controlled by a "principal," an absolute

* "Races of Mankind," Vol. I., p. 228.

autocrat, with an irrefragable veto; however, all the tribes join in a general council, presided over by the Superintendent of Indian Affairs for the Southern Superintendency; but at this meeting only the rendition of criminals, the joint action in regard to land, and

A CHEYENNE INDIAN CHIEF.

similar matters, which are of comity between the nations, are legislated upon. When the Civil War broke out, the greater number of the Indians sided with the Confederates. Hitherto they had known next to nothing of the Northern people, or of Northern politics, while, on the other hand, most of these were slave-owners, some of the Cherokees owning from 200 to 300 slaves, and negroes who had settled among the Indians also held slaves.

LOADING A COTTON STEAMER.

The Cherokees sent to the war one regiment, under the command of General Stand Weatie, a full-blooded Indian. Meantime, the Indian territory was occupied alternately by Northern and Southern troops, and plundered by both. The loyal Indians removed into Kansas, while those who had adopted the Southern cause emigrated with their families into Texas. Many of the Indians also enrolled themselves in the service of the North, and now that they have returned to their homes the feud is still, to some extent, kept

CREEK INDIANS

up, and will not die until this generation has passed away. Before the war many of the Indians were rich. It was not uncommon for a well-to-do stock-raiser to own 15,000 head of cattle. He was a poor Indian who had not twenty. During the war they lost all this property, but the general Government having appropriated money for the purchase of new stock, the tribes now possess nearly as much as before.

Though education—even high-class—and religion are spreading among the Indians, yet their manners are still very aboriginal, and will doubtless long remain so. They have all the native American's taste for strong waters and for patent medicines, of which alcohol forms an ingredient. Hospitality is unbounded. No sooner does an Indian take a wife than all her relatives come to live on him, and remain for life, or until they have impoverished him. Mothers-in-law are in the Indian territory a tyranny undreamt of by the Benedicts of other lands. At present the Indians live very peaceably together. The chief feud is the land question; one party wishes to allow white men to settle in the territory while the majority scornfully reject any such proposition. It has been remarked that an Indian suspected of wishing to dispose of land to any white personage speedily dies, while a Caucasian who shows rather too strong an aptitude to ingratiate himself with his red-skinned neighbours runs a risk of sudden and mysterious disappearance. "This superb country," writes Mr. King, to whom we have been indebted for these facts, "unquestionably one of the most fertile on the globe, is a constant source of torment to the white men of the border, in whom the spirit of speculation is very strong. The hardy citizen of the South-West bears no ill-will towards the Indian Southern tribes, but it irritates him to see such vast tracts of land lying idle. He aches to be admitted to the territory with the same privileges granted to Indian citizens, viz., the right to occupy and possess all the land that they may fence in, and to claim all that remains unfenced within a quarter of a mile on either side of the fenced lots. He is crazed with visions of the far-spreading, flower-bespangled prairies, the fertile foot-hills, the rich quarries, mines, and valley lands. He burns to course at free-will over the grassy regions, where even the Indians raise such fine stock. And now that the railroad has entered a protest against continued exclusiveness on the part of the Indians, he thunders at the northern and southern entrances of the territory, and will not be quiet."* And, of course, he will not thunder in vain. Voters never do in the United States, or, for the matter of that, anywhere else. Politicians who value place more than the national good faith will arise, and perjure the Government for the sake of popularity, as they have done elsewhere. Time after time have reserves been "secured" to the Indians further west, who have no right to the Indian territory proper, which solely belongs to the tribes to whom it was originally secured, or to those whom the possessors may choose to admit, and again and again they have been deprived of it, and made to remove elsewhere. And so it will be here, and with the usual consequences.

LOUISIANA.

Louisiana is at once one of the richest, and, in its history, one of the most romantic of the States of the Union, but, at the same time, since the war, one of the saddest pictures of bad government and civil discord in the history of any country. To-day, it is to me the simile of one of its modern historians, Paradise lost. In twenty years it may be Paradise regained. Its possibilities are unlimited. Its bayou-penetrated soil, its rich uplands and vast prairies, comprise everything necessary for a great commonwealth. But a gigantic struggle is in progress within its bounds—a battle of race with race, "of the picturesque and unjust civilisation of the past, with the prosaic and levelling civilisation of the present. For a century and a half it was coveted by all nations; sought by those great colonisers of America—the French, the English, the Spaniards. It has in turn been the plaything of monarchs and the butt of adventurers. Its history and traditions are leagued

* "Southern States," p. 200.

with all that is romantic in Europe and in the Eastern Continent in the eighteenth century. From its immense limits outsprang the noble sisterhood of South-Western States, whose inexhaustible domain affords an ample refuge for the poor of all the world." How romantic was this history, from the time when De Soto explored until the day when the First Napoleon, terrified lest New Orleans should fall into the hands of the English, sold, in 1813–14, the "Earthly Paradise" to the United States for fifty million francs, and how full of romance, let other works relate.* It is enough for us to jot down a few particulars regarding its modern condition.

With the exception of Florida and Texas, it is the most southern of the United States. It measures, from north to south, 200 miles, and from east to west, at its widest part, 290 miles, with a total area of 40,790 square miles, or 26,105,600 acres. Situated at the mouth of the greatest river on the continent, it contains within its limits the delta of the river, which is intersected by numberless lesser rivers and bayous, and filled with lakes. Yet, contrary to the popular idea on the subject, even in America, we must warn the reader against hastily concluding that it is, throughout its entire extent, a low, wet, swampy region. Much of it is, no doubt, a great plain of wonderful fertility, with an indefinite succession of dense jungle, tangled swamps, marshes, lakes, sloughs, cane, and cypress brakes. To those whose only knowledge of the State has been derived from sailing through it by way of the Mississippi, such ideas are very natural. However, if we penetrate back from the river, our ideas will speedily alter. The whole surface may be divided, according to Colonel Lockett, into two grand areas, the hilly and the level country. The hilly parts may be again subdivided into three regions, different from each other in configuration of the surface, in soil, in forest growth, and in fertility. These divisions have been named the good uplands, the pine hills, and the bluff lands; while the level country may be subdivided into pine flat, prairies, arable alluvial lands, wooded swamps, and the coast marsh. The alluvial region can be best seen from the deck of a Mississippi steamer when the river is "well up." The panorama which spreads out before the eye as the boat speeds along is one which is apt to cling for a long time to the traveller's memory. For years afterwards there will rise up before him the stately residences of planters, half hidden in groves of magnolia, pecan, and live oak, the massive sugar-houses with their tall chimneys; the neat villages of negro cabins whitewashed, and arranged in prosaicly systematic order; the broad fields of cotton, Indian corn, and cane sweeping back in green waves to the blue line of distant woods; the sleek, fat cattle and horses grazing on the level or embankment; and the verdant meadows stretching down at places almost to the water's edge. The scenery is not majestic, but it is, nevertheless, pleasing. If we penetrate one of the bayous, overhung with moss-covered cypress, willows, and live oak, it will give another idea of Louisiana. "He must ramble," writes the enthusiastic Louisianan, on whose stores of information we are drawing, "along the clear, quiet lakes, whose polished surfaces reflect with perfect fidelity everything above and around them, save where float the broad leaves and bright flowers of the *graineaurelli*; he must penetrate the tangled swamps with their primeval forests standing as the repre-

---

* Gayarre: "History of Louisiana," &c.

sentatives of past ages, with their dense jungle of luxuriant cane, with the ponds, sloughs, and morass where the wild fowl nestle among the water-lilies; and if he has anything of an artist's eye, he will everywhere see new and peculiar beauties." The coast marshes are composed, when the soil is at all firm, of rich black mould, but the greater portion of these are liable to be overflown by the river, or by the tidal-overflow, and often the green meadow-like covering is only a treacherous crust concealing the unknown depth of water and oozy mud below. The Great Prairies are pleasanter spots. They are of the rolling type (p. 74), being waved like the billows of the sea. In fact, it is difficult to ride through these prairies without being reminded of this resemblance, and the grass moved by the wind ripples like the bosom of the ocean, while the dark blue border of woods are like the distant shores, the projecting spurs like capes and promontories, the "coves" like bays and gulfs, while the clumps of trees that here and there diversify the grassy surface look like islands in the sea. The population of the Louisiana prairies are mostly of Acadian origin, and are usually very thrifty. The people are kind and hospitable, fond of little junketings and "socialities"—as such merrymakings are termed in that quarter—but shy of strangers, especially if he speak no French. The prairie region is healthy, and altogether, perhaps, the best part of the State. The bluff lands of the Mississippi we have already alluded to, while little need be said of the pine flats, except that, like the pine hills, the soil is thin, sandy, and poor, and the surface a perfect level. The woods are so open that in travelling through them a herd of deer may be often seen half a mile ahead, and the surface is covered with grass, and contains many wet oozy places called "bay galls," from the clumps of bay-trees (magnolias) always found in them. In the pine hills are peculiar tracts called "hogwallow lands," characterised by a stiff, sticky, calcareous soil, which in wet weather is terribly muddy. The inhabitants of the pine woods are proverbially poor, but as a set-off to this unhappy characteristic, are honest, moral, virtuous, simple-hearted, and hospitable— not invariable corollaries to a lack of goods. Louisiana is a region of great wealth, but until there is a great infusion of new blood by immigration it can never, under the altered state of affairs, do much good. The negro, for the time being, is rejoicing in his freedom, while the white man is too often allowing his natural indolence to gain the mastery of him, or to relapse into the despair which was begotten of the ruin wrought in the sad struggle into which the State plunged in the dreary years of 1861–65. Even election day (p. 133), once so busy an event to New Orleans, fails, except spasmodically, to excite him. A better day is, however, beginning to dawn.

New Orleans is Louisiana on a condensed scale—so far as the people, the chief industries, and the ways of thinking are concerned. The "Crescent City," as the inhabitants love to style it, still retains much of its old French character—in its manners and customs, in the veracity of its inhabitants. "Business here, as in foreign cities, has usurped only half the domain; the shopkeepers live over their shops, and communicate to their commerce somewhat of the aroma of home. The dainty saloon, where the ladies' hairdresser holds sway, has its doorway enlivened by the baby; the grocer and his wife, the milliner and her daughter, are behind the counters in their respective shops. Here you pass a *café*, with the awning down, and peering in can distinguish half a dozen bald, rotund old boys, drinking their evening absinthe, and playing picquet and *vingt-et-un*, exactly as in France. Here, perhaps,

AN ELECTION DAY IN NEW ORLEANS.

is a touch of Americanism: a lazy negro recumbent in a cart, with his eyes languidly closed, and one dirty foot sprawled on the side walk. No! even he responds to your question in French, which he speaks poorly, though fluently. French signs abound; there is a warehouse for wines and brandies from the heart of Southern France. Here lives a group of French negroes, the buxom girls dressed with a certain grace, and with gaily-coloured handkerchiefs wound about an unpardonable luxuriance of wool. Their cavaliers are clothed mainly in antiquated garments rapidly approaching the level of rags, and their *patois* resounds for half a dozen blocks. Turning into a side-street leading off from Royal, or Chartres, or Bourgoyne, or Dauphin, or Rampart Street, you come upon an old little wine-shop, where the cobbler sits at his work in the shadow of a grand old Spanish arch; or upon a nest of curly-headed negro babies ensconced on a tailor's bench at the window of a fine ancient mansion; or you look into a narrow room, glass-fronted, and see a long and well-spread table, surrounded by twenty Frenchmen and Frenchwomen, all talking at once over their eleven o'clock breakfast; or you may enter aristocratic restaurants where the immaculate floors are only surpassed in cleanliness by the spotless linen of the tables; where a solemn dignity, as befits the refined pleasure of dinner, prevails, and where the waiter gives you the names of the dishes in both languages, and bestows on you a napkin large enough to serve you as a shroud, if this strange *mélange* of French and Southern cookery should give you a festal indigestion. The French families of position usually dine at four, as the theatre begins promptly at seven, both on Sundays and week-days. There is the play-bill, in French, of course, and there are the typical Creole ladies, stopping for a moment to glance at it as they wend their way homeward. For it is the shopping hour; from eleven to two the streets of the old quarter are alive with elegantly, yet soberly-attired ladies, always in couples, as French etiquette exacts that the unmarried lady shall never promenade without her maid or her mother," and so on. The French quarter is not so highly educated, perhaps, as the American portion of the city; but it has decidedly more of what the Americans style "elegance." The Spanish and French residents never attempt to set the fashion in New Orleans. They live quietly among themselves, match-making and marrying, dining and giving dinners, church-going, shopping, and calling upon each other in simple unaffected fashion. The average American in New Orleans knows little of his French neighbours, and does not always appreciate them. You cannot talk for five minutes to the go-ahead New Orleans business man of the dominant race without his telling you that "we have a non-progressive element amongst us; it will not be converted." At the same time he will laud the many virtues of his French neighbour, though at the same time finding it difficult to forgive him for taking so little interest in public affairs, and in the daily whirl of life, which is the normal existence of the Anglo-Saxon. The older families, Mr. King tells us, still speak with bated breath and touching pride of their "ancestors who came over with Bienville," the founder of the city, or with such and such Spanish governor, and many a name has descended untarnished to its present possessor through centuries of valour and adventurous achievement. Yet the grooves in which Louisianan society once ran have been so broken up by the Civil War, that old residents declare that since "the surrender"—from which all things are dated in the South—four hundred years seem to have passed over the State. "The Italy of Augustus was not more dissimilar to the Italy of to-day than is the Louisiana of to-day to the Louisiana before the war. There was no longer the spirit to maintain the grand unbounded hospitality once so characteristic of the South. Formerly the guest would have been presented to planters, who would have entertained him for days in royal style, and who would have sent him forward in their own carriages, commending him to the hospitality of their neighbours. Now these same planters are living on corn and pork." Most of these people have now vanished from their old homes; some—happier still—are beneath its soil. Ladies of culture and refinement, whose incomes were gigantic before the war, are now washing clothes for their daily bread. The misery, the despair of hundreds, are beyond belief. Every other white face bears a look of sadness indescribable, though subdued it may be by manly courage, even by hope. But it is still there. For the time being the negro has got the upper hand; and having learned only too well the art of the political plunderer, is now impoverishing the State still more. However, that day is likely soon to pass away. A better state of affairs is dawning; and though Louisiana can never attain the property it did under the old artificial state of things, with negro slavery as the coping-stone of the social system, it may in time recover from the wreck and ruin which have overtaken it.

Cotton is the staple trade of New Orleans. In the American quarter cotton is the only subject of conversation during certain hours of the day. "The pavements of all the principal avenues in the vicinity of the Exchange are crowded with smartly-dressed gentlemen, who eagerly discuss crops and values, and who have a perfect mania for preparing and comparing the estimates, as the basis of all speculations in the favourite staple; with young Englishmen, whose mouths are filled with the slang of the Liverpool market, and with the skippers of the steamers from all parts of the West and South-West, each worshipping at the shrine of the same god. From high noon until dark, the planter, the factor, the speculator, flit feverishly to and from the portals of the Exchange, and nothing can be heard above the excited hum of these conversations except the sharp voice of the clerk reading the latest telegrams."

In 1824–25 the cotton crop of the United States was 569,249 bales; in 1859-60, it was 4,861,292 bales. In 1861 came the war. After the war was over of course there came a temporary lull in cotton produce; but it may be mentioned, as proving that under free labour, with all its drawbacks, in reality more cotton was produced than under the slave system, that in 1870–71, notwithstanding the depression in trade, 4,362,317 bales were produced. In fact, under slavery, the planters left everything to their overseers, and did not obtain all their lands were capable of yielding. A large proportion of all the cotton product passes through New Orleans, and is piled up in its *levées* and wharves. Most of the estates are now worked by the freed men on shares, and on the whole is working well. The sugar interest was at one time more important than even the cotton trade to New Orleans. But that time is past. The *levée* system of the Mississippi is well seen at New Orleans. These *levées*, or embankments, are necessities for the river-side towns, to prevent inundations. Accordingly, the people of Louisiana have stuck to the maintenance of their *levées* with all the pertinacity of the Dutch for their dykes, and for the same reason. They have built and endeavoured to maintain more than 1,500 miles, or 51,000,000 cubic feet, of these works within the State. The cost of the present system was about 17,000,000 dollars, but it is estimated that as much more will be requisite before it can be completed and perfected. On page 125 a typical specimen of one of them is sketched, and by way of contrast on page 124 another view of the river more than 1,700 miles from the Gulf of Mexico. Next to cotton comes sugar, as the of the Louisianan industries, then rice, wheat, and cattle-rearing. In 1874 the sugar product of the State was estimated at 134,504,691 lbs., while in 1861 it was 528,321,500 lbs. In 1874 it was estimated that 101,963 barrels of rice, each barrel weighing 230 lbs., were yielded by the State.* There are orange orchards in the State producing 3,000,000 oranges annually. The "moss," or *Tillandsia usneoides*, is another vegetable product of Louisiana. About 20,000 bales are annually gathered, for the purpose of stuffing

A STREET IN NEW ORLEANS.

mattresses, chairs, &c. Tobacco and other tropical crops are also reared. Finally, rock-salt and crystallised sulphur may be mentioned as among the mineral riches of Louisiana. Iron is also scattered over the State; coal abounds in certain regions; and petroleum wells are found in one or two counties.

In 1810 the population of the State was 76,555. In 1870 it was 726,915, of whom the whites were by about 2,000 in the minority to the negroes. New Orleans, in 1722, had 100 wooden houses and about 200 inhabitants. In 1800 it had 8,000 people within its limits; in 1860, 168,675; and in 1870, 191,418.

* Bennett: "Louisiana as it is," 1876, p. 225.

CHAPTER VIII.

THE UNITED STATES; TEXAS; MARYLAND; VIRGINIA.

STILL taking Mr. King, one of the latest and certainly the best of American students of the ex-Slave States as our guide, we descend the Mississippi River on our way

PICKING COTTON

to Texas. We might, it is true, have taken the railroad route to the Gulf by the Missouri, Kansas, and Texas Railway; but the old steamboat journey will enable us the better to cover the ground we must traverse in the course of this section. Within fifty or sixty miles of the mouth the river banks become too low for cultivation. No longer do you see the great sugar plantations and the negroes busily cutting the cane, or boiling the expressed juice, while the air is dark with the piles of *bagasse* burning. The river now broadens until, on reaching the "Heads of the Passes," it separates into several streams. Through these channels the Mississippi passes into the Gulf of Mexico (p. 141). Across the mouths of these passes bars of sand are formed, owing, no doubt, partially to the sea-water pressing back the sediment-laden current. The deposition of the mud is also due to the fact that when salt water is mixed with fresh water holding any sediment in suspension the precipitation takes place much more speedily. This is a question which has been determined by experiment, and certainly has an important bearing on the formation of these bars at the mouths of rivers. When the water is high the current of the Mississippi can be seen as far as fifteen miles out to sea, the fresh water being as sharply defined as if a torrent of oil had been poured on the surface of the gulf. "Sometimes," we are told, "when a steamer is running through a dense pea-soup-coloured water on top the paddle-wheels will displace it sufficiently to enable one to see clear gulf water rushing up to fill the displacement. The flood-tide runs up underneath the water for a long distance, and at extraordinary high tides is distinctly visible as far as New Orleans, 110 miles above." The bar of the Mississippi is peculiar in this respect, that it is not always of the same depth. Sometimes "mud lumps" will form in the shape of cone-like elevations at the bottom, often in the course of a few hours, so that the pilot may one day find ample depth for the largest ship, and the next ground with one of much less draught. At the present time elaborate works are in progress with a view to improving the mouth of the Mississippi, and all kinds of apparatus have been fixed with the object of effecting a permanent deepening of the entrance to the river. The principle of most of these is either to lift up the mud with an ordinary dredger, or to stir it up so that the current will sweep it out to sea after being so loosened. Of late Mr. Eades has endeavoured to concentrate an increased flow of water in the South Pass as well as across the shoal bars at each end, firstly, by means of jetties, funnel-shaped at the head of the pass and parallel at its lower end; secondly, by closing the Grand Bayou, which diverted a portion of the water after it had entered the South Pass, and prevented it reaching the lower end; and, thirdly, by regulating the proportion of water to be admitted into the Pass à l'Outre and the South-west Pass. The result has been, so far (1877), that vessels of twenty-two feet draught had passed through the South Pass to New Orleans during the month of April, and it may be safely concluded that before long New Orleans will be a port open to vessels of the largest sizes now used in commerce,* and most probably in the end a canal will solve the problem effectually. At the mouth of the river are a few woe-begone-looking villages—the homes of pilots, or of a few "damp and discouraged fishermen," though in early days, before vessels could securely reach New Orleans, Belize and Pilot Towns were important places. Once out of the Mississippi—past its swampy wilderness of shrivelled cypress and stagnant waters—we sight a country full of splendid sugar lands, and immense groves, from the boughs of which the Spanish moss, or *Tillandsia*, hang in long beards, giving a sombre appearance to the long aisles or vistas which can be seen through the depths of the forest. This moss is collected for stuffing (p. 156), but also acts as a kind of scavenger to the close, foul air of these sunless thickets. The traveller is now passing along the shores of

TEXAS,

the coast-line of which, "bordering upon the Gulf of Mexico, from Sabine Pass to the Rio Grande, from the Louisiana boundary to the hybrid, picturesque territory where the American and the Mexican civilisations meet and conflict, is richly indented and studded with charming keys. Trinity, Galveston, West, Matagorda, Espiritu Santu, Aransas, and Corpus Christi

* Eades: "Transactions of the British Association," Plymouth Meeting, 1877.

Harbours, each and all offer possibilities for future commerce. The whole coast, extending several hundred miles, is also bordered by a series of islands and peninsulas, long and narrow in form, which protect the inner low-lying banks from the high seas. The plains, extending back from the coast in the valleys of the Sabine, the San Jacinto, and the Colorado, seem in past centuries to have formed a vast delta, whose summit was probably near the Colorado, and where angles were formed by the Sabine and the Nueces. Great horizons, apparently boundless as the sea, characterise these plains. The wanderer on the gulf sees only the illimitable expanse of wave and alluvial; the eye is fatigued by the immensity, and seeks rest upon the lines of ancient forest which covers the borders of the Colorado and Nueces. Beyond these plains comes the zone of the prairies, whose lightly undulating surface extends inland as far as the Red River, while the mountains on the north-west crown the fertile knolls of rolling country. These mountains are portions of the Sierra Madre (p. 15), which is itself but a spur from the Grand Andean range. Running to the north-west is the State of Coahuila (once a portion of Texas). The Sierra Madre spur bifurcates to enter the Texas of the present, and continues in a north-westerly direction, under the name of San Seba, in whose breast are locked the rich minerals which the Spaniard, during his period of domination, so often and so vainly tried to unearth. The Texan coast sweeps downward and outward by a wide curve to the Mexican boundary. Approaching it from the sea, the eye encounters only a low-lying level of white sand, with which, however, at all hours, the deep colours of the gulf are admirably contrasted.*

The State of Texas was at one time part of the Spanish possessions in America. It then passed into the hands of Mexico; but in 1835 the American settlers, under "Sam Houston," drove the Mexicans out of the country and formed an independent government. In 1845, the "Lone Star Republic" joined the United States. In 1861, it joined the Confederate States, but since the war it has received immense accessions to its population from the Southern, South-Western, Western, and even Northern States, so that of all the seceding States it had suffered least by the Civil War. According to the census of 1870,† there were 2,964,836 acres under cultivation, producing 20,554,538 bushels of Indian corn, 415,112 bushels of wheat, &c. The live stock at that date included 574,611 horses, and 3,990,158 cattle; while the chief manufactures were salt, iron, and woollens; and the principal exports cotton, sugar, tobacco, cattle, and wool. In 1870, there were 818,579 people in the State, and the whole value of the assessed property was 149,732,929 dollars. These figures must, however, be now greatly increased, if we are to have a just view of the enormous strides in prosperity which Texas has made during the last few years. Indeed, any account of the State dating prior to, or, indeed, immediately after, the Civil War, would give a most imperfect, and, indeed, erroneous impression of the "Lone Star State." For this reason, even at the cost of having to slightly abridge the account of some of the States which follow, we shall burden the account of the Great

---

* "The Southern States," p. 161.

† Here, as elsewhere throughout this work, I have given the latest statistics accessible to me. In some cases, however, these are not very material to the accuracy of the account, as they are not average figures, but vary from year to year. Hence it sometimes happens that statistics of an older date give a clearer view of the material and moral progress of the country under description.

Republic with a rather fuller description of things Texan than would have otherwise been called for. One of the largest of the States of the American Union, it is usually spoken of by its inhabitants as divisible into eight sections—Northern, Eastern, Middle, Western, Extreme South-western, and North-western Texas, the Mineral Region, and the "Pan Handle," a section comprising more than 20,000 square miles, at present almost entirely inhabited by Indians. The Mineral Region is so called from a belief that it is, more than the rest of Texas, rich in iron and copper ores, and is in area 50,000 square miles. Between San Antonio River and the Rio Grande, as well as the 700 miles stretched between San Antonio and El Paso, is a vast section given entirely over to grazing herds of cattle, horses, and sheep, or to the predatory Indian, who takes tithes of them. "Across the plains," wrote Mr. Edward King, "runs the famous 'San Antonio Road,' which for 150 years has been the most romantic route upon the Western Continent. The highway between Texas and Mexico, what expeditions of war, of plunder, of savage revenge, have traversed it! What heroic soldiers of liberty have lost their lives upon it! What mean and brutal massacres have been perpetrated along its dusty stretches! What ghostly processions of friar and arquebusier, of sandaled Mexican soldier and tawny Comanche; of broad-hatted, buckskin-breeched volunteer for Texan liberty; of gaunt emigrant, or fugitive from justice, with pistol at his belt and Winchester at his saddle; of Confederate grey, and Union blue, seem to dance before our eyes as we ride over it! The romance of the road and of its tributaries is by no means finished. There is every opportunity for the adventurous to throw themselves into the midst of danger, even within forty miles of 'San Antón,' as the Texans lovingly call the old town; and sometimes, in the shape of mounted Indians, the danger comes galloping into the very suburbs of San Antonio itself."

San Antonio, of all American towns, is the one that has the most distinctly Old World look about it. In some of its quarters the traveller might suppose himself in some country town in Spain, so that it is even more mediævally European than most of the bustling towns of France or Germany. Indeed, this European aspect is typical of nearly all the old Spanish towns of America. They have, to the man wearied with the push and the noise, the dollar-worshipping vulgarity, and the "go-ahead" ways, in which he has no part or place, a calmness wondrously pleasant. In San Antonio, encased amid the trees and flowers of Flores Street, or any of the lovely avenues which lead from it into the beautiful country around, "there seems a barrier let down to shut out the outer world: the United States is as a strange land. In San Antonio, too, as in Nantucket [or in California and the Pacific region generally], you may hear people talking of 'going to the States,' 'the news from the States,' &c., with the utmost gravity and good faith. The interests of this section are not so identified with those of the country to which it belongs as to lead to the same intense curiosity about American affairs that one finds manifested in Chicago, St. Louis, and even in Galveston. People talk here more about the cattle trade, the Mexican thieving question, the invasion of Mexico by the French, the prospect of opening up the silver mines, than of the rise and fall of the political mercury; and the general government comes in for consideration and criticism only when the frontier defences or the Mexican boundaries are discussed." But every day brings San Antonio nearer to the busy world. As yet, however, the railroad is far off, and we can almost sympathise with the old Mexican inhabitants—the "apparently immortal old men and women who are preserved in Chili pepper"—at their dread of the advancing iron horse. The climate in San Antonio is delightful. The enthusiastic inhabitants, with a logic a little "mixed," perhaps, declare that "if you wish to die here you must go

VIEW ON THE SHORE OF THE GULF OF MEXICO.

somewhere else;" and when one sees, month after month, consumptives on their last legs creeping into "San Antōn" to die, and day by day finding renewed life and vigour, it is difficult not to allow that there is not a *fond de vérité* in the cheerful saying.

Galveston is another Texan city, but of quite another type. It is a pushing commercial centre—"a city in the sands"—by the shore, and yet one where yellow fever is the set-off for the wealth which cotton and railways pour into it. Still, externally, it is a pretty town, fragrant with orange and myrtle, with oleander and roses, and other rich-hued blossoms of a semi-tropical country; while to eyes accustomed to the cold twilight of the

North, the evenings are charming beyond words. The thirty-one miles of beach are ever being laved by the restless water beyond. The town itself is built on an island, and the heat is never disagreeably intense, a cool breeze blowing over it day and night, so that the wonder always is, where the fever comes from. Once the home of the gulf pirate, Lafitte, whose followers numbered 1,000, all refugees from justice, and afterwards a noted *depôt* for the sale of negroes from the Louisiana plantations, Galveston is now fast becoming the *entrepôt* for the cotton crop of Northern Texas, and looks forward to a day when it will compete with New Orleans for the honours of being the "gossypeopolis" of America. In Texas there are 20,000,000 acres of cotton-bearing lands, all yet destined to be brought under cultivation by the freed negroes, who, throughout the State, are a much more industrious and prosperous class than elsewhere in the South.

The country in the vicinity of Galveston and the San Jacinto Bay is as well adapted for growing sea-coloured cotton, worth from 200 to 300 dollars per ton (in gold, for, as on the Pacific coast, paper, now almost equal in value to coin, is little seen). In 1873, the export of the ordinary upland cotton from Galveston was 353,502 bales, worth 32,125,806 dollars, while the value of the imports in the latest year I can obtain accurate returns of (1872) was 4,910,292 dollars. From 700 to 1,100 vessels annually enter Galveston harbour, though so shallow is the bay that the small vessels often unload into cars which drive alongside them, while the wharves themselves look strange, owing to the multitudes of long jetties necessary owing to the shallowness of the water in shore. Beef is also exported to a great amount, chiefly to the West Indies, and so the cheery "land crabs," as the Houston people call the Galvestonians, manage to live and prosper. Its present population may be about 31,000, but like nearly every Southern city since the war—where the people have not thoroughly given themselves over to indolence and despair—Galveston is infinitely more prosperous than it could ever have been under the old *régime*.

Houston, the city of "mud turtles," as the Galvestonians, in memory of its all-abounding mud during the wet season take a good-natured revenge in calling the inhabitants, is one of the most promising of Texan towns. Cotton and wheat are also its staples, and as the wheat region of Texas comprehends 40,000 square miles, it must be long ere Malthusian fear of hungry men and no bread become here an accomplished reality; and as Texas could put its wheat into the market two months ahead of the Western States, the prospects of the grain-growing South-West are great indeed. The abundance of food-growing land in Texas will have also an effect of stimulating all kinds of manufactures, owing to cheap food making wages sufficiently low to render goods sufficiently cheaply produced to enter into competition with the same class of wares in the Old World. When Texas was a Republic, Houston was for a time the capital. Audubon, the naturalist, has left us a curious memorial of the city in those days. The residence of the President, "Sam Houston," was a common log cabin. The ornithologist found the head of the Lone Star Government, and the founder, indeed, of its fortunes, "dressed in a fancy velvet coat and trousers, trimmed with broad gold lace, and was at once invited to take a drink with him. All the surroundings were uncouth and dirty in Audubon's eyes; but he did not fail to recognise that the stern man who had planted a liberty-pole in that desolate prairie in memory of the battle of San Jacinto would make Texas an autonomy. They did their rough work in their rough way; but it will stand for all time." Sam Houston was certainly one of the most remarkable men whom America has ever produced, and one who, though much talked about and written about, was, in the contending passions to which his actions gave rise, never fully appreciated at his true value, for good or bad. He was a Virginian, born in 1792, near Lexington. In the war of 1812 he served as a private soldier with a courage that won the life-long esteem of "Old Hickory" (Andrew Jackson). In 1825 he was elected a member of Congress for Tennessee, and in 1829 married the daughter of an ex-Governor, and became himself Governor of the State. Hitherto Houston's life was that of the ordinary prosperous Southern gentleman. Now commenced his wild and romantic career. A few months after his marriage, from causes which were never clearly known, he deserted his young wife, and abandoning country, civilisation, and career, joined the Cherokee Indian nation, was adopted as a son by the chief, and in due time became himself one of the chiefs of that people (p. 126). In 1832 he was in Washington, making complaints against several rascally Indian agents, and fighting duels with their friends. In due time he would no doubt have met the fate of the fire-eater "Chiv." of those days, had not at that period the Texan War opportunely—for him—broken out. This was a new field for his ambition—a fresh opening for his restless, reckless energy. Houston was a man after the "Ranger's" own heart, and soon they elected him commander-in-chief. After various reverses and successes, he fought, on the 21st of April, 1836, the decisive battle of San Jacinto, which at one blow annihilated the Mexican army under Santa Anna, and won the independence of the State. In 1837 he became the Second President, and in 1841 was re-elected. When Texas united itself to the Great Republic, Houston went to Washington as Senator, and there remained until 1859, when he was elected Governor. He opposed secession, but finding his influence powerless to stem the current, he retired into private life, and died in 1863, in the midst of the turmoil of the struggle, his death not creating that noise in the world which in calmer times it would undoubtedly have done. He was a dignified man, yet of scanty book-learning. At school he was not allowed to learn Greek, which he anxiously desired to do after reading a translation of the "Iliad." He then swore he would never recite another lesson, and he kept his word. Of Sam Houston many anecdotes are related, but as this parenthesis, though too short, has yet been too long for the space at our disposal, I most unwillingly spare the reader the narration of these illustrative traits. He was a man fitted for the times and the men, and though he had often difficulty in restraining the impetuosity of his wild "mustangs," every one of whom thought himself as good as the President or anybody else, there were few other men who could have accomplished the task as he did; and, as years roll on, General Houston's name will undoubtedly keep its place as the greatest in the early annals of Texas.

Houston is built on prairie land, and is often exposed to a "norther," a potent breeze, which "comes raving and tearing over the town" at intervals, especially after the prevalence of a series of calm, sultry days. It is exhilarating, but icy. "Suddenly clouds vanished, only a thin mist remained, and after a brief reign of a brace of hours, the 'norther' was over. He is the physician of malarious districts; from time to time purging them thoroughly. Sometimes he blows down houses, trees, and fences, forcing the beasts on the plains to huddle together for safety; rarely, however, in his coldest and most blustering moods bringing the mercury of the thermometer below 25°." Houston is a well-laid-out, pleasant town, neat and spruce, like most Southern towns where the Germans are found in any great numbers, and with a "coloured" population, peaceable and well-to-do. Society is simple, frank, and cordial. The people are hospitable and courteous, proud—as all the Texans are, of their country—in their intense praise of the beautiful State, overlooked by "stars, which Northern skies have never known," a feeling which in Europe we are accustomed almost solely to associate with France or Frenchmen. In the vicinity of Houston can be seen that familiar feature of Texas, the cattle-herd and his lasso, or cord, the loop of which he skilfully throws over the horns, head, or under almost any portion of a four-footed or other animal. It is, however, a too well-known implement in all the open parts of both North and South America to need any detailed description here (p. 115). A Texan on horseback looks like a centaur, so easy and firm is his seat. He is indeed so rarely seen afoot, that the cattle run at a pedestrian, supposing him to be some strange wild animal with designs against them.

In the Wichita region in North-western Texas are magnificent copper deposits; but there the "Indian difficulty" is found in a very pronounced form, though Indians never yet in the history of America were allowed to impede mineral development.

Waco, now a fine town on the Brazo River, with many manufacturing establishments, was once an Indian village, and not long ago the scene of a great battle with the Wacos and Cherokees. Dallas, another town, "grows like an enchanted castle in a fairy tale," while everywhere, in regions once only cattle-runs, fields of cotton, maize, and sugar-cane are springing up. The farmers are in most cases in this part of the country composed of settlers from the old non-slave States. These people are also the most intelligent and cultured of the rural Texans. The northern and middle counties are types of such regions, while the eastern corner of Northern Texas was before the war treaty all held by emigrants from Alabama, Georgia, and Mississippi. Then slavery flourished, but the post-"surrender" times have neither improved them nor their "negroes," and both are gradually going to the wall before the Northerners.

Austin, the Texan capital, is prettily situated in an amphitheatre of hills, the landscape being bordered by the blue Colorado range. Its population is from 9,000 to 12,000, but, like most of the cities selected as State capitals in America, does not seem to rapidly increase in size or citizens. The Legislature, of course, meets here, and is a very free-and-easy assembly. A visitor in 1874 notes that "there were no objections apparently to the enjoyment of his cigar by any honourable senator on the floor of the senate if the session were not actually in progress. The senators sat with their feet upon their desks, and the friendly spittoon handy; but these are eccentricities which prevail in many a State beside Texas."—even it may be added, in the National Congress, not a hundred years ago. Texas, in that process of reconstruction which fell to the lot of all the seceded states after the war closed, had many troubles. For a time lawless men, "equal to anything," and who made the State notorious in the old times, were rampant. Between 1865 and 1868

CATCHING WILD HORSES ON THE PRAIRIES WITH THE LASSO

it is said that there were 900 homicides, while the "Conservative" and "Radical" sections—in other words, the ex-secessionists and the loyalists—fought a battle not always confined to words, for the political mastery. Affairs are now tolerably quiet, though the reconstruction troubles are not yet over, the ex-rebels, having waxed fat, being inclined to kick against the new order of things, which is, however, in all its main features, irreversible.

In Austin are preserved many curiosities of the time when the State was one of the independent governments of the earth, and made treaties, and war, and peace (when they could, though that was rare), like any other sovereign power. In Austin, when it succeeded Houston as the capital, the European governments had their representatives, while the congressional halls swarmed with adventurers, and the city generally abounded with that unsavoury mass known as the scum of the earth. Once in a while there was a great diplomatic muddle in Austin, which threatened serious consequence for a time. Some pigs having been killed for encroaching on the French minister's grounds, the owner used somewhat heated language to his Excellency. The diplomat was grossly insulted, and his master, Louis Phillippe, to show his displeasure at the conduct of the proprietor of the slaughtered swine, prevented the Republic, of which the offender was a citizen, obtaining what was then known as the "French Loan," whereby Texas was nearly ruined. Altogether there were four Presidents of the Texan Republic—Burnet, Houston, Lamar, and Jones—and their history is the history of a stirring and instructive epoch in the lives of rude men, groping after freedom, of bold ones fighting for it in the readiest way they found to their hand, and of adventurers intriguing for power through rascality of a peculiarly Texan type.

In Austin the negro and the Mexican are both familiar figures. The "coloured man" is doing very well in his free state, though, as might be expected, never having formerly known what it was to be the possessor of coin, he is extravagant and improvident to a degree bordering on the condition of a sailor ashore. Sometimes he undertakes long journeys, without the slightest idea of where he is going, and then, finding that he has not money to get back again, "locates" anew. Food and raiment—much of the former and little of the latter—are the articles on which the freed man chiefly spends his money. Swindling prize lotteries and sweetmeats are baits which Sambo can rarely pass by on the other side. Education and journalism flourish in this State, and though there are still within the limits of the commonwealth gentlemen whose manners might bear great improvement without rendering them singular for virtue, yet Texas bears signs of, in time, losing much of its ancient evil reputation. Everybody, of course, knows the old stories, some of which are undoubtedly true, but others so greatly spiced with exaggeration as to be to all intents and purposes lies. When the judge went to Texas he applied to the chief lawyer in Austin to show him a copy of the State laws. "Certainly," was the polite reply, and the attorney produced from a drawer a two-foot bowie-knife. Not very many years ago I was assured by a traveller that sitting one evening in the common room of a Texan hotel he got into friendly conversation with a resident. On parting for the night, his new acquaintance sidled up to him, and, with a suspicious glance around, inquired, "Stranger, what mout have been your name befo' you left the States?" It was only then that my friend became aware that he was in the midst of a community where it was considered an act of prudence to change one's name with the sky! The truth is, that for long the West will always be an asylum for lawless characters and outlaws generally. They congregate in new towns. They haunt the termini of new railways, and appear like ravens over a carcase in the desert when a new digging or other haunt over which the strong arm of the law has not been able to extend itself springs into life; but in time they as mysteriously disappear. They get shot off, die, are hanged; or reform, and at least live cleanly enough to be lost in the crowd of respectable people around them. Texas has been peculiarly unfortunately situated. In its independent condition it was an asylum for many fugitives from justice. It has, in addition, a good deal of the floating desperadoism of the West attracted to it, while, being one of the Southern States, the rowdyism which always hung and still hangs about the old slave regions fell to its share in a disproportionate degree. The people who are ready to employ a revolver on scant provocation are not yet treated with due rigour in Texas. There is a toleration for them as men of spirit—ready to back their opinion with a pistol. The great mass of the people are, however, law-abiding, and determined to drive ruffianism out of the State. Yet, for at least two years after the war, and during the whole time of its continuance, "society" was thoroughly disorganised. "Road agents" and murderers infested the highways, robbing remote hamlets, and effecting jail deliveries in the most unconventional manner, and in some cases doing their evil deeds with impunity. Yet, as a rule, the murdered were of much the same type as the murderers—professional scoundrels and gamblers, whose exit from the world was a good riddance to it. They "died with their boots on," and were rather proud of the circumstance, and, such being the case, assuredly no one could find fault with their self-satisfaction. Mr. King considered that the present condition of Texas may be summed up in a very few words: A commonwealth of unlimited resources, and with unrivalled climate, inhabited by a brave, impulsive, usually courteous people, by no means especially bitter on account of the war, who comprise all grades of society, from the accomplished and polished scholar, ambassador, and man of large means, to the rough, unkempt, semi-barbaric tiller of the soil or herder of cattle, who is content with bitter coffee and coarse pork for his sustenance, and with a low cabin surrounded with a scraggy rail fence for his home. The rough side of Texas always clings to the imagination of men, just as the same aspect of California is the one that is invariably associated with the Golden State.

Texas has, however, its peculiarities, as all American States have, notwithstanding the migratory character of the Americans. In the northern and extreme southern parts of the State the difference between the townsmen and those of the Northern and Western States is not great; yet, in the remote districts, Mr. King—to whom once for all we must acknowledge ourselves indebted for the greater part of these notes—confesses that there may be found more ignorance and less idea of comfort than he could have thought possible in America. An army of cooks is probably more required than anything else for the civilisation and amelioration of Texas; salt pork in the culinary condition called "fried" is not a toothsome article of diet. Rude, uncultivated, rich men of the old school, who rear their cattle by the thousand, are still met with, but they are fast disappearing, and finding that they must either yield or acquiesce, sullenly tolerate frame, or board houses, railroad stations, and hotels as necessary evils. In old times, unthrift and slovenliness were the characteristics of a Texan family. They counted their wealth in "niggers," much in the same manner as the Russian landowner did until the emancipation of the serfs; and though boasting that he was worth a hundred thousand dollars, the possessor of this wealth might be living more meanly than the poorest labourer in the North. "The only amusement of the paterfamilias was a hunt, or a ride to the country seat in court time, where, in days when every one carried arms, there was usually some exciting event to disturb the monotony of existence—perhaps to disturb existence itself. There was no market, no railroad within hundreds of miles, no newspaper, no school, save, perhaps, some private institution, miles from the farm or plantation, and no intellectual life or culture whatever."

The rich slave-owner was a kind of patriarchal savage, proud of his dirt and ignorance. The heroic epoch of the struggle for independence being over, thousands of persons settled down to such a life as this, and thought it vastly fine. What a magnificent awakening has come to them! The mass of people in the interior have still a hearty scorn for anything good to eat. The bitter coffee, and the greasy pork, or "bacon," as it is always called, still adorns the table of most farmers. A railroad president, inspecting a route in Northern Texas, stopped at a little house for dinner. The old lady of the homestead, wishing to treat her guest with becoming dignity, inquired in the kindest manner, after having spread the usual food before him, "Won't ye have a little bacon-fat to wallop your corn dodgers in now, won't ye?" This was the *acme* of hospitality in that region. Now and then in these days of immigration a housewife will venture a timid "Reckon ye don't think much of our home-made fare, do ye?" when the visitor is a stranger; and, indeed, he shows up on his face his wonder that a well-to-do farmer's stout sons and pretty daughters are satisfied with pork and molasses, and clammy "biscuits,*" with no vegetables whatever. The negro is responsible for the oceans of grease which form such a feature in Texan cookery. The black cook liked it, and his easy, indolent "owners" accepted his taste, as they accepted certain peculiarities in his dialect. To this day a Texan countryman will say "dat 'ar" and "dis yer," and say "furder" for further. One phrase will always mark out a Texan among a crowd of South-Western people. This is, "I reckon *so*," the accent being put on the last word. "Two sights and a look" is another Texan phrase, though neither so common or so intelligible. Profanity is a characteristic of the whole West. The English vocabulary of reputable words is not sufficient to express the force of the Occidental's feelings. He supplements them by what he calls "swear words," and the rest of the world oaths. Mr. King tells us in Western Texas, owing perhaps to the people's long contact with the Spaniard and Mexican, the profanity is more highly coloured, vivid, and dramatic than in the rest of the State. In parts of Northern Texas, however, the force of language exceeds anything that the reader is likely to encounter elsewhere. In Western Texas it is fantastic, almost playful. "I once travelled from Galveston to Houston in the same car with a horse-drover who will

---

* A "biscuit" is in America what would be called a "roll" in England, though of a very solid description, and generally yellow with saleratus.

serve as an example. This man was a splendid specimen of the Texan of the plains, robust, and well formed. There was a certain chivalrous grace and freedom about all his movements which wonderfully impressed me. His clean-cut face was framed in a dark, shapely beard and moustache, which seemed as if blown backward by the wind. He wore a broad hat, with a silver cord around it, and I felt impelled to look for his sword, his doublet, and his spurs, and to fancy that he had just slipped out of some Mexican romance.

VIEW ON THE BALTIMORE AND OHIO RAILROAD (MARYLAND).

His conversation was upon horses, his clear voice ringing high above the noise of the car wheels, as he laughingly recounted anecdotes of adventures in ranches in the West, every third word being an oath. He caressingly cursed, he playfully damned, he cheerfully invoked all the evil spirits that be, he profaned the sacred name, dwelling on the syllables as if it were a pet transgression, and as if he feared it would be too brief. Even in bidding good-bye he cursed as heartily as an English boatswain in a storm, but always with the same cheeriness, and wound up by walking off, lightly laughing, and murmuring blasphemous assent to his friend's last proposition."

Texas since the war has almost increased in property as greatly as she did after her annexation to the United States. In 1845 the Lone Star Republic had 150,000 inhabitants. It has now considerably more than 1,000,000. But the wide plains are destined to support many millions more. At present cattle are the chief denizens of these seas of grass. It is not uncommon in the great plains between San Antonio and the Rio Grande for a single individual to own 200,000 head. A Texan likes to talk of a "purchase of 60,000 head." In 1872 there were 150,000 driven overland from Western Texas to Kansas, through the Indian territory, and in 1871 as many as 700,000 were driven across. The Kickapoo and Comanche Indians, as well as the Mexicans, are, however, harrying the beeves of Texas. The end of this will be swift vengeance. Things have almost already come to a head, and a Texan is not the most patient of mortals under wrong. The inevitable destiny of Mexico is to be annexed to the United States. This event will come sooner or later, and, all things considered, perhaps better sooner than later.

We have digressed concerning Texas more as a matter of convenience, for physically it might have been better joined in the description of Kansas and the Indian Territory, or in that of Mexico, in a future chapter. For the same reason we shall now leave the Mississippi Valley, and, at one jump, land ourselves under the shadow of the Alleghany, or Appalachian Mountains, which may be said to separate the northern States, so called, from the southern ones. We shall then work southwards from Maryland to Mississippi, and northward by the lakes, until we pass through the eastern States in a hurried journey, and so bid farewell to the great Republic of North America.

MARYLAND,

as a State, must occupy some space, as it comprises cities of the importance of Baltimore, and unless we consider Washington as in a territory of its own—the district of Columbia is—it has within its borders the capital of the United States. Maryland was one of the old colonies, and was founded by George Calvert, the first Lord Baltimore, in disappointment at his efforts to form settlements in Newfoundland. It was at first intended to call it Crescentia, but Charles I., when the charter of incorporation was presented for his signature, struck out Crescentia, and inserted Maryland, to do honour to his queen, Henrietta Maria of France. Lord Baltimore was made absolute proprietor of the colony, his quit rent being simply two Indian arrows annually, and one-fifth of all the gold and silver found within the limits of the territory. In 1659 Baltimore was built, in the midst of a rude region almost entirely occupied by Indians. In 1752 it had but twenty-four houses, and three years later we find the Indians coming within eight miles of the little city, and creating such terror that palisades were erected to defend the place, while the women and children were placed in vessels in the harbour. The population of the city is now 350,000, and the trade of the town has grown so enormously since the war that few of those who then knew Baltimore would have recognised it. The State has equally prospered since slavery was abolished. It has an area of 9,500 miles of land, and the waters of the Chesapeake covers 1,000 more. These waters, from the enormous quantity of fine oysters and fish living in them, and the flocks of wild ducks, and multitudes of terrapins and crabs obtained, are quite equal in productiveness to the land, while the shores of the beautiful bay are crowded with market-gardens, which supply Baltimore, Philadelphia,

Washington, and New York with their finest vegetables. Wheat and Indian corn prosper, and along the eastern shore, where the grape prospers, wine is made in considerable quantities, and of excellent quality. Tobacco is also a profitable crop, though an exhausting one. Here, under the thriftless culture of the old times before the war, the soil in many places was "worn out," but it is now, under a better system of agriculture, getting re-fertilised by marl and other manures. In the latter part of the seventeenth century as many as 100 ships would sail annually to England and the West Indies with tobacco, "raised" in Prince George's County alone. On the southern border of Montgomery County are the great falls of the Potomac, one of the best water-powers in the United States. The mountain district of the State is fertile, and interspersed with lovely valleys, well adapted for stock-raising, and the culture of wheat. Maple sugar is also made here to the extent of many thousand pounds weight per annum. The iron ores of Cumberland, and the coal mines of Alleghany County, are among the mineral resources of Maryland. At the close of the war 89,000 slaves were emancipated in the State, and, as has been the rule throughout most of the Southern States, flocked to the towns. Accordingly, the old planters finding it impossible, or at best very difficult, to procure labour, are now anxious to break up their great estates into smaller farms, for sale, thus affording an opportunity of settlers acquiring rich lands at a low price, and adding one more element to the revolutionary faction which the Civil War introduced into the old Slave States.

Baltimore was the first city in the United States to inaugurate a railway. It was laid on the 4th of July, 1828, by Charles Caroll, of Carollton, the last surviving signer of the Declaration of Independence. Within half a century Baltimore population has increased nearly 300,000, and the railway, after long struggles, is now one of the most successful in America. At first the cars were drawn by horses, and, indeed, even after steam-engines were introduced, it was not uncommon for the train to be beaten in speed by a fast-trotting horse, galloping alongside the line. In 1873, 7,250,000 bushels of grain were exported from Baltimore, chiefly owing to the impetus the trade received from the elevators erected at Locust Point. In 1873, the Baltimore and Ohio Railroad brought 2,752,178 tons of coal to Baltimore, while the coffee and flour trades of the State are also important. As a sugar market, Baltimore nearly ranks second to New York, and in timber, corn, cotton, and petroleum, the commerce of this active hive of industry is also rapidly increasing. This oil is chiefly received from Western Virginia and Pennsylvania, and the total exports in 1873 were 3,170,995 gallons. Copper, pigs, oysters, and whiskey are among the other multifarious means which bring wealth to the Baltimoreans. The capital invested in whiskey is alone 5,000,000 dollars, and the receipts from the sales 6,000,000 dollars. Twenty thousand people find the means of livelihood in dredging and tinning the "gentle oyster," and making lime out of its shells, or in printing the labels for the tins in which it is exported. "When the spring comes, and the great army of *employés* who have been occupied with the oysters during the winter would otherwise be idle, the fleet of schooners and boats penetrate all the streams flowing into the Chesapeake, and their crews purchase from the orchards and market-gardens along those streams thousands of tons of fruit and vegetables. The oyster packeries are transformed into manufactories of savoury conserves. Peaches, pears, apples, berries, tomatoes, pickles of every imaginable kind, are so prepared that they can be exported to any part of the world. Large kegs are annually sent to Hindostan, to China, to Japan, and throughout Middle and Western Europe." Maryland altogether draws an annual revenue from the oyster trade of between 50,000 and 60,000 dollars. Added to this, the clubs—famous for luxurious terrapin—the picture galleries, and the Academy of Music, give Baltimore a claim to be considered in a somewhat higher light than a mere hive of money-grubbers. The monuments of Baltimore have led it to be called the "monumental city," while the schools and universities

VIEW OF JEFFERSON'S ROCK, CEMETERY HILL, HARPER'S FERRY, MARYLAND.

—particularly that known as the John Hopkins—are famous even in a country where education is so flourishing as in the United States.

Maryland has within its limits some fine scenery. The railway traverses a beautiful valley—the Monocacy—between the Monocacy River and the Catoctin Mountains. It traverses the battle-field at South Mountain, running at the foot of a precipice for three or four miles (p. 119), and passing Haggerston Junction, enters the celebrated gorge at the village of Harper's Ferry. Harper's Ferry will ever be famous in American history as that spot just on the borders of Maryland and Western Virginia where John Brown made his heroic attempt to bring freedom to the slaves, and whose blood may be said to have fertilised the seeds which eventually sprung into fulness of life in President Lincoln's Emancipation Proclamation. But long before that date it was a place to which the reverent American pilgrimed. Here is the rock on which Thomas Jefferson is said to have sat when he wrote his "Notes on Virginia." It commands a fine view of the junction of

THE BRIDGE ACROSS THE POTOMAC AT HARPER'S FERRY FROM THE HILL.

the Potomac and Shenandoah, in the Gorge, which is noted as one of the most picturesque bits of Southern scenery (p. 152). Before the war Harper's Ferry contained 3,000 inhabitants. Its population at present comprises about 1,600 whites and 700 negroes. The Potomac, which runs in Western Virginia, and rushes down through the Alleghany Mountains, traverses the northern part of Western Virginia, and divides that State from Maryland. At Harper's Ferry, the Baltimore and Ohio Railroad crosses the river over a fine bridge (p. 153). The village was named after Mr. Robert Harper, a native of Oxford, who established the first ferry over the Potomac, and who was one of the earliest adherents of the revolutionary party in the United States. Before the war it was the site of a national armoury, and where the little engine-house still stands John Brown made his defence against the Orleans and the Virginian militia, when he was planning his raid for the purpose of freeing the slaves of Virginia; and not far from Harper's Ferry stands the hamlet of Charlestown, where that heroic, high-minded, though indiscreet man was in 1859 executed under the laws of the State of Virginia.

Harper's Ferry was a famous spot during the Civil War. The Confederates took possession of it just after the Federal troops had destroyed the armoury and 15,000 stand of arms to prevent their falling into the "rebels'" hands, and for long it was the base of supplies for Bank's and Fremont's armies when they were "operating" against Stonewall Jackson. The population of Maryland in 1870 was 605,497 whites and 175,391 blacks, and its debt at the date of writing, over and above its assets, 6,219,172 dollars. The Chesapeake and Ohio Canal Company, however, owe the State some 20,000,000 dollars, which is at present a non-interest-paying debt. As a specimen of the proper feeling which prevails through the State to the "freedman" we may add that the last Legislature voted 50,000 dollars for the establishment of schools for the coloured children.

The *Federal District of Columbia* was established so as to put the country immediately surrounding the capital out of the turmoil of State politics, though territorially it is within Maryland. It consists of sixty square miles, its chief towns being Washington and Georgetown. Its population in 1870 was 131,700, of whom 82,278 were whites and 43,404 coloured. Up to 1846 the District of Columbia was 100 square miles in area, but in that year Alexandria County was incorporated with Virginia. For long the District sent no representative to Congress. Now, however, like the rest of the territories, it sends one delegate who can speak, but is not allowed to vote. The capital, Washington, like all cities built to order, is as yet rather raw and unknit together, "a city of magnificent distances," as it has been sometimes styled. It is situated on the left bank of the Potomac, and is distant thirty-nine miles from Baltimore, 136 from Philadelphia, 226 from New York, 120 from Richmond, 1,203 from New Orleans, 2,000 from San Francisco, and 300 from the Capes of the Chesapeake. At the city the Potomac is one mile broad, and deep enough to admit the largest vessels, though its trade is small. Its population was, in 1870, 109,199, but the Federal capital owes its chief importance to the numerous public buildings, nearly all of a very handsome and tasteful description, the non-official portion being, as a rule, mean and scattered.

CHAPTER IX.

THE UNITED STATES: VIRGINIA; NORTH AND SOUTH CAROLINA.

THE "Old Dominion," as the Virginians proudly style their State, was one of the oldest of the British colonies in America, the most loyal while it was under the English crown, the most gallant in resisting the encroachment of the old country, and in the whirligig of events the most determinedly bitter of all the States which in 1861 seceded from the Union, and fought the fight, which ended at Appomatox Court House, in the same State. Virginia, "mother of presidents," never does anything by halves. First sighted by Sebastian Cabot, in 1498, and again explored along its shores by Sir Walter Raleigh, it was colonised in 1607 by a party of "gentlemen of no occupation, without family," a few labourers and mechanics. In honour of their sovereign, the "Virgin Queen," Elizabeth, they named the colony Virginia. Then disaster overtook the colonists, who were saved from total ruin by the famous Captain John Smith, whose name is so indissolubly mixed up with the early history of the country. Next came Lord Delaware with supplies and emigrants. One of the latter, John Rolfe, married Pocahontas, the daughter of Powhatan, the principal chief, and so gained the friendship of the Indians. To this day the Randolphs in Virginia, with what truth may be doubted, trace their descent from this couple, whose romantic history has often been told, though it must be confessed that when robbed of the picturesque surroundings, it bears a much more prosaic aspect than we are usually led to associate with it.* However, there must be no scandal about the "Princess Pocahontas," whose swarthy sire, "the Emperor of Virginia," became an English peer, under the title of Lord Roanoke. Still later the aristocratic colonists had their number diluted by an infusion of convicts— a free shoot for their moral refuse being one of the uses to which the English Government until comparatively recently put the fairest portion of the lands which they annexed. In 1671 the population was 40,000, and the Governor, Lord William Berkley, said he was thankful that they had no free schools or printing, which he considered the greatest evil a State could labour under. In 1773 Thomas Jefferson prepared the Declaration of Independence, and drafted the document as it now stands; and up to 1825 four out of the five presidents had been Virginians. Finally, when the Southern States seceded, Richmond, the chief city of Virginia (p. 160), became the capital of the Confederacy, and after a gallant struggle was restored to the family of States on the 20th January, 1870. The history of Virginia is thus the history of the Union. To this day the English origin of the great Transatlantic Republic is better seen, in the manners, customs, and ways of thought of the people, than in any other portion of the United States. A Virginian has ever been the proudest and most aristocratic of men, and as the State is again beginning to be colonised by English settlers, it is likely that its Old World feeling will continue, now that the curse of slavery has passed away, without apparently affecting the State so greatly as it has done some of the regions further to the South, where the white man finds it more difficult to toil, or where the climate affords an easier livelihood to the lazy, indolent negro. Virginia is sometimes

* De Vere: "Romance of American History," p. 69.

called one of the Middle Atlantic States.* The longest line in the State, from the Atlantic south-west to Kentucky, is 276 miles, while from north to south the length is 192 miles. Its area comprises, according to the census of 1870—in this instance, it is said, not very accurately—38,352 square miles, or 24,545,280 acres, divided into ninety counties. Taking official publications of the State,† we find it divided into six great natural districts, belts of country extending across the State from north-east to south-west, in a general direction parallel to each other, and corresponding to the bend of the Atlantic coast on the east, and of the range of the Appalachian system of mountains on the north-west. The first division is the *Tidewater Country*, which comprises 11,350 square miles, and a population in 1870 of 316,305. It is mainly an alluvial country, composed of clays and sands deposited by the tidal rivers. Little of it is as much as 100 feet above the sea, and much of it is marshy and malarious. It is estimated that 30,000,000 bushels of oysters are annually obtained in the waters of this region. Indeed, the State collects a tax on 20,000,000. The *Middle Country* is a wide, undulating plain, crossed by many rivers, that have cut their channels to a considerable depth, and are bordered by alluvial bottom lands. The *Piedmont Region* is more diversified, with many broken ranges of hills and mountains, enclosing valleys of many forms, the hills generally rounded, but embracing in places extensive plains. The *Blue Ridge* is a mountain range, stretching into plateaus, and rising into domes, and forming one of the most prominent features of the State over which it extends. *The Valley* is a portion of the great Central Appalachian Valley, that extends for hundreds of miles, from Canada to Alabama, a broad belt of rolling country, enclosed between lofty mountain ranges, diversified by hills and valleys, with many winding streams of water. The Blue Ridge is on the east, and the Ketatenny, or "Endless Mountains," on the west. The best natural division of Virginia is the *Appalachian Country*, made of numbers of parallel mountain chains, with trough-like valleys between them, the mountains often running for fifty or more miles, as an unbroken single, lofty ridge, with an equally uniform valley alongside. Goshen Pass, in the Rockbridge County (p. 157), may be engraved as a characteristic picturesque portion of this region. Sometimes the mountains die out, and the valleys widen. Some of the mountains and valleys are of sandstone, some of slates and shale, and others of limestone, so that here is "great variety of surface." Indeed, few States of the Union have greater diversity of surface, the altitude of the country extending from a little over the sea level to nearly 4,000 feet, which is the altitude of the Alleghany range in some places. In the Valley region is the celebrated Shenandoah Valley, one of the most fertile and wealthy portions of the State—indeed, the "pride of Virginia"—embracing, as it does, 30,000 acres of fine farming and grazing land, margined on either side by inexhaustible deposits of hæmatite iron ore. Its grain and grass-growing capabilities are especially celebrated. In 1866 it produced three million pounds of tobacco, and five and a half million bushels of Indian corn. It is now beginning to be a point of attraction to English settlers, who have purchased some of the estates of the planters ruined by the war. This lovely valley felt the

---

* Guyot classes New York, New Jersey, Pennsylvania, Delaware, Maryland, Virginia, and West Virginia, as Middle Atlantic States.

† Hotchkiss: "Virginia: A Geographical and Political Summary" (Richmond, 1876).

VIEW OF GOSHEN PASS, ROCKBRIDGE COUNTY, VIRGINIA.

shock of the civil struggle as keenly as any portion of the seceded States. Both Northern and Southern armies overran and ruined it by their repeated foraging expeditions, so that at the close of the war the inhabitants were harried to about the extremity of their woe.

The Alleghany, or Appalachian Mountains, traverse the whole length of Appalachian Virginia. This may be considered, according to Major Hotchkiss, as a series of comparatively long parallel valleys, running north-east and south-west, separated from each other by mountain ranges, that are generally equally narrow, long, and parallel, and rather elevated. "In crossing this section to the north-west, at right angles to its mountains and valleys, in fifty miles one will cross from six to ten of these mountain ranges, and as many valleys. A strip of this region is embraced in the valley countries, and they include the two or three forest ranges that have drainage into the valley, so that some 900 square miles of Appalachia are politically classed with this valley." This region is in Virginia "an irregular belt of country, 360 miles long, varying in width from ten to fifty miles. Its waters, generally, flow north-east and south-west; but it has basins that drain north and north-west, and south and south-east. The head of the valleys are generally from 2,000 to 2,800 feet above tide, and the waters often flow from each way to a central depression—that is, from 600 to 1,200 feet above the sea level—before they unite and break through the enclosing ranges." The Alleghanies divide the State into *East* and *West Virginia*.

The latter division, for our purposes, we have classed as simply a part of the old State; but on the 20th April, 1863, it was erected into a separate government, the people having disagreed with those of the eastern portion of Virginia on the question of secession, preferring to remain loyal to the Union. One of the fairest portions of the domain, it comprises fifty-three counties, nearly 300,000 people, and a mineral region all but unequalled. Charleston, a little town of 3,000 or 4,000 inhabitants, is the capital. The view on p. 161 is a characteristic sketch of the Alleghanies.

The *climate* of Virginia is a mean of extremes between the great heat of the Southern Mississippi Valley States and the extreme cold of the North. However, the varied character of the surface renders any account of the climate difficult to be stated in general terms. From the circumstances of the country comprising level tracts hardly raised above the sea, to long valleys 2,000 feet above its level, and ranges of mountains, running across the entire breadth of the State, and attaining a height of 3,000 to 4,000 feet, all possible exposures may be found. It accordingly follows that Virginia comprises a variety of climates, "temperature, winds, moisture, rain, and snow-fall, beginning and ending of seasons, and all the periodical phenomena. In Virginia the total amount of improved land is 9,091,691 acres, while the woodland is stated at 11,428,958, the cash value of the whole being 273,236,274 dollars. Virginia has thus more cleared land than any other State of the Union, except Illinois, New York, and Pennsylvania. There were, at the date of the statistics I have access to, 1,067 farms, over 30 and under 10 acres; 2,590 over 10 and under 20; 10,538 over 20 and under 50; 13,800 over 50 and under 100; 2,684 over 500 and under 1,000; while of the whole 57,188, 577 contained 1,000 acres and over. On an average each 100 persons have twenty cows, and each cow gives on an average 10.4 lbs. of butter. This is under the average of some of the other States. For instance, in New England 75 lbs. are yielded by the average cow, while in the Middle States it rises to 87. However, in the Southern States, it falls to 22, while in the Pacific States it is only 15. In England we have only an average of nine cows to every 100 persons. In Virginia there are 590,955 sheep, an average of 6.48 to every 100 persons. Honey-bees, swine (1,262,707, an average of 105 to each 100 persons), cattle (422,645), working oxen (79,165), &c., are among its other animal riches. The fisheries—shad, herring, rock-perch, sturgeon, sheepshead, ban, chub, spots, hoglish, trout, tailor, Spanish mackerel, crabs, lobsters, oysters, terrapins, &c.—are also productive, to the extent of over 1,000,000 dollars worth of those named being annually caught. The delicious canvas-back duck, so gastronomically associated with Maryland and Virginia, mallard, bald-face owl, and other wild fowl, are common; quail, pigeons, wild doves, "grouse," and wild turkey, also afford good sport; while wild deer are still plentiful in all portions of the State, especially in Tidewater, Middle, and Mountain sections. Among the other animal resources of the State may be enumerated 201,933 horses, and 39,444 mules and asses. Notwithstanding the ravages of the war, the live stock of Virginia was, in 1870, valued at 43,029,030 dollars, a considerable rise over the estimate made in 1860. Wheat, rye, Indian corn, barley, oats, and buckwheat, are among the grain crops "raised," while peas, beans, potatoes, sweet and "Irish," sugar-cane, maple sugar, sorghum, or Chinese cane, beetroot, wine, hops, clover, cotton, hemp, flax, &c., are among the other vegetable resources of the State. Tobacco is, of course, the staple. In 1850 it produced about one-third of the United States' crop, making about 100 lbs. to each of its inhabitants. The State crop was then 121,787,646 lbs. Weaving—especially at home—has always been a Virginian industry, and it shares in all the other manufactures of the Middle States.

Its mines of coal and iron are celebrated, though the native Virginians are more addicted to cultivating the earth above than digging into it beneath the surface. The first is ever congenial to their solid practical character, as befits the descendants of the country gentleman who founded the Commonwealth. In 1870, Virginia (exclusive of West Virginia) had 712,089 whites and 512,841 blacks, a decrease of 15,000 compared with 1860. There were also a few Indians and Chinese living within its bounds. It also appears that in 1870, 988 out of every 1,000 people were born in the United States, so that twelve in each 1,000 were foreign born. The State has 3,105 persons to the square mile, so that taking the census of 1870 as the basis, the United States having at that date 38,558,371 people, Virginia was the tenth State in point of population. Including the territories, the population over the whole United States was 10.70 to the square mile, or, omitting the territories, 19.21. The centre of population, according to the "Statistical Atlas," was, in the United States in 1870, sixteen miles north of Woodstock in Virginia. In 1830 it had passed nineteen miles west-south-west of Moorefield; in 1840, to sixteen miles south of Clarksburg; in 1850, to twenty-three miles south-east of Parkersburg; and in 1870, to forty-eight miles east by north of Cincinnati. The land surface area of the United States, including territories, is

3,603,884 square miles, and that of land and water about 1,000,000, so that Virginia has about the eightieth of the area of the whole country. Of the foreign population of Virginia, Ireland furnishes nearly one-half, Germany one-third, England one-sixth, and Scotland one-twentieth. About forty-nine per cent. of the foreign-born population are found in Tidewater, and these chiefly in the seaport towns. Over twenty-nine per cent. live in the Middle Country, eleven per cent. in the Valley, while Piedmont has only eight per cent. Of the blacks only 9,124 were born in other States, and these chiefly in the

VIEW OF THE PORT OF RICHMOND, VIRGINIA.

neighbouring States, showing that buying and selling of negroes was never, as in the more Southern States, a practice of the Virginian planters. The females were considerably in a majority compared with the males, except in the case of foreigners, where, as might be expected, the men had the majority. In 1872-73 there were 253,441 white children and 83,297 black ones in the public schools; and the sceptics in centenarianism may note the "fact"—if fact it be—that there were reported, according to the census of 1870, to be 230 people—65 men and 165 women—100 years or more of age. It might be desirable to see the birth certificates of some of these ancient "aunts" and "uncles," who are, of course, chiefly found among the negroes. Richmond had, in 1870, 51,038 people, while Petersburg, Alexandria, and Norfolk had all over 14,000. The others fell below this, most of them being under 3,000. In 1870 there were 3,890 paupers

VIEW OF BALCONY FALLS, JAMES RIVER, VIRGINIA.

in the State, each supported at an average cost of seventy-seven dollars ninety-one cents. Of these twenty-six were foreigners. Crime is not high; the receptions into the Virginia Penitentiary, the only great prison in the State, having been, for 1872-73, 211 (55

VIEW OF THE ALLEGHANY MOUNTAINS, VIRGINIA.

whites and 181 negroes), or about one in 5,000 of the population. These statistics are given in order that the many inquirers after Virginia in this country may obtain a more perfect idea of some of its features than can be supplied in any other form. We may now visit more at random some of the typical localities of this—in the opinion of many—most interesting portion of the United States.

Lynchburg—"Old Lynchburg," the inhabitants love to call it, veneration for what is ancient being a distinguishing feature of the more cultivated classes in the United States as elsewhere—lies among the mountains on the southern bank of the James River, in the centre of the Piedmont District, and not far from the base of the Blue Ridge. It was once one of the wealthiest towns of America. Fortunes have been amassed here in tobacco, and as it is now becoming a railway centre, it is likely that in time it will rise from being a "little city" of 12,000 people to become once more a great wealth-producing hive of industry. In that world which lives among dictionaries it is famous as having given a new word to the English language. Colonel Lynch, the Irish emigrant, whose name has been applied to the town, was a noted soldier in the Revolutionary War. This hot-headed Hibernian, when he caught a "Tory," punished the individual whose chief crime was that he did not think as Colonel Lynch thought, and did as he did, after so summary a fashion, that in time speedy "justice" of a somewhat equivocal type became known all the world over as "Lynch Law." Coal is found in the immediate neighbourhood in abundance, while the tobacco manufactories are prosperous. "Old Lynchburg" still preserves some of the features of its colonial days, and in this part of Virginia bits of dialect and phrase, smacking of old English and Scotch manners, still linger among the people. This shows how little the community has been altered within the last hundred years by the changes—and even revolutions—which have so transformed other sections of America. Yet, when one experiences the courtesy and graceful hospitality of the Piedmontese-Virginians, there cannot be much regret that the world has for so long passed by them on the other side. "South-west Virginia," writes Mr. King, "is a region which will in time be overrun by tourists and land speculators. The massive ramparts of the Alleghanies (p. 161) are pierced here and there by cuts through which crawls the line of the Atlantic, Mississippi, and Ohio Railroad, and towns are springing up with almost Western rapidity. Stores of coal and iron are daily brought to light, and the farmer of the old *régime* stares with wonder, not wholly unmixed with jealousy, at the smart new-comers, who are agitating the subject of branch railroads, and the searching into the very entrails of the hills. . . . The traveller who hurries through Lynchburg, repelled by the uncouth and prosaic surroundings of the railway station, will lose real pleasure. A residence of a few days in the old town will show him much that is novel and interesting. He may wander along the beautiful banks of the James below Lynchburg; by the canal whereon the gaily-painted boats slip merrily to their destination; or he may climb the steep hills behind the town, and get a glimpse of the winding stream which looks like a silver thread among the blue mountains. At noon-tide he may hear the mellow notes of the horn by which buyers are summoned to a tobacco sale; and at sunset he may watch the curious groups of negroes returning from their labours, singing and chattering, or noisily disputing some momentous political issue." One of the most famous of Virginian natural curiosities is the Natural Bridge, which is situated in the same part of the country as Lynchburg, two miles from the mouth of Cedar Creek. Perhaps we could best describe this great monumental wonder in the language of a native Virginian. "The first view of the bridge is obtained half a mile from it, at a turn on the stage road. It is revealed with the suddenness of an apparition. Raised a hundred feet above the highest trees of the forest, and revealed against the purple side of a distant mountain, a whitish-grey arch is seen, in the effect of distance, as perfect and clean cut an arch as its Egyptian inventor could have defined. The tops of trees are waving in the interval, the upper half of which we only see, and the stupendous arch that spans the upper air is relieved from the first impression that it is man's masonry, the work of art, by the fifteen or twenty feet of soil that it supports, in which trees and shrubbery are firmly imbedded, the verdant crown and testimony of Nature's great work. And here we are divested of an imagination which we believe is popular, that the bridge is merely a huge slab of rock thrown across a chasm, or some such hasty and violent arrangement. It is no such thing. The arch and the whole interval are contained in one solid rock. The average width of that which makes the bridge is eighty feet, and beyond this the rock extends for 100 feet or so in mural precipices, divided by only a single fissure, that makes a natural pier on the upper side of the bridge, and up which climb the hardy firs, ascending, step by step, on the noble rockwork till they overshadow you. The mighty rock in the earth side, of which even what appears is stupendous, is of limestone, covered to the depth of from four to six feet with alluvial and clayey earth. The span of the arch runs from forty-five to sixty feet wide, and its height to the under line is 196 feet, and to the head 215 feet. The form of the arch approaches the elliptical. The stage-road which passes over the bridge runs from north to south, with an acclivity of thirty-five degrees, and the arch is carried over a diagonal line, the very line of all others most difficult for the architect to realise, and that best calculated for picturesque effect."

In the Piedmont section farms may yet be had on very easy terms, owing to the ruin wrought by the war; and in Virginia, on account of the land having so generally descended from father to son for long generations, it is easier to get a consecutive and secure title than in most other parts of the United States. The Alleghany springs in Montgomery County, near the Roanoke River, at the eastern foot of the Alleghany Mountains, have long been a favourite summer resort of the Virginians. The scenery is very beautiful, and may be imagined from the sketch of a waterfall in the close vicinity to "the springs" (p. 165). The routine at these resorts is always much the same. There is a huge hotel, with a large dining-room, a large "bar," a large ball-room, and endless corridors of tiny bed-rooms. Here, however, there is some variety. The "guests" are not lodged in one monster barn, but the caravanserai is divided off in a number of pretty one-storeyed cottages, where the travellers sleep. "Society" is "good," and those who love to dance through half the night can indulge in such athletics to the music of a band of negro fiddlers, and have for partners no "carpet-bagger Yankee," but the scions of dilapidated "fust families," who have accepted the inevitable, and commenced the world anew. If they are rabid they will talk bitterly of the lost cause, and if moderate men and philosophers, will, in all likelihood, refer euphemistically to the fratricidal war in which they shared as "the late unpleasantness." Still, everything in the South smacks yet, and will long smack, of the "old times," which look to those who knew them as of yesterday, but in the light of the results as something that was in ages that are past. Hosts of polite "coloured" serving-men—ex-valets, ex-nurses, ex-maids—of the "before the war" epoch, who will tell you with pride, "I used to belong to ole Mars' ——," mentioning some name famous in the annals of slave proprietorship, are familiar features of the place. "There is no gambling, save an innocent whist party by some sleepy old boys, who lurk in the porches, keeping out of the strong morning sun. There is no Saratogian route of carriage and drag; no crowded street, with ultra style predominating in every costume; nothing but simplicity, sociable enjoyment, and excellent taste. In the sunny mornings the ladies and their cavaliers wander about the mountain pathways; dress does not exact homage until dinner time, and the children join with their parents in the strolls and promenades, followed by the venerable 'aunties,' black and fat, who seem indispensable appendages to every Southern family having young children."

The Montgomery White Sulphur Springs and the Yellow Sulphur Springs—near the head of the Roankoe—are also favourite holiday residences for Virginians and the Southern people generally. They afford a pleasant retreat for families, cheap, healthy, and good, while the waters of the last named are in great request among the ladies on account of the reputation—deserved or not—of curing children's diseases, and of their imparting a rare purity of complexion to their mammas.

Balcony Falls (Plate XVI.) is another sight of Virginia, a State abounding in natural beauties of this nature, while the Brine Springs at Saltville are interesting, as supplying a great portion of the salt used in Virginia and elsewhere. Weir's Cave, the Natural Tunnel, seventy feet high; the Hawk's Nest, a pillar 1,000 feet high; ebbing and flowing springs, and the Blowing Cave, which sends out a blast of cold air in summer, and draws in air in winter, are among other sights of Virginia, which we can only mention. Eggleton Springs, on the New River, form a point whence the traveller may easily diverge from his route to examine a remarkable natural curiosity of the Virginian mountains. This is a pond or lake, with no visible source of supply, sunk in a kind of earthen cup 1,500 feet below the level of the sea. It has been forming and enlarging until it is now about three-quarters of a mile long by one-third of a mile wide. Submerged trees may be seen below the surface, and so deep is it that a line hundreds of feet long will not reach the bottom in the middle. It is, of course, fed by springs, the overplus of which is not sufficiently carried off by the outlets. The resources of South-western Virginia, in agriculture and mines, are so extensive that General Lee was justified in declaring that theoretically he "could carry on the war for twenty years from those western mountains." Only he showed his wisdom (which is always practical) by not doing so. All Virginia is sacred with memories —memories of the gallant men who fought its earliest fights, and of the swashbuckling cavaliers who followed them. Its streets, to the student of history, resound with the far-distant echo of arquebusiers, firing at the "salvages," and with the clatter of men in mail, or the jingle of spurs on the jack-boots of the buff-coated cavaliers of James and the Charleses. George Washington and his revolutionaries are here, and in about, and about and over all, are memories of Robert Lee, and of the gallant, though mistaken, men who obeyed his fighting orders. The journey from Lynchburg to Petersburg calls up many memories. Not very many years ago the mad rush of desperate and final battle swept across it. From the log and earth parapets of Five Forks, where Pickett's forces met their doom at the hand of Sheridan; from the Appomattox, and from Hatcher's Run; from Fort Gregg, where the splendid Mississippians held on against hope and fate, until nearly all of them had perished; from the entrenchments of deserted Petersburg; from Burkesville;

from the road to Jeters-ville, over which Sheridan and the "Fifth" went clattering; from

VIEW OF STILES FALLS (FOUR MILES FROM ALLEGHANY SPRINGS), VIRGINIA.

Amelia Court House and from Sailor's Creek; from the High Bridge, and from Cumberland Church, near Farmville, where Mahone made his heroic stand, and would not be driven; from all the bloody and memorable fields which stretch, sunlit and peaceful now; from the hills around Petersburg, to the village of Appomattox Court House, come echoes which recall to us some faint impressions of "the splendour and grandeur of that last resistance of the broken army of Northern Virginia."* Appomattox Court House lies silently hidden in groves and gardens, "as if frightened by the notoriety it attained." Few signs of the great struggle—material, terrestrial, I mean—remain. Here and there in a field the railway passenger, as he leans out of the window of the "cars," may notice a green grass-covered entrenchment, but even these are disappearing, and mile after mile of cotton, corn, and tobacco covers the places around which only a decade or so ago the wild whirl of battle swept. The once much-beleaguered towns are now again getting alive with the rattle "of spindles and the ring of hammers on tobacco hogsheads."

In Norfolk County is the "Great Dismal Swamp" celebrated in the anti-slavery literature of the days "before the war." It is a succession of weird and apparently irreclaimable marshes, "through which run black currents of water, and in the midst of which spring up thousands of dead tree trunks," many of them charred by recent fires.

The city of Norfolk has quite an English aspect, and might at first sight be mistaken for one of the ancient towns on the south-eastern coast of England, but with a life rather brisker than the latter. The negroes are, however, struggling to get the upper hand at every municipal election, and it is just possible that as they are a large element in the "vote" they may succeed in ruining the old town for a time.

Richmond, the capital of Virginia (p. 160), and so long as it lasted, of the Confederacy, is a pleasant town, the first view of which, from the James River, is really very striking, and gives the city the appearance of being much greater than it really is. It has now nearly 60,000 inhabitants, and the total assessed value of its real estate and personal property amounts to 37,000,000 dollars. Its exports amounted, in 1875, to 2,243,716 dollars, and, like most Southern towns, its trade has taken an upward turn since the war. Tobacco, coffee, and flour are its staples, and there is little doubt but that it will become in time one of the most important of the Southern iron manufacturing centres. The days of "land and niggers" are past, and railroads are teaching the Virginians that there are other things quite as profitable. In twenty years, manufacturers will become in Virginia what, in the days "before the war," planters were—the aristocrats of the country. The old Virginian hated cities. In the spread of the manufactories, as Mr. King well remarks, he saw the symbol of the decay of the society which produced him and his. Cities are democratic, and accordingly the aristocratic planter disliked them, their corruption, and the ambitious populations that never showed him sufficient respect. In the planter's eye the rich manufacturer was just as much a *parvenu* as the Manchester man is according to the ideas of the long-pedigreed English squire. To the mind of the ex-Virginian slave-owner, the lordly agriculturist was, and is still, the only fitting type of the real "gentleman." "He shudders when he sees the youths of the new school joining in commerce, buying and selling mines, talking of opening new railroad routes, and building cotton-mills. He flies to the furthest corner

* King: "Southern States," p. 597.

of the lands that have been spared out of the wrecks caused by the war, and strives to forget the present and to live as he did 'before the surrender,' like a country squire in England two hundred years ago." Charlottesville is another lovely town, perhaps one of the loveliest of all the many beautiful Virginian towns. It is, of course, small compared with Richmond and Norfolk, but being the seat of the oldest of the Southern Universities it has airs of learned leisure which befit its reputation and surroundings.

Lexington is one of the Meccas—for they have several—of Virginians. From the Military Institute some of the best talent of "the war" went out. Three of its professors and 123 of its alumni were killed, and 350 of its graduates were wounded in the struggle. Here is the grave of Stonewall Jackson, who was once a professor in this institution, and in the Washington and Lee College is the tomb of General Lee, who died as principal of the establishment which now shelters his body.

With all its magnificent resources the negro difficulty—though not so prominent in Virginia as elsewhere—is yet of sufficient importance to seriously impede the prosperity of the State. The African has taken here, as elsewhere, to politics and its pelf with singular avidity. The Conservatives declare that in trickery they are no match for the coloured politician. Among other things—in Petersburg, at least, where the scandal has its *locale*—it is said that there is great difficulty in securing the burial records of the negro population, since it is their custom "to make a dead voter renew his life in the person of one of his friends." The negro is, however, showing himself anxious for knowledge. If he would but add this to a moderate anxiety for work, he would soon become a blessing instead of, as in too many cases, a nuisance to the State. Schools for his education are being established throughout the country, though as yet the ex-slave-owners, from mistaken ideas regarding the dangerous character of "book-learned niggers," or from mere apathy, are not showing much desire to forward these praiseworthy exertions of private philanthropists and the Government. A little too great inclination to "take liberties" is always the characteristic of the Ethiopian, bond or free, and though no doubt the law gives him his "rights," yet it cannot be expected that the proud planters will always be inclined to grant them when they run to the indiscreet length of insisting on first-class seats in railway carriages. Still we must say, that, taking the State as a whole, there has been an almost studied desire to do nothing ungenerous towards the "freedman," which presages great things for the "Old Dominion." The farmers also seem more anxious to make up for lost time by hard work in reconstructing their shattered fortunes than in spending their time in the empty wind-grinding of politics. Now and then some extreme Secessionists—probably those who were the readiest at first to take the oath of allegiance to the United States—will flout a few Confederate flags. But these mementoes of a lost cause excite little enthusiasm among a people only too anxious to accept the inevitable, and make the best of what has been left them after the "late unpleasantness."

North Carolina.

To the ordinary reader we have little doubt that this description of the United States —State by State—brief as the account must necessarily be, is wearisome. However, it is impossible to give a just idea of the Great Republic in any other way. One State may not differ very widely from the one next to it in physical geography. But it is often another country so far as prejudices, feelings, interests, and the peculiar American feeling of State pride is concerned. In the North this is marked, but it is in the South that it attains its maximum of development, and the intensity with which "the State" is loved, apart from the fact of its being one of the component members of the Commonwealth of Governments, had a weighty influence in bringing about the Civil War. When one Southerner meets another, almost the first question he asks is in regard to "his State." Two gentlemen will meet in a railway car or on a steamboat, and will introduce each other by a query as to their respective States. The ice being thus broken, they will proceed to discuss politics or things in general. But a knowledge of each other's respective States is essential, probably for the simple reason that until they know this it is impossible for them to meet on common ground, or avoid the well-known prejudices or raw points of each other. Every traveller in America must have noticed—more particularly in the Southern States—how anxious travellers in a public conveyance are to know if there is any one from "their State." If there happens to be such an individual, then the states-men instantly fraternise, and for the rest of the journey are on the most amicable terms. County, and even parish pride are characteristics of some parts of England, but though we have counties exceeding in population all but the largest of the American States, yet nothing like the State pride of America is witnessed amongst us. One traveller assures us that on a certain night the train in one of the Southern States halted at a little side station in the middle of the pine woods to pick up a solitary traveller. Before taking his seat he shouted into the carriage, "Is thar' any one heah from Tennessee?" Obtaining no response, he repeated the question in the next carriage, and his states-men being apparently scarce in that train, for the whole length of it was heard out of the darkness the monotonously plaintive cry, "Is thar' any one heah from Tennessee?"

North Carolina must, therefore, be described apart from South Carolina. The former State is about 450 miles long and about 180 broad, with an area of about 50,704 square miles. In 1870 there were 678,470 whites, 391,650 negroes, and 1,241 civilised Indians, in all 1,071,361 people in the State. The part of the State lying on the Atlantic seaboard is swampy in many places, and flat and alluvial throughout; but to the west it rises into undulating hills, and is traversed by ridges of the Alleghanies, which culminate in Mount Mitchell, elevated 6,582 feet above the sea. There is between these ridges a table-land about 2,000 feet high. Pitch pine is one of the chief products of the low lands, but lead, copper, iron, and coal, and even gold, are mined. In 1870, 12,824,695 dollars worth of materials were used in manufacturing 19,021,327 dollars worth of goods. Raleigh is the capital, but Wilmington and Fayetville are also considerable towns. North Carolina was the first State to demand separation from Great Britain, and the taste for secession being an acquired one in the State, it promptly joined the Confederacy in 1861. The flat, swampy country extends about 100 miles back from the coast, and is inhabited by "a low and almost worthless population." The next region supports wheat, tobacco, and cotton, while the mountains grows nothing in particular, but is rich

VIEW OF GRANDFATHER MOUNTAIN 5897 FEET, NORTH CAROLINA.

in mines. The mountaineers of North Carolina seem a different race from those of the lowlands, so different, indeed, that they have sometimes tried hard to secure a division of the State, adding to their mountains those of Tennessee. The Lowlanders know little of the Highlanders, and to the latter have almost always belonged the men in whom the State takes any pride. Seen from the sea, the North Carolina coast is flat and uninteresting. "There is an aspect of wild desolation about the swamps and marshes which one may at first find picturesque, but which finally wearies and annoys the eye. But the coast is cut up into a network of navigable sounds, rivers, and creeks, where the best of fish abounds, and where trade may some day flow in. The shad and herring fisheries in these inlets are already sources of much profit. The future export of pine and cypress timber, taken from the mighty forests, will yield an immense revenue. The swamps or dry tracts along the coast are all capable of producing a bale of cotton to the acre. They give most astonishing returns for the culture of the sweet potato, the classic peanut or 'guber,' the grape, and many kinds of vegetables. Malarial fevers will of course seize on the inhabitant of this region who does not pay proper attention to the drainage all about him. It is believed that along this coast great numbers of vineyards will in time be established, for there are unrivalled advantages for wine growing." Carolina suffered greatly by the war; and in that State may be probably found men more bitter and desponding than in any of the other Southern States. North Carolina was always one of the chief pro-slavery communities. When the war broke out it held 350,000 negroes, and comprised an aristocracy of men-owners, whose fortunes have been greatly shattered. The negroes, moreover, are of rather a low type, and have only too completely demonstrated their unfitness to be trusted with political power, by the wild orgie of plunder and oppression which, aided by a few scoundrelly "carpet-baggers" from the North, and, indeed, from quarters nearer home, they inaugurated on the whites. The carnival of robbery and maladministration burdened the State with a debt of between 36,000,000 and 40,000,000 dollars, most of which went into the pockets of the black and white thieves, who in the turmoil had put themselves into power. Out of 16,000,000 dollars voted by the Legislature for public works only half a million has ever been devoted to that purpose!

We have engraved some specimens of the scenery of North Carolina. Grandfather Mountain, in the Alleghanies (p. 169), is one of the chief points seen from the summit of Mountain Mitchell, looking towards the north-east, Table Rock and Hawk Bill, twin mountains, being in front of it. More pleasing bits of scenery are the sketches given on pp. 172 and 173, which require no description. North Carolina is shockingly illiterate. Out of 350,000 pupil-children there were, in 1873, only 150,000 at school, and in the same year it was estimated that within the borders of the State were 350,000 people who could neither read nor write.

SOUTH CAROLINA

has an area of 34,000 square miles, and a population, according to the last census, of 705,606, of whom 289,667 were whites, 415,814 negroes, and 125 civilised Indians. Carolina had, until its reconstruction in 1868, an essentially aristocratic Constitution, in so far that its Presidential electors and its executive were not elected by popular

suffrage, but by the joint votes of the two Houses of Legislature. It is, however, democratic enough now. The negro dominates it almost without check. Physically, South Carolina is only a continuation of North Carolina, so that it is unnecessary to repeat what we have previously said regarding the topography of the State. Even bad government of the superlatively evil type of that which has for so long afflicted South Carolina has been unable to crush all life out of it. Its lands are too rich for that. Accordingly, this State, like most others, has made some progress since the war, its railways, and consequently its cotton products, having greatly increased. The planters' lands have passed, in many cases, into the hands of the negroes, by the Act of Forfeiture, a scandalous piece of legislation unsurpassed by anything of the kind with which we are familiar. Hence the black reigns supreme, especially in the cities, where they herd, and are in an immense majority over the whites. Rice is one of the staple crops of South Carolina, and on rice and cotton Charleston, the capital, grew rich. In 1870 the product was 32,304,825 lbs., against 119,100,524 lbs. in 1860. It is, however, now recovering the shock it sustained during the war, when the rice-fields were deserted. Charleston is, however, an exception to the general rule. It is prosperous mainly because it has not depended solely on cotton and rice, but has established manufactories, and her business men have not allowed them to incontinently wilt under their temporary reverse of fortune. Every year since the war the cotton receipts at Charleston have increased, and the same may be said of rice, timber, and other of its materials of trade. Charleston is one of the oldest of American towns, and one of the loveliest. Of these characteristics the monstrous corruption and spoliation which have prevailed here as elsewhere in the State since the advent of black government could not deprive it. In the work I have so frequently quoted, for the reason that it supplies the best and most impartial view of the Southern States which we have met with, full details of the changed condition of affairs are supplied. The following description of Beaufort, once one of the centres of sea-island cotton cultivation in South Carolina, may give in a few words an idea of the pass things have been brought to:—
"If the planter of the days when the royal colony of South Carolina was in the height of its glory could return now and wander through the streets of moss-grown Beaufort, he would be amazed; but no more so would the planter of 1850 or 1860, if he too might return. For it would be found that in a decade and a half one of the most remarkable revolutions ever recorded in history has occurred. A wealthy and highly prosperous community has been reduced to beggary; its vassals have become its lords, and dispose of the present and pledge the future resources of the State. In ten years the total valuation of the commonwealth has been reduced from nearly 500,000,000 dollars to barely 15,000,000 dollars at the present time; the banking capital of Charleston from 13,000,000 dollars to 5,000,000 dollars; the insurance capital is nearly destroyed. The taxes have been increased from 392,000 dollars in 1860 to 2,000,000 dollars in 1870. Slaves valued at 174,000,000 dollars have been freed, and set to learn the arts of self-government and civilisation. More than 400,000 blacks now inhabit the State, and their number is constantly increasing. Thousands of planters have been so utterly ruined that they can never hope to attain comfortable circumstances again. Opposite an elegant mansion in one of the main streets of Beaufort is a small unambitious structure, in which the former

occupant of the grand mansion is selling goods at retail. He returned after the capture of the town to find himself stripped of everything, and has been living in view of his former splendour ever since. His fields are held by strangers, his house is converted into offices. In a day, as it were, he and thousands of others were reduced to complete dependence, and compelled to live under the government of the ignorant slaves, whose

VIEW NEAR MORGANTON, NORTH CAROLINA.

labour they had grown rich on."* It may be poetic justice, but still, apart from all ideas about slavery, on which we believe the world is now tolerably well agreed, it seems cruel to the last degree. The same picture might be drawn of almost any portion of the Southern States where the negro is in power. To use the words of his own savage song:—

"De bottom rail's on de top,
An' we's gwine to keep it dar."

But perhaps he won't. Already there is a dawn of better things. Education will not make

* King: "Southern States" (1875), p. 428.

men—black or white—honest; but it may teach them that if they wish to retain some share of power, robbery of the public purse is not the best policy. The negro is seeking knowledge,

VIEW OF WATAUGA FALLS, WESTERN NORTH CAROLINA.

and filling the benches in the universities deserted by the white students, whose racial prejudices are keener than their love of learning. Brain is in the end stronger than muscle, and even in South Carolina may not prove synonymous with right.

CHAPTER X.

THE UNITED STATES: GEORGIA; FLORIDA; ALABAMA; MISSISSIPPI; TENNESSEE; KENTUCKY.

GEORGIA has an extreme length from north to south of 320 miles, and a breadth, where it is widest, of 254 miles, the whole area being 58,000 square miles. In 1870 it contained 1,200,609 inhabitants, of whom 545,142 were "coloured." The physical features of the State are varied. Much of it along the sea-shore consists of low alluvial lands and swamps, from which it rises into an undulating and hilly country, culminating in the Blue Ridge Mountains in the north and north-western portion of the State. The chief rivers are the Savannah, and Chattahoochee, and all of the principal streams run towards the south and south-east. About a fifth of the country is under cultivation, and the products are as varied as the soils. The islands fringing the coast grow the famous Sea Island cotton, while the river bottoms produce rice, cotton, maize, sugar, and the pine barrens abundance of timber, and could be easily made to yield other crops. The central region contains soil which is now exhausted, owing to the wasteful system of agriculture under the slave system, but the "Cherokee country," though long cultivated by the Indians, is yet so rich that it will yield fifty to seventy-six bushels of grain to the acre. Gold was at one time mined in considerable abundance, as are silver, copper, iron, lead, marble, and precious stones still to some extent. Passing from Aitken in South Carolina to Augusta in Georgia, the traveller passes through the region known as the Sandhills. He sees busy manufacturing villages, hears the whirl of spindles, and on every hand witnesses the signs of progressive industry. Yet all this of a date subsequent to the war, and leads one to reflect on the lost opportunities of their Southern States, for more than two centuries. This Sandhill region extends from the north-eastern border of South Carolina to the south-eastern border of Georgia. Its climate is wonderfully revivifying, and its soil excellently suited for rearing valuable fruits unknown in the bleaker north. The upper limits of the Sandhills in South Carolina are close to the rivers, and very clearly defined. They are supposed to have been ancient sandbanks not far from the sea-shore, and are now usually clothed with aromatic forests of the yellow and "short-leaved" pine, the Spanish and water oak, the red maple, the sweet gum, the haw, the persimmon, the wild orange, the China tree, the lovely *Kalmia*—the spoonwood, or calico bush of the South—the flaming azalea, the honeysuckle, the white locust, the China burr, and other evergreens, while under their shade flourish the iris, the phlox, and the silk grass. Japonicas grow ten feet high in the open air, and blossom late in the winter, and the "fringe tree" and *Lagerstrœmia Indica* "dot the lawns with a dense array of blossoms." The "un-stimulated soil" will not produce cotton or cereals more than two years in succession, but it supports thickets of peaches, apricots, pomegranates, figs, pears, all kinds of berries, and grape-vines of extraordinary luxuriance, in addition to Northern vegetables, which, however, here ripen in the months of April and May. To this pleasant land and kindly climate came, in the halcyon days "befo' the wah," alike the planter from the lowlands, and the merchant from Boston and New York. The latter still arrives with the first hints of winter to occupy his pretty cottage in Aitkin. But he has now the little town all to himself. The planter comes no more: his splendour and spendthrift profusion are things of the past. Still, Aitkin, like many other places, is recovering from the depression that the fatal struggle brought on it. Traces of the war are disappearing, and when prosperity comes once more back to the Southerners, it will no doubt rise from its ruin an infinitely more flourishing, if an entirely "reconstructed" city. Nothing can ever make it otherwise than one of the loveliest of Southern towns.

"Reconstruction" in Georgia was a failure. Accordingly the negro in that part of the world holds the white men, and especially that mystic portion of them leagued as the avenging Ku-Klux faction, in profound respect. The African is here more inclined to "take a back seat" than in most other quarters of America, to herd together in little villages with his kith and kin p. 176), and, above all, is particularly careful never to be obtrusive in quarrels with white men, as we are assured by an eye-witness that the rural Caucasian has a kind of subdued thirst for negro gore, which, when once aroused, is not readily quenched. In Atlanta, outwardly, at least, nearly all the old scars inflicted by the war have healed. Its streets have a smart appearance, and its handsome residences are in agreeable contrast with the many tumble-down, unpainted mansions one sees throughout the ex-Slave States.

Savannah—the "forest city"—is a town of another type. The contrast between it and Atlanta is so marked that the visitor on arriving seems to have come into another country, and in a few hours to have crossed the sea. It is still a pretty town, but its loveliness is that of the sombre and voluptuous semi-tropical lowlands, and its atmosphere no longer the bracing air of the uplands, but the sluggish climate peculiar to the coast regions from Cape Hatteras to Florida. The Savannah River, up which you sail, returning from Florida some radiant morning, seems to you to have no affinity with the Savannah, which far among the Northern Mountains you have seen born of the frolicsome or riotous streamlets, for ever leaping and roaring in the passes or over mighty waterfalls. Here it is broad, and deep, and strong, and near the bluff on which the city stands it is freighted with ships from European ports, and from the northern cities of the American coast. The moss-hung oaks, the magnolias, the orange trees, the bays, the palmettoes, the oleanders, the pomegranates, the lovely japonicas, astonish the eyes which have learned to consider a more northern foliage as Georgian. Very grand in their way were the forests of pine, with their sombre aisles, and the mournful whispers of the breeze stealing through them; but here is the charm of the odorous tropical South, which no one can explain. Yet it is not here that one must look for the greatest wealth of the State; for middle Georgia is, perhaps, the richest agricultural region in the commonwealth, and the hundreds of farms along the western boundary are notable instances of thorough and profitable culture (King). It was at Savannah that the existence of Georgia began, for it was here that Oglethorpe planted his tiny colony a century and a half ago. When the traveller walks through its sylvan streets, shaded by a wealth of foliage, and yet with all the conveniences of a great commercial centre, it is difficult to believe that in 1734 the inhabitants had to lock themselves up in their cabins at night "because the alligators strolled through the town seeking whom they might devour," and that up to a much more recent period they were kept in constant dread of the neighbouring Indians scalping them off the face of the aboriginal earth on which they were encroaching. Savannah is also a city of health to which used-up folks much resort. It has a temperature resembling that of Gibraltar, Palermo, Bermuda, Sydney, or Shanghai, and is about equally proud of its share in all the struggles which America has gone through, and of the remarkable progress which it, like all the States, has experienced since "the war." Savannah is the rival of Charleston in cotton (p. 177),

A NEGRO VILLAGE IN GEORGIA.

and more than its rival as a city to which the light-hearted Southerner who has "reconstructed" himself, and prospered under the operation, resorts to have "a good time generally" when his harvest is housed. About the end of November the city is filled with gaiety. Agricultural shows, races, reviews, and balls are plentiful; wassail resounds, and money seems to flow almost as plentifully as before "the late unpleasantness." The laugh of the tall planter mingles with the cough of the Northern invalid in the halls of the hotels; elegant equipages dash along the roads leading to "Thunderbolt"—a pretty riverside resort—or to the sombre "Bonaventure" cemetery, once the site of the home where the

Tatnall family lived in almost regal splendour. Savannah is a well-conducted town. Brawls are few, and larcenies, save when they arise from "the undeveloped moral consciousness" of the freedman, are few. But the day of the negro was of short duration in the "Forest City." He has now no longer a voice in city affairs; only about 100 vote, and they are at present unrepresented in the city council. Education is much more

LOADING COTTON AT SAVANNAH.

advanced here than in the "back country," but the black children are still to a great extent excluded from the advantages of the schools, an absurd and very mischievous prejudice against raising Sambo out of the slough of his old ignorance still prevailing among the majority of the citizens of the *ancien régime*. Georgia is not likely to soon rapidly increase in population. The "poor mean white" is in that State the poorest of his poverty-stricken order, and, unable to make a living in the more sterile districts out of the exhausted soil, is moving in great numbers into Texas, as indeed he is from Alabama and most of the cotton States. Yet the wealth of Georgia is on the increase, though the labourers have decreased. Improved agriculture, the use of fertilisers, as well as that energetic spirit of the Georgians which distinguishes them above the people of the neighbouring States—have accomplished great things, and will yet accomplish still greater. Before the war, the cotton States were dependent on the North for almost every manufactured article. As a Georgian journalist remarked a few years ago—"A Georgian farmer uses a Northern axe-helve and axe to cut up the hickory; chops out his cotton with a New England hoe; gins his cotton upon a Boston gin; hoops it with Pennsylvania iron; hauls it to market in a Concord wagon, while the little grain that he raises is cut and prepared for sale with Yankee implements. We find the Georgian housewife cooking with an Albany stove; and even the food, especially the luxuries, are imported from the North. Georgia's fair daughters are clothed in Yankee muslins, decked in Massachusetts ribbons and Rhode Island jewelry." This is still true, though the creation of manufacturing cities at Columbus, Macon, Albany, Thomaston, Augusta, Atlanta, Marietta, Athens, and Dalton promise in time to make this reproach a something of the past. The number of small farms is increasing, and the negroes have got into their possession much good cotton land, which, with an utter recklessness for the future, they are rapidly ruining. The black man is lazy, just as the white one was (and is), and just as fond of hunting, fishing, and lounging through the beautiful woods, and along the noble streams, as was his "owner" in the bad old times. In the lowlands of the State the whites are shiftless, indolent, uneducated, and always complaining. It was in this section that an old woman explained to a thirsty traveller that they could not give him any milk "because the dog was dead." It appears on further inquiry that the defunct animal was in the habit of driving up the cows to be milked at eventide, and that since his untimely decease none of the family had felt inclined to go in search of the errant kine! Salt pork and whiskey form the staple diet of the inhabitants of a region where the finest of oxen and sheep could be fattened, while butter and milk are articles rarely seen on the farmer's table. The "Georgia cracker" is a pessimist, who will never allow that he is well, but only "tollable"—a lean, sallow people "of dry fibre and coarse existence, yet not devoid of wit and good sense." He seems, according to Mr. King, to have been born with his hands in his pockets, his back curved, and his slouched hat crushed over his eyes, and he does his best to maintain this attitude for ever. "Quarrels, as usual among the lower classes throughout the South, grow into feuds, cherished for years, until some day at the cross roads, or the country tavern, a pistol or a knife puts an often fatal end to the difficulty. There is, in all the sparsely-settled agricultural portions of Georgia, too much popular vengeance, too much taking the law into one's own hands; but there is a gradual growth of opinion against this, and even now it is by no means so pronounced as in Kentucky and some other more Northward States." Still, Georgia is a goodly country, and the Georgian one of the pleasantest of men, when he has emerged from the "poor white" stage of existence, or still better if he has never been in it. The ladies are occasionally a little bitter at "the Yankees;" the men are, however, inclined to sink politics and attend to business, so far as this is compatible with the Government of their State being one "for white men." They are particularly on the alert when the African begins to be loquaciously political.

FLORIDA.

There is so much to be said of the "Flowery State"—the most southern of the American United Commonwealths—that it is perhaps better not to enter upon this hopeless task, but merely whet the reader's appetite for fuller accounts* by mentioning a few historical facts regarding it. It is for the most part a peninsula about 400 miles long, and at its broadest 360 miles wide, its area being 60,000 square miles. The St. John River—the *Illaka*, the stream that "has its own way" of the Seminoles—runs north-east through the peninsula for 300 miles (pp. 181, 184), and among others it is drained by the Suwannee, the Appalachicola, the Chattahoochee, Escambia, and Perdido. Tallahassee is the seat of Government, while St. Mark's, Pensacola, and the charming St. Augustine the oldest settlement in Anglo-Saxon America, are the chief towns. Its population was, in 1870, 96,057 whites, and 91,689 blacks. A census might now show rather over a quarter of a million, all told. In 1870, 2,373,541 acres of its 35,000,000 were in farms, but only 736,172 acres were improved. There are now over 10,000 farms, and the acreage improved has increased, but not so rapidly as it ought. These lands produced in the year mentioned 2,225,056 bushels of maize, and 39,789 bales of cotton, beside other crops. The introduction of railways is, however, greatly altering the country, though it has been rather stubborn in "reconstructing" itself, after the close of the Civil War, into which it rushed on the Confederate side. Much of this State is sandy and marshy along the seaboard, forming as it does part of the belt of that description which stretches from the Potomac to the Mississippi. Even the interior has "water privileges" over many. The inlets carry the tide to within fifty miles of any point, and the endless lakes, streams, and springs (some of which are 250 fathoms in depth), make many parts of the interior almost amphibious. Indeed, that immense district known as the Everglades is inundated over a considerable portion of its extent. Tillage accordingly is not suited to much of the soil, though the inherent fertility of parts of the State, and the stimulating influences of heat and moisture, enable it to bear large crops of sugar, cotton, and rice. Its growth of ship-building timber is all but inexhaustible, while the rivers and coasts swarm with fish (p. 185), which the salt-encrusted "keys" furnish the means of curing. The climate is comparatively salubrious, and some parts of the coast along the shore of the Gulf of Mexico render it particularly favourable for those affected by chest diseases. Florida accordingly has become to the North American what the "Consumptive Coast," from Nice to Spezzia is to phthisical Europe. Jacksonville is one of the favourite health and pleasure resorts of the peninsula. There the climate is delightful. In the early days of December the mercury will frequently range from 79° to 80° and at night sink to 70°, though a cool breeze from the river produces a delicious tempering of the warm air. Over the older-settled portions of Florida there still lingers much of the *dolce far niente* repose which is the characteristic of all the old and most of the modern haunts of the Spaniard. But the busy "Yankee" is here too, and in a few years

---

* Lanier: "Florida: its Scenery and Climate." Philadelphia, 1876. Hadock: "Camp Life in Florida." New York; Fairbank: "History of Florida." Philadelphia, &c.

will transform this State, as he has transformed nearly every quarter into which his steam-engines, his tramcars, his cotton-mills, and overflowing energy and industry have penetrated. The beauty of the semi-tropical scenery p. 189, and the softness of its semi-tropical climate, will, however, always be the chief charm of that State, which Ponce de Leon, early in the sixteenth century, called by* its pleasant name, when on Palm Sunday he landed here in search of the fabled " Fountain of Youth." But were we to speak of the banana gardens, the orange groves, the palmettoes, the cacti, the Seminoles, the "roughs," the pleasant people, and the all-abounding "coloured man," we should cover more pages than we have lines to devote to this fair, if not particularly wealthy commonwealth.

ALABAMA

was first penetrated by the Spaniards under De Soto, though the present site of Mobile was occupied by the French as early as 1711. But in 1763, with the rest of the Gallic territories east of the Mississippi, it passed into the hands of England, and in due course slipped from under our rule, with the rest of our rebellious children's farms in that part of the world. In 1819, after having been the scene of many ineffectual struggles by the Creek Indians to stop the progress of the whites, it entered the Union as a State. It seceded in 1861, and since "the surrender" has been undergoing the painful operation known as "reconstruction." In 1870 its total population was 996,992, of whom 475,501 were coloured, and 98 Indians. It is not an educated State. Of persons ten years old and upwards, 349,771 were returned in the census quoted as unable to read, and 383,012 as unable to write. Among 175,000 voters, there is only a newspaper circulation of 10,000. In 1873, all the schools, except those in the large cities, were closed on account of the inability of the State to pay the teachers! In length it is 330 miles, and in average breadth 154, its whole area being 50,772 square miles. Though the Alleghany Mountains stretch into the State, the elevation is nowhere great. The centre of the country is hilly and broken, but the southern portion of it, for nearly sixty miles inland, is flat, and indeed raised very little above the sea-level. The Alabama is navigable from Mobile to Wetumpka on the Coosa branch, 460 miles from the Gulf of Mexico, and its tributaries also for some distance. The climate is semi-tropical, the temperature ranging from 82° to 48° Fah. during the winter, and from 105° to 60° during the summer, the mean temperature of the year being a little over 60° Fah. The forests have been felled; hence the winds from the snow-covered Rocky Mountains sweep more uninterruptedly than formerly over the country. It also follows that hence the temperature has increased in severity of late years. The uplands are healthy, but the people of the lowlands are subject to intermittent, bilious, and congestive fevers. In coal and iron Alabama is rich, while granite, marble, flagstones, roofing slate, lime, and porcelain clay, with a little gold, are among its other mineral resources. Much of the soil is worthless for agriculture, and much of the central and northern parts are covered with forests of oak, poplars, cedars, chestnuts, pines, hickories, mulberries, elms, and cypresses. There are few manufactures. Cotton is the principal

---

* In Spanish, *Pascua Florida*.

VIEW ON ONE OF THE TRIBUTARIES OF THE ST. JOHN RIVER, FLORIDA.

article of export. In 1874 287,974, and in 1872 137,997 bales—the greater part of which came to this country—were "raised" by the State for the foreign market. Alabama is a State of magnificent resources, the development of which has been retarded by the curse of slavery. It is said that within a century she expended 200,000,000 dollars on the purchase of slaves, a sum which, if devoted to the cultivation of other elements of wealth, would have yielded a very different result to the impoverished State. The soil is inexhaustibly rich, but it has been most recklessly abused. Good lands which were once worth fifty dollars an acre can now be had for five; yet, before the break up of the slave system, the State has been known to produce a million bales of cotton *per annum*. Everywhere ruin has come on Alabama. Montgomery, the capital—and at first the seat of the Confederate Government—is a pretty town of 14,000 inhabitants. But its streets are now filled with black and white idlers. There is little work for anybody, and that little nobody does. Politics and plunder are popular trades among the negroes, while the white people are afflicted with a kind of political stagnation, which allows them to consent mutely to almost any misfortune which may overtake them. The Alabamians are noted for their frankness and generosity, and their women are celebrated for beauty, among a people not the most homely-faced in the world. Yet they have their drawbacks. "One even now and then sees among the degraded 'poor whites' who 'dip snuff' and talk the most outrageous dialect, some lovely creature, who looks as poetic as a heathen goddess until one hears her speak, or she pulls from her pocket a pine stick, with an old rag saturated in snuff around it, and inserts it between her dainty lips." Some of the men are veritable Titans—giants, alongside whom even the tall backwoodsmen from Maine and Minnesota look dwarfed, but nearly all of them know little outside their State, and though easily led, are difficult to drive against the bent of their ignorance and prejudice. Mobile, the chief city of Alabama, is a charming place, with a sleepy quiet about it, inexpressibly pleasant to those accustomed to the noise, and hurry and scurry of the great towns of the inner world. Its inhabitants number about 36,000, but its harbour is tranquil and free from commercial bustle. On Government Street are numbers of fine mansions; many of the gardens are luxuriant, while superb oak-trees shade the same street, as well as the public squares between Dauphin and St. Francis Streets. There are fine shops with few buyers, and altogether the Mobile of to-day is about as quiet, and as dull, but hardly so rich as some of the old fishing villages that one lights upon along the shores of Massachusetts—when the fishermen are "on the banks." Mobile was a famous place during the Civil War. Block-running prospered briskly here, until Farragut's fleet forced its way into the harbour. Even then the town held out for another twelve months, only yielding little by little as Spanish Fort, Blakely, Hager, and Tracey were "invested, besieged, and taken." Even yet it is the home of many Southern celebrities, and a place where "good society" of the "befo' the surrendah" type may be found in as great perfection as anywhere else in the Southern States. An opulent future is in store for Alabama. But meantime it wants capital, farmers, and enterprise, and could get along with a few less lazy whites, a great number fewer "political" negroes, and altogether with the cotton worm.* For what the "coloured man" and the "carpet bagger" have left, that the dismal moth devours.

* The caterpillar of *Aletia argillacea*, a species of moth.

Mississippi.

The State of Mississippi is 332 miles in length, and from 78 to 148 in breadth, from east to west, containing an area of 47,156 square miles, and in addition to the mainland portion includes those islands in the Gulf of Mexico of which the principal are Horn, Deer, and Ship Islands. Altogether the State has sixty counties, and eighty-eight miles of sea-coast, but no good harbours. The general contour of the country is undulating, and its soil fertile, especially in the bottoms of the Yazoo, Black, Sunflower, and other tributaries of the Mississippi River, and of the Pearl, Pascagoula, and smaller streams flowing into the Gulf. The sea-coast, unlike much of the shore north of it, is not swampy but sandy, and well wooded with live oak, magnolia, and pine. It also bears the reputation of being one of the healthiest regions in America. For fifty miles the State borders the Mississippi. Hence it can dispense with sea-harbours, the river affording an outlet for its semi-tropical harvests of cotton, sugar, maize, tobacco, hemp, flax, sweet potatoes, figs, oranges, &c., as well as wheat and peaches, which grow luxuriantly. In 1870, it had a population of 827,922 (444,201 of whom were blacks), 564,938 bales of cotton, and 15,637,316 bushels of maize. Mississippi must always be an agricultural State, as it has no minerals on which to found a future prosperity. The climate, though warm, is yet sufficiently varied to allow the apple to flourish at one end of the State, and fig and orange groves at the other. But Mississippi reconstructs slowly. The ruin brought by the war was great, and the people recuperate slowly. Property has fallen ruinously both here and in the neighbouring State of Alabama. In 1860, the latter State boasted a valuation in real estate and personal property of nearly 150,000,000 dollars; in 1870, 155,000,000 dollars. Mississippi, when the war broke out, had a valuation of 509,472,912 dollars, and in 1870, 151,455,527 dollars. The cotton product was also in 1870 less than one-half of what it was in 1860. During the Civil War, Mississippi was one great camping ground, and the tracks of the contending armies are still visible in devastated timber and waste lands. The State could readily support on its 35,000,000 acres 12,000,000 of people, but over its whole extent there are not more than half-a-dozen towns of any considerable size. The chief of these are Vicksburg, Natchez, Jackson, and Columbus. The rest of the "cities" are mere villages—trading-places for the surrounding country. Yet the good lands are accessible, and the whole State is intersected by railways. Jackson, a quiet, pretty town of 5,000 or 6,000 people, is the capital. The negro is here predominant, and wields his power with that self-consciousness, which, if not the peculiar attribute of the black politician, is yet common to him with other *parvenus*. Jews appear to monopolise the trade, Hebraic names predominating on the signboards. The State debt is about 5,000,000 dollars, and as Mississippi robbed her creditors some years ago by repudiating her honest indebtedness, it is not likely that it will greatly increase. Altogether, the financial state of Mississippi is not very wholesome, the vicious system of issuing State warrants having been pursued until a late date, with the effect of reducing the State paper sometimes as much as forty per cent. below par. These warrants are now, however, funded, and taxes have to be paid in greenbacks. The negroes are, if anything, rather more intelligent than those in the neighbouring States, and are likely soon to be greatly improved.

as the majority of the black children attend school. The Ku-Klux-Klan and other anti-negro associations flourished for a time in the Sate. But though the system of "regulating" the sentiments of speakers at public meetings by the controlling influence of a shot-gun is not yet extinct, and Southern Mississippi bears but an indifferent reputation, yet, take the country as a whole, life and property are probably as safe as in any other Southern State. There is, however, but little property to save, and life in Mississippi is

VIEW ON THE ST. JOHN RIVER, FLORIDA.

a dull, cheerless existence. The negroes emigrate to Texas and Lousiania in search of work, while the planter at home complains of the scarcity of labour. Duelling is still common in the State. Hence respect for life, especially in Vicksburg, which has attained an unenviable notoriety as a town where "shooting at sight" is a popular method of vengeance, is reduced to a minimum. It is not, indeed, duelling, but cold-blooded murder. The authorities do their best to expunge this blot on the State's fair fame. But the ultra-Conservatives afford them little aid in securing the duellists, on the ground that the (unblessed) "Yankees want to do away with duelling so as to make their own heads safe." At Natchez the

negro has generally the upper hand, and in most of the parts of the State has his
fair share in the control of public affairs. Yet on the plantations, though in many cases
insolent and inclined to exercise his newly-acquired liberty to a degree bordering on license,
he still, in the majority of cases, keeps up almost instinctively his respect for his old
"owners"—now his "employers." "Mas'r," "Massa," and "Sah," still come readily to
his lips, and the old planters have generally little trouble in exercising a moral control

BAITING IN A DRUM FISH (*Pogonias chromis*) OFF THE COAST OF FLORIDA.

over their quondam slaves. The negro's deference to the white man is instinctive; his
politeness, a habit rather than a desire. Still the planters almost universally acknowledge
that though the "free nigger" is not a desirable personage, he is more profitable as a
labourer than he was as a bondsman.

TENNESSEE

has an area of 45,600 square miles. In addition to the Mississippi, which bounds it on the
west, it is drained by the Cumberland, the Tennessee, the Obion, Hatchee, and other rivers

which afford "water privileges" of an excellent character to the State. Eastern Tennessee is mountainous, owing to several ridges of the Alleghanies crossing it; the middle region is hilly, but the west is level. In the middle region, between the Cumberland and Tennessee Rivers, are extensive iron mines, and in the Cumberland Mountains are many unexplored limestone caverns, some of which are 400 feet deep, and several miles in extent. In one—at a depth of 400 feet—is a river, while another, opening perpendicularly into a mountain, is, as yet, unexplored; in others fossil remains of animals are abundant. In these cavern districts hundreds of acres have, in places, sunk to a depth of 400 feet. Traces of ancient mounds and fortifications of a date most probably prior to the advent of the present American aborigines are found in some places. Unless in the river bottoms the climate is healthy, and the soil, except in the mountainous eastern section, fertile. Cotton, tobacco, maize, figs, peaches, grapes, wheat, and most of the other products of Southern temperate regions, grow luxuriantly, while woods of pine, oak, hickory, walnut, sugar-maple, cedar (juniper), black walnut, and other trees abound. Raccoons, bears, deer, opossums, foxes, &c., are found in the forests, and the same may be said of the swine, for though not *ferae naturae*, yet, as they roam in great herds in the woods, feeding on acorns and nuts, they may, without any great stretch of language, be ranked as such. In 1870 there were 936,119 whites and 322,331 blacks in the State. In the same year the farms, averaging 166 acres in size, covered 19,581,214 acres, though less than one-third of their area was improved. At the same date the live stock was valued at 55,081,075 dollars, while the crops consisted, *inter alia*, of 41,343,614 bushels of Indian corn, 6,188,916 bushels of wheat, 21,465,452 lbs. of tobacco, and 181,842 bales of cotton *per annum*. In 1870 the State valuation was 253,782,464 dollars, and its indebtedness 18,827,494 dollars; but at present its debt is not one-half of that amount, and it is yearly being decreased. The State seceded in 1861, but it was only after a struggle that the "rebels" succeeded in severing themselves from the North, the first rebellious proposals being defeated. Even after the war entered upon the loyal men in Tennessee raised five Union regiments against the fifty Confederate ones, which, in ten months, took the field. At Chattanooga and Knoxville some of the most important operations of the Civil War were carried on, and for years after "the surrender" many portions of the State were in a very disorganised condition. The negro has not, however, managed to gain much power, owing to the fact of his being still in a great minority to the whites. Though Nashville is the capital, yet Memphis is the chief city. It is noted for its broad, regular streets, lined with handsome buildings, and since 1873, when yellow fever decimated the inhabitants, is tolerably well drained and cleaned. To-day it has about 67,000 inhabitants, who, though acknowledging that the city is not a sanatorium—and this in an American is a dangerously liberal concession—indignantly deny that, next to Prague and Valparaiso, it is the most unhealthy city in the world. It is perhaps not worse than other towns of the central valley of the Mississippi, but owing to the cemetery on the Chicksaw Bluff receiving the dead not only of the city, but also of the "migratory multitudes who toil up and down the currents of the half-dozen great streams which bring trade and people to Memphis," it has, therefore, been done some injustice—but perhaps not a great deal—in the returns of statisticians. It is in the centre of the cotton belt, and of the cotton trade of the surrounding country.

Its commerce in this one product must represent an annual return of something like 40,000,000 dollars, chiefly received from Western Tennessee and Northern and Central Alabama, Mississippi, and Arkansas, as well as from the south-eastern section of Missouri. Altogether the trade of the city is believed to average about 62,000,000 dollars yearly. The yellow fever ravages of 1855, 1867, and 1873 were primarily due to steamers from "down river" bringing the infection. In the last-named year two men arrived ill on board a New Orleans steamer. They were landed at "Happy Hollow," a low, marshy place, "which the genius of Dickens would have delighted to picture, filled with shanties and flat boats, with old hulks drifting up during high water, and then adopted by wretched long-shore men as their habitations. One of the two men died before he could be taken to the hospital, the other shortly after reaching it, and the physicians hinted that they thought the disease was yellow fever. For three weeks it was kept in 'Happy Hollow'; then it moved northward, through the navy yard, and suddenly several deaths in Promenade Street, one of the principal avenues, were announced. The authorities then went at their work; but it was too late, except to cleanse and disinfect the city. The deaths grew daily more numerous; funerals blocked the way; the stampede began. Tens of thousands of people fled; others, not daring to sleep in the plague-smitten town, left Memphis nightly to return in the day. From September to November hardly 10,000 people slept in town over night. The streets were almost deserted, save by the funeral trains. Heroism of the noblest kind was freely shown. Catholic and Protestant clergymen and physicians ran untold risks, and men and women freely laid down their lives in charitable services. Twenty-five hundred persons died in the period between August and November. This thriving city had become a charnel house. But one day there came a frost, and though suffering too severely to be wild in their rejoicings, the people knew that the plague itself was doomed. They assembled and adopted an effective sanitary code, appointing a fine Board of Health, and cleansed the town. Memphis is to-day in far less danger of a repetition of the dreadful scenes of 1873 than are Vicksburg or New Orleans, or half-a-dozen other Southern cities. Half a million dollars contributed by other States were expended on the burial of the dead and the needed medical attendance during the reign of the plague. The terrible visitation did not, however, prevent Memphis from holding her annual carnival, and repeating in the street so lately filled with funerals the gorgeous pageants of the mysterious Memphi— such as the Egyptians gazed on 2,000 years before Christ was born—the pretty theatres being filled with glittering costumes and echoes of delicious music. The carnival is now so firmly rooted in the affections of the citizens of Memphis that nothing can unsettle it." (King.)

A revolution is in progress around Memphis, as in every other portion of the Southern re-United States. Never was there a more wasteful, thriftless set of people than the old slaveholders. Cotton was their staple, and to that they devoted all their energies. When the planter made his annual settlement with his agent in town, he drew what surplus might be to his credit, and invested it in land and "nigroes." Everything—all articles of daily consumption—he imported from the North. This can be done no longer. He has had to reconstruct and commence the world anew on borrowed capital; for his money went where his negroes went. All was lost. He now finds it impossible to cultivate great tracts of country, as year after year the labour question becomes more and more a "burning one." The negroes have deserted the worn-out lands; hence cotton culture shows year by year a tendency to move further west—to the virgin lands on the other side of the Mississippi. The negroes, moreover, have a fondness for herding together in communities, and mingling less and less with the whites. It is, indeed, the belief of far-sighted observers, that the rich bottom lands of the Mississippi, where the white man cannot labour, and where even to live would be difficult for him, will be eventually the home of the black man in America. The great estates in the temperate regions must be cut up into workable farms, cultivated by that white labour which never comes, or if it does come, speedily leaves under the present aspect of affairs. The planters of the new school are now beginning to be alive to this fact, and to the wastefulness of the system under which "supplies" were drawn from a distance. This might work in the "old times," but nowadays it will not. If the crop fails, and his credit along with it, he gets crushed under an overwhelming load of debt, and is forced to gather up the wreck of his fortunes, pack the Lares and Penates into an emigrant wagon, and sullenly start afresh in Texas. At present in the vicinity of Natchez, and a score of other places, there is a magnificent farming country. And yet the people depends on the West and North-West for every barrel of flour which they use; for the bacon on which they cultivate indigestion, while they could eat abundance of beef and mutton, had they thrift enough to "raise" them; for the clothes on their backs, the shoes on their feet, and for the very vegetables which in every other part of America the poorest country labourer grows at his own door. All come hundreds of miles, by steamer and by rail, and before they reach the buyer are taxed for their profits by shipper, carrier, wharfinger, re-shipper—if they have gone far back from the river or line of rail—and by the local shopkeeper.

The result is, that if a few worms creep over the cotton leaves, the planter is in despair, and if the neighbouring streams rise he is in the midst of a swamp physically and financially—in a slough of debt and misery which he will not get the better of for at least two years. If you ask them the reason of all this, they have rarely any other answer than that if they grow these things the negroes would steal them, and that the "South is ruined for ever." Efforts are being made to utilise the negro. In some cases he "works on shares," in other cases he leases the land, and pays a rent of so many bales an acre, while others prefer the wage system, or to work the land by "squads," who get advances secured by liens on the squad's share of the crop, and on the horses and mules which they own. The negroes live in little villages, and on that account are less inclined to roam than they were when they first rejoiced in their newly-acquired freedom. They feed on pork, cornmeal, and molasses, all brought hundreds, or it may be thousands of miles, though with a little care they could grow enough to feed themselves and their families on the land which they cultivate. But if his "ma-'r" is thriftless, the ex-slave is more so, and in addition is lazy and improvident, and fond of holidays and junketings. "The planter," according to Mr. King, "always feels that the negro is irresponsible, and must be taken care of. If he settles on a small tract of land of his own, as so many thousands do nowadays, he becomes almost a cumberer of the ground, caring for nothing save to get a living, and raising only a bale of cotton or so wherewith to get 'supplies.' For the rest he can hunt and fish. He doesn't care to become a scientific farmer. Thrift has no charms for him. He has never been educated to care for himself; how should he suddenly leap forth, a

FOREST SCENERY IN FLORIDA.

new man, into the changed order of things?" Yet, in justice to the negro, it must be allowed that when he is prosperous he is much more ready to pay his advances to the merchants than a white is in the same circumstances. If, however, his crop fails, he is exceedingly unwilling to meet his liabilities, even when he has the means to do so. The Jews who, throughout the South almost monopolise the commerce, quite understand him. The negro likes to be treated with consideration when he visits the "store." Accordingly, the profuse Oriental manner and conversation of the Hebrew merchant are much to his mind. Hence Messieurs Shadrach and Abednego secure his custom, at the comfortable profit of 100 per cent. They, however, watch both black and white planters narrowly, and regulate their advances in accordance with the prospects of the crop, to such an extent that the Israelites in some sections are becoming at once task-masters, arbiters, and guardians of the destinies of the planters—an utter spoiling of the Egyptians, which is, however, on the whole, vastly to the eventual benefit of these improvident debt-contracting individuals. The negro is as extravagant as the white. If the former secures a little money, the front of his cabin soon becomes strewn with sardine tins and whiskey bottles. If the latter is fortunate, the surplus is apt to be expended on a "barbecue" and a case of champagne, in which for an evening to drown his woes, re-fight his battles, and curse "the Yankees."

Nashville, the capital of Tennessee, was, and is yet, renowned for the wit and beauty of its men and women. At present it possesses between 40,000 and 50,000 inhabitants, and an annual cotton trade of more than 130,000 bales. It also supplies the South with provisions—including Nashville whiskey, which is a much-esteemed beverage in "Dixie's Land "*—and is exported from the city sometimes to the extent of 100,000 barrels per annum. Its prosperity is likely soon to be further increased by the rapid development of the coal trade and the manufacturing interests of the city and surrounding country. Tennessee has much good land, well suited for emigrants from temperate regions. Mr. Killebrew, in his work on the resources of the State, advises settlers to "locate" in colonies, so as to have neighbours whose habits and modes of thinking are similar to their own, though considerably different from those of the native population, which is, as a rule, rather ignorant, and prone to be prejudiced against all innovation. This is, however, chiefly in the rougher mountain regions, where a labouring man is apt to be despised, and a Northerner especially to be regarded as a natural enemy.

KENTUCKY.

"Ole Kantuck," as the negroes love to call it, has an area of 37,680 square miles, or 24,115,200 acres, divided into 110 counties. As a rule the country is rolling, hilly in places, and in the south-east mountainous. The soil is rich, and raises some of the finest wheat, maize, cotton, hemp, tobacco, and fruit crops in the warmer temperate regions of America. Cattle-grazing is extensively followed, and in the woods millions of swine fatten on the nuts and acorns. The coal-beds, some of which are rich in oil, stretch nearly

---

* The real "Dixie's Land" was, however, in Manhattan Island, where the eminent Mr. Dixie was a slave-owner until he was forced to remove South, to the regret of his slaves, who have celebrated the charms of their old Utopian home in the familiar ditty.

across the State, while the lead, iron, marble, and salt deposits are rich. In 1870 the State assessment was 409,544,294 dollars, and the value of the agricultural produce 87,177,574. Kentucky resisted all persuasives to induce her to secede, and though many of the inhabitants joined the rebels, yet the State, as a whole, remained faithful to the Union. In 1870 the population was composed of 1,098,692 whites and 222,210 blacks.

A voyager down the Ohio River is rather impressed with the difference of the two sides of the river. On the Ohio shore all is bustle. Fine manufacturing towns, iron furnaces, and the clatter of hammers, give evidence of thrift and industry. On the Kentucky shore the towns from Huntingdon to Cincinnati (Ohio) are few but substantial, though not bustling. "Around the various taverns in each of them is grouped the regulation number of tall, gaunt men, with hands in pockets and slouched hats drawn over their eyes. A vagrant pig roots here and there in the customary sewer. A few cavaliers lightly mount the rough road leading into the unimposing hills; a few negroes slouch sullenly on a log at the foot of the *levée*, and on a wharf-boat half a hundred white and black urchins stare open-mouthed, as if they had never seen steamboats or strangers before." Log cabins sprinkle the shore at intervals, and altogether the voyage down *La Belle Rivière*—as the early French explorers called it—is one of considerable interest, even though the romance of the great American rivers has either ceased, as in this section of the Union, or is rapidly fading into the region of shadows. The trade of the Ohio is enormous. It drains 214,000 square miles, and may yet be the high road for the transportation of the commerce of fifty millions of people.

Louisville is the chief Kentuckian town, and though life there is not so vivacious as in New Orleans, Savannah, or Charleston, is yet as brisk as in most other Southern or South-Western towns. It boasts the best hotel in the United States, which fact speaks volumes—a whole library—for it, and the Louisville *Courier Journal*, once famous for the cutting wit of Prentice, and the duels—actual and threatened—which its strictures provoked. The population numbers 130,000 or more, celebrated for their frankness yet high-bred courtesy, and fine character generally.*

The Mammoth Cave of Kentucky is the great lion of the State, albeit there are several other similar caverns, though on an infinitely smaller scale. It is situated in Edmondson County, near Green River, 130 miles south-west of Lexington, and in reality is not one cavern but a series of caverns, which have been explored for at least ten miles (p. 192). A stream called the Echo River flows through it, and in this, which is shut off from the daylight, is a fish without eyes,† and a crayfish which is blind. The river finds its outlet beyond the cave in Green River, and is here and there wide and deep enough to float a large steamer. One of the earliest guides to the cave was a negro named Stephen. At that time the Stygian Echo River was unexplored, and for years the ambitious African urged a white man, living near the cave, to build a boat with which to examine it. At length the boat was built and a voyage under the arches decided on; but Stephen's courage failed him, and it was only at the pistol's mouth that he could be induced to enter the boat and proceed. The voyage was a daring one, for neither of the explorers could have predicted that

---

* Casseday's "History of Louisville." † *A. Amblyopsis.*

the river would find an outlet beyond the cave in Green River (p. 193). The temperature of the cave is very equable; hence it has been recommended as a residence

INTERIOR OF THE MAMMOTH CAVE OF KENTUCKY.

for consumptive and asthmatic people. The cave is now resorted to by numerous visitors every year, and in accordance with custom, every part of it has received fanciful names, while its roof is frescoed with the ignoble patronymics of ambitious tourists, traced with the smoke of a lamp fixed to the end of a pole. The various passages of the cave

VIEW OF LAKE GEORGE, NEW YORK STATE.

have a total length of more than 200 miles, though there are numerous caves far down—dead seas into which the visitor may peer from overhanging precipices—as yet only vaguely known. The cave is said to have been discovered in 1809 by a hunter, who accidentally strayed into it in pursuit of a bear which had taken refuge there. At present it belongs to nine heirs, who receive each about £200 yearly income from it. The cave with 200 acres of land was bought for forty dollars. A company has talked of buying it,

EXPLORING THE ECHO RIVER, MAMMOTH CAVE, KENTUCKY.

and by increasing the facilities for reaching it, and providing better accommodation, it is believed that it could yield excellent interest on the half a million dollars demanded for it.*

Frankfort, the capital, on the Kentucky River, is one of the prettiest spots in the whole State. It lies in a deep valley, surrounded by sharply-defined hills, while the river flows between banks of that limestone which supplies the excellent building-stone of which the town is partially constructed. The Frankfortians are being roused from their pleasant dream of life by manufactories which are being established, but they seem

* Foxwood: "The Mammoth Cave: an Historical and Descriptive Account thereof." Philadelphia, 1874.

neither to care nor to seek for them. The Ku-Klux—that murderous Vehmgerichte of the days succeeding the collapse of the Confederacy—until comparatively recently instituted a reign of terror in the neighbouring country, and the ignorance and general carelessness which are yet only too prevalent in the back country of Kentucky lent that aid to their operations, without which they would have been early crushed.

The State is famous not only for its lovely women, tall, clear-complexioned, and courteous men, but also for its horses reared on the Blue Grass region of Middle Kentucky. This part of the country is also noted for its Bourbon whiskey, which, in the head of a "Bourbon democrat," becomes a potent factor about election time. Lexington is another wealthy and beautiful Kentuckian town, built on ancient Indian ruins and fortifications of great extent and magnificence. In this vicinity is also an extensive cemetery of the pre-historic inhabitants of the country. The coloured portion of the population at present gives little trouble in the State. The public debt is small, and the authorities are chary in contracting more, so that the State may be said to be prosperous.

CHAPTER XI.

THE UNITED STATES: INDIANA; MICHIGAN; OHIO; PENNSYLVANIA; NEW YORK; NEW JERSEY AND DELAWARE.

WE have now—still reserving a visit to Delaware (which never seceded from the Union)—left the Southern States, the home of slavery, and the scene of that terrible retribution which man-owning and man-selling brought upon them. We are again into a freer atmosphere—among farmers who toil—among freemen who eat the bread that they have gained by the sweat of their brow, in a land which, though possibly not rich, nor abounding in the gorgeous vegetation ripened under the mellow skies of the semi-tropical South, has yet charms of its own, perhaps more attractive than those which we have for some time past been among. Its people may not be so polished, as in exceptional communities and cases you may find in the South. But as a rule, the majority of them are infinitely superior to the greater part of the inhabitants of the ex-Slave States, and their life, thoughts, and expressions a thousand times more wholesome, and more in keeping with nineteenth-century civilisation, and that working-day world which lies on the other side of the Atlantic. Let us, therefore, before finally quitting the United States, say a few words about these middle and northern States of the Great American Republic.

INDIANA.

This State is 276 miles long, 140 miles broad, and has an area of 33,809 square miles, and a population of 1,680,637, a large number of whom are from Germany, Ireland, or from other States of the Union. There are immense coal-fields, and the fertile soil, though the climate is colder than that of France in winter, yields bountiful crops of wheat, maize, tobacco, fruit, &c. The hills on the Ohio River are covered with vineyards, the plains with cattle, and the bush with pigs. More than 3,800 miles of railway intersect the State, while the Wabash and Erie Canal, 467 miles long, connects the Ohio River with Lake Erie. The capital is Indianapolis (80,000 inhabitants), but Evansville, New Albany, Madison, Vincennes, Terre Haute, Lafayette, and Fort Wayne, are all towns of some importance. The contour of the country shows that it is in general level, or gently rolling, consisting of great prairies drained by sluggish rivers, such as the Wabash and its tributaries. Altogether, Indiana is one of the finest of the American States.

MICHIGAN.

The State of Michigan has a land area of 56,451 square miles, and had a population in 1874 of 1,333,861, including at that date 4,926 Indians, one Chinaman, and one Japanese. The *tribal* Indians numbered, in 1874, 8,923, chiefly Ottawas, Chippewas, and Pottawattomies, who depend chiefly on hunting and fishing, or are engaged in agriculture. The chief town is Detroit, with a population in 1874 of 101,255, though Lansing, the capital, has not 8,000. The general aspect of the northern peninsula is rugged and picturesque. That portion east of the meridian of Marquette is an undulating plateau, "sinking gradually towards the south, and more rapidly towards the north, the watershed being much nearer Lake Superior than Lakes Huron and Michigan." Over this plateau numerous lakes and marshes are scattered. Except where fires have destroyed the timber the surface is covered with forests, and where it has the region is simply a desert. West of the plateau the country is "irregularly mountainous, interspersed with swamps and lakes," some of the peaks attaining a height of 1,500 to 2,000 feet above Lake Superior. The copper or mineral region occupies the north-west extremity of the peninsula. This region contains most of the mineral wealth of the State, but the soil is for the most part sterile. The southern peninsula is in every respect different from the northern one. The surface is generally level, though in the south there is an irregular cluster of low conical hills. The soil is luxuriantly fertile, except on the northern part, and is underlain by beds of bituminous coal, covering an area of about 12,000 square miles. The coal, however, from the difficulty of working, is only shipped to a small extent. Michigan abounds with objects of natural and antiquarian interest. "Among the former the most noteworthy are the 'Pictured Rocks' (p. 113) on the shores of Lake Superior, about thirty miles west of Sault St. Marie. There are sandstone bluffs of various colours, worn by the action of the waters into grotesque forms resembling castles, temples, arches, colonnades, &c., which, from a steamer on the lake, have the appearance of a gorgeous picture." These rocks extend along the shore for about twelve miles, and rise from 200 to 300 feet above the water. Sometimes the cascades shoot over the precipice, so that a vessel may sail between the descending waters and the natural wall of rock. "The line of cliffs," to use the words of an American writer, "terminates at the eastern end in what is known as 'The Grand Chapel.' It stands about fifty feet above the present level of the lake, and its roof, which is arched, is supported by two gigantic and beautiful columns, that appear to have been hewn and placed where they are by skilful hands. The backward reach of the roof rests upon the main cliff, and within the chapel is the base of a broken column that is strongly suggestive of a pulpit. The roof is crowned with a growth of fir-trees that maintain a terrible struggle for life with the storms that are so frequent here, and to which they are always exposed." p. 197. In the northern peninsula and on Isle Royale there are the remains of very ancient copper mines and mining tools, and it is evident that a race well advanced in civilisation occupied the country at some distant period in the past, of which the Indians found in possession by the early explorers from Canada could give no account.* The climate of Michigan, though tempered by the proximity of the lakes, is one of extremes; that of the winter, especially in the northern peninsula, being arctic, while in summer it is disagreeably hot. Good crops are, however, reared, especially in the southern peninsula, where there is a deep soil of a dark loam, often mixed with gravel and clay, and very fertile. Apples are grown in great quantities. Peaches form one of the crops grown successfully on the shores of Lake Michigan, while most of the fruits of temperate climates flourish, and even the vine is cultivated on the shores of Lakes Michigan and Erie. The fisheries of the lakes and rivers are a great source of wealth, while as regards agriculture, Michigan was in 1870 tenth among the States in the value of its agricultural produce, and ninth in the value of its manufactures. When the last assessment was made, the taxable value of property was 630,000,000 dollars. The total value of sawmills in Michigan was greater than in any other State. In 1870, 304,054,000 laths, 2,251,613,000 feet of lumber, 658,741,000 shingles, and staves, &c., to the value of 1,332,922 dollars, were among the products.

Michigan appears to be derived from the Chippeway words *mitchi* (great) and *sawgyegan* (lake), and was formerly applied to both Lakes Huron and Michigan.† Lake Michigan is one of the five great lakes of North America, but the only one which is entirely included in the United States. Its length is 320 miles, its mean breadth 70 miles; its mean depth 1,000 feet, its elevation above the sea 589 feet, and its area 22,400 square miles, thus exceeding the area of Lake Huron by 2,000 square miles. The shores are for the most part low and sandy, and here and there are hillocks formed of the sand which has blown inland. These wind-formed mounds are sometimes 150 feet in height. It is more than probable that at one time the water of Lake Michigan found an outlet through the valley of the Illinois River into the Mississippi. There is only a low watershed separating the lake from the Illinois River, and the valley through which the latter flows is of greater extent, and looks as if it had been worn to its present dimensions by some great river coursing through for unnumbered ages. The lake is believed to be just now working westward, gradually encroaching on the shores of Wisconsin, and leaving those of Michigan. There is a lunar tidal wave on the lake. From observations made at Chicago it was found that high water occurred there half an hour after the meridian passage or sinking of the moon, and that the difference of elevation of the lake surface between high and low water was

---

* See also Foster and Whitney: "Executive Document," No. 69, 31st Congress, 1st Session; Agassiz: "Lake Superior;" Strickland; "Old Mackinaw," and the works of Schoolcraft, Daniel Wilson, Kohl, &c.

† Houghton: "Michigan Geological Survey," 1858-43, and Brooks, Pumpelly, and Rominger: "Geological Survey of Michigan" (1873).

VIEW OF "THE GRAND CHAPEL" ROCKS, LAKE SUPERIOR.

155-thousandths of a foot, while spring tides gave a difference of over three inches. The lake has few harbours, and accordingly is not very safe to navigate, more especially as it is subject to severe storms at different seasons. (Hawes.)

Lake Huron is another of the five great American lakes. It has been estimated to contain 20,000 square miles. Its surface is raised 578 feet above the level of the sea, and though its average depth is 1,000 feet, yet soundings have been taken in it as deep as 1,800 feet without touching the bottom. From the clearness and purity of its waters, it received the name of *Mer Douce*, or Fresh Sea, from the French fur-traders who first visited it. It is said to contain 3,000 islands. One of these—the Great Manitoulin, or Sacred Island—runs parallel to almost the whole Northern Rocky Coast.

Lake Superior is the largest body of fresh water in the world. Its greatest length is 335 miles, its greatest breadth 160 miles; its area 32,000 square miles—or, in other words, it is a lake equal in size to Iceland. Its mean depth is 1,000 feet, and its elevation above the sea 600 feet, or 22 feet above that of Lakes Huron and Michigan. Its shores on the British side are as yet only inhabited by the fishermen and miners, who extract the silver ore from the veins on Silver Island, near Thunder Bay. But on the American side there are several little towns, chiefly connected with the copper mines.

The presence of the wonderful city of Chicago (in the State of Illinois, p. 115)—pronounced *Shek-argo*[*]—the name being derived from an Indian word signifying "wild onion,"[†] will always give Lake Michigan a position over most of the other American lakes, notwithstanding its poor harbours. Its growth has been even more rapid than that of San Francisco, and though St. Louis has outstripped it in the race, yet the great city of Missouri has had a longer period of probation than has its rival in the trade of the West. There are men yet living who remember when the site of Chicago was a swamp, and could have been bought for a few gallons of rum or a pack of beaver pelts. Land bought forty years ago from the government for 1¼ dollars an acre is now worth 10,000 dollars an acre. Business property is worth, on an average, 25 dollars per square foot. Its nucleus was Fort Dearborn, built in 1803, but long after that date it was a mere frontier fort, frequented by hunters, trappers, and backwoodsmen. Indeed, the writer was well acquainted with a man who, some fifty years ago, encamped in a cotton tent on what is now the chief street of Chicago, for the simple reason that there was then no house to give him shelter. His chief regret to the day of his death was that he had ever "lifted his stakes" from the swampy ground, the price of which is now estimated in more dollars per rood than he could then have bought it per square mile. In 1830 the town was organised, but two years afterwards the population was only 1,000. In 1850 it had increased to 28,269, in 1860 to 110,973, in 1870 to 298,977, in 1873 (p. 115) to 367,396, while in 1877 the population was certainly over 450,000, and probably near 500,000. On the 8th October, 1871, the city was almost entirely destroyed by fire, the loss of property being estimated at 196,000,000

---

[*] This is among Americans something of a shibboleth and sibboleth to distinguish the stranger and the resident, the former almost invariably calling the Illinois city *Che-kargo*.

[†] Others, however, will have it that the name of the city is derived from the Chicago River, which divides the town into three districts. In the tongue of the neighbouring Indians the river was *Checaqua*, which means thunder.

dollars; but it has now been entirely rebuilt in a much better style than it was previously. Indeed, fires in a new town in America seem essential to its prosperity. It is the test of its security. If the town is a substantial one it is soon rebuilt, with many improvements which would otherwise have taken time to have been introduced, thus giving both the citizen and the stranger confidence in it. If not the fact is taken as a proof that it was originally built under mistaken views of its requirements, thus illustrating in the history of cities the law of the "survival of the fittest." Chicago is the great storehouse for the grain crop of the West, its annual export of wheat being over 15,000,000 bushels, of maize 27,000,000 bushels, besides large quantities of rye and barley. Its pork and beef packing, brewing, distilling, iron and steel, boot, shoe, and other establishments, are almost as celebrated as those of Cincinnati, while its colleges, schools, and journals are scarcely less prosperous than its manufactures of machinery, carriages, &c. In 1875 the taxable value of the city property was given at 293,188,950 dollars. The Michigan Government, considering that the first duty of the State is to educate its citizens, has established a university at Ann Arbor, which has thirty-four professors and an endowment of a million acres of land. Yet the only fees exacted from the student are ten dollars on admission and five dollars per annum until they graduate.

OHIO.

We are now approaching commonwealths, not only great in size, but populous, prosperous, and polished. The State of Ohio is one of these; albeit, not very long ago it was looked upon as being almost in "the West." But where the West is one cannot always be certain in America. It is like the North in Britain.

> "Ask where's the North? At York 'tis on the Tweed,
> In Scotland at the Orcades, and there
> At Greenland, Zembla, or the Lord knows where;
> No creature owns it in the first degree."

And so it is with the West. At one time all New England considered it at Ohio, Kentucky, and certainly at Chicago. There for a time it remained. Then it shifted to Missouri, and for a time came to a standstill at St. Louis; but "westward the course of empire" took its way, and then, among the trappers and rowdies of Omaha, "St. Joe's," and Independence, it was customary to fix this movable point. But the great plains offered little in the way of obstacles to the roaming emigrant, and he talked of going West, where he "calc'lated" to trap in the Rocky Mountains, or hunt buffalo in the plains of what is now Dakota. But there was still a Further West, and that was Oregon; and in Oregon people talked of going West, until the Pacific stopped all further progress. I suppose they still speak in the same way; but Japan and China must be now their Furthest West.

Famous among the burghers of Rouen, two hundred years ago, was the rich family of the Caveliers. Though not nobles, they held distinguished positions, and were courtiers and diplomatists. But of the noble deeds and skilful word-fencing history has failed to record much of any of them, except of Robert—better known by the designation of La Salle, the name of the estate of which he was Sieur—born in 1643. In Canada, in 1666, we find the young Sieur—a lad of ardent imagination, bright intellect, and profound piety—according to the piety of the Jesuits. He wearied to make conquests for the Church, and longed, with the longing of three-and-twenty, for the dim mysterious lands that lay "west." Like all the explorers of the time, he thought of finding a passage to the South Sea; and when the Seneca Indians told him of a river called the Ohio rising in their country and flowing into the sea, but at such a distance that its mouth could only be reached in eight or nine months, he immediately came to the conclusion that it and the Mississippi were one, and that it must needs flow into the "Vermilion Sea," as the Gulf of California was then called. This, then, was the long-sought-for water-way to the Pacific; and if not, the populous Indian tribes on its banks would afford a sure source of profit. La Salle's imagination was on fire. He sold his property, and with the proceeds fitted out four canoes, manned by fourteen men, and set off to explore this great river of the West. How he fared, what crosses, what toils, what dangers he encountered, are now matters of history, which will ever accord to the young Jesuit Sieur the honour of first sighting the Ohio, and probably also the Illinois; but that he discovered the Mississippi has not been proved.* It was not, however, until 1787 that a company of forty-seven emigrants reached the site of Marietta, and began the first settlement, and about the same time Congress began to exercise jurisdiction over the territory north-west of the Ohio. But the troubles of the settlers began. The Indians fell on them, and, indeed, in 1791, the Miamis defeated General St. Clair. In 1803 Ohio was admitted into the Union, though Connecticut reserved 3,666,921 acres to the north-east, along the shores of Lake Erie. This has since been known as the "Western Reservation." The State, as at present bounded, is 228 miles long in its extreme length, and 220 miles broad, with an area of 39,964 square miles, or 25,576,960 acres. Most of it is in the form of a great plain, descending from the Alleghanies towards the Mississippi Valley, the highest point, namely, between the Scioto and the Miami, being only 1,540 feet above the sea. Cincinnati, for instance, is 525 feet above the tide-water, Cleveland 685 feet, and Hudson 1,137 feet. Between Lake Erie and the Ohio River, a low chain of hills constitutes the watershed, and along the lake shore are cliffs which sometimes attain a height of 750 feet above the surface of the water. The south-eastern section of the State is undulating, with precipitous hills along the banks of the Ohio. In this section are found the famous mounds of the Ohio Valley. These fortifications and tumuli appear to have been the work of a people who lived long prior to the advent of the present race of Indians, and to have been the site of extensive towns and settlements. When the whites first came into the Ohio Valley, they found the Red Indians roaming through the forest, unacquainted with any of the arts of civilisation, and with not even a tradition of the nation who built these great earthworks and lived here. It is believed by some that they may have been the same people who worked the copper mines of Lake Superior, where their stone tools and weapons, left behind in the workings, abundantly attest their numbers and the skill with which they mined the masses of virgin metal. But whether they were the old Aztecs of Mexico, who only visited this

* Parkman: "Discovery of the Great West," p. 25.

part of the country in the summer, or whom, we can only guess at, for there is no certainty whatever in the matter. Ohio is peculiarly favoured in "water privileges." In addition to the Ohio—which, opposite Cincinnati, is 1,600 yards broad—there are Muskingum, the Scioto, the Great Miami, and the Little Miami, all navigable tributaries of the river from which the State takes its name, while emptying into Lake Erie are the Maumee,

AN OHIO FARMHOUSE.

Sandusky, Cuyahoga, and Chagrin Rivers. Lake Erie itself affords a frontage on the north of 230 miles of navigable water, while there are many other lakes which, though not utilised for water carriage, abound in cat-fish, sturgeon, pike, perch, shad, &c. Among the trees may be mentioned seven species of maple, eleven of walnut, and twenty-six of oak, though, owing to its more rapid growth, most of the timber of the Western country is softer than that of the Eastern States. The soil is mostly clay, but in the "Reserve" there is gravel and sand. The climate is very variable, the temperature in the winter falling far below zero, and in the summer rising many degrees above it. As an agricultural region, Ohio ranks, so far as the production of Indian corn is concerned, third among the States, fourth in oats, fifth in barley, and seventh in wheat. In 1870, there were 195,953 farms, of which sixty-nine contained more than 1,000 acres each, while the average size was 111 acres.* In the amount of improved land it was, in 1870, only excelled by New York and Illinois, having 14,469,133 acres more or less under cultivation. The value of its farms, farm implements, and live stock, was 1,200,158,511 dollars; while the value of farm products, "including betterments, &c.," was 198,256,907 dollars. In 1874, there were within the State 738,000 horses, 22,500 mules, 882,000 oxen and other cattle, thus, in this respect, ranking next to Texas and Illinois; 778,500 cows, second to New York; 4,639,000 sheep, next to California; and 2,017,100 hogs. Fruit grows in great abundance, over 350,000 acres being devoted to orchards. In 1872 (an exceptionally good year), the apple crop was 23,000,000 bushels, and the peach crop 105,619 bushels. As a manufacturing State it ranks third; in the fabrication of agricultural implements, first; and in pork curing, next to Illinois and Missouri. In the winter of 1874–75, there were 874,736 hogs, of the value of 16,597,190 dollars, packed. The establishments where this operation is chiefly carried on are in Cincinnati, which, on that account, has been often styled Porkopolis. Bituminous coal is profitably mined in thirty counties, iron in twenty counties, and salt and petroleum in many parts of the State. Its shipping is considerable. Vessels load in Lake Erie—573 feet above the sea—and sail directly into the Atlantic by way of the River St. Lawrence; and as the 455 miles of the Ohio River are connected through the Mississippi with the Gulf of Mexico, Ohio has the advantage of foreign vessels visiting it. In all, during 1874, 1,362 vessels entered, and 1,388 cleared in the foreign trade; 8,417 entered, and 8,160 cleared in the coastwise trade. Of these, 219 vessels belonged to Cincinnati, and 609 to the Lake ports. Twenty-eight vessels were built upon the Lake, and forty, of which nineteen were steamers, upon the Ohio. In 1870, the fisheries were valued at 383,424 dollars, thus giving Ohio the fifth rank among the piscatorial States. Nor does this great and prosperous Commonwealth neglect education in its eagerness in the pursuit of wealth, as may be seen from the fact that in 1873–74 there were 11,688 school-houses, 22,575 teachers, and 985,917 children being taught. There are thirty-two colleges—one of which, Oberlin, has 1,330 students—twelve schools of theology, eleven of medicine, four of science, and three of law. The State contains thirty-one "cities." The State capital is Columbus, with (in 1870) 31,274 inhabitants; but the greatest town in Ohio is Cincinnati—the "queen city of the West"—with a water frontage of ten miles on the Ohio, and a census, in 1870, of 216,239, of whom rather more than one-half were natives of the United States. In 1873, the valuation of property was 185,645,710 dollars. Thirteen railroads enter the city, while eight lines of river packets moor their vessels alongside the wharves of this prosperous town, which has been made more familiar to English readers as the

---

* For the statistics, &c., of Ohio and the remaining United States, I have drawn my material from the official census of 1870, from a few of the State or local censuses, and from other documents, published and unpublished, which have been supplied me by the courtesy of many private and official personages, who are so numerous that I can only thank them collectively. For much information also which I could not obtain otherwise, I am indebted to Burley's "United States Gazetteer," edited by Mr. C. H. Kidder, and more particularly to the topographical articles, by the Rev. M. B. Williams, in that excellent work. (Philadelphia, 1876.)

anathematised home of Mrs. Trollope, who published such an unflattering account of it, than as a great commercial and even educational centre. Cleveland (92,829), "the forest city" of Lake Erie; Toledo (51,580), on the Maumee River; Dayton (30,745), Sandusky (13,000), Springfield (12,652), Hamilton (10,081), and Akron (10,006), are the principal towns, the others having, in 1870, less than 10,000 inhabitants. The growth of Ohio has been remarkable. At the beginning of this century it was eighteenth among the States in point of population; in 1840 it was third, and it has since retained this position. In 1870, the population numbered 2,665,260, the increase during the last decade being 13.92 per cent. Ohio is, however, the source from which Indiana, Illinois, and Iowa have drawn much of their population, 70,000 persons having gone to Iowa in seven years from Ohio.

The town of Chillicothe, which has now over 9,000 inhabitants, is about the only "city" in the United States which was once a considerable town of the Indians. Here lived the Shawnees—the *chats sauvages*, or wild cats, of the French—and so large was their village, that when Daniel Boone was brought as a captive to the place in 1778, he saw 450 fully armed warriors setting out on a foray on the settlements of the whites. However, there were at least three Chillicothes in existence, though all three were in Ohio. The town of Erie bears, in a fragmentary state, the name of a tribe of Indians now wholly extinct. Thomas Morton calls the lake, in 1632, "the Lake of Erocoise," by which he, no doubt, means the French "Iriquois." Father Hennepin speaks of it as Lake Erike, "that is to say, the Lake of the Cat, as the Hurons call it." Afterwards Charlevoix calls the Indians "the Nation of the Eriez, or the Cats"—hence the present name of the lake and town; though the tribe who gave both their name were totally exterminated by the Iriquois in the year 1654. And here we may say a few words on the names of American towns generally.

The early New England settlements were generally named after the town or village in England from whence the settlers had originally come, though in many cases the New World namesake surpassed its Old World godfather. We need only recall New York and Boston to the reader's mind, though, at the same time, New Edinburgh and New London have scarcely shot ahead of Old Edinburgh and Old London, nor are they likely to do so. Then came a period when the colonists were particularly loyal, and accordingly, especially in Virginia and the other Southern States, the princes and other notabilities of the House of Stuart—the Hanoverians were not so popular—were honoured by having their names attached to various towns. With the Revolution came an end of this. Then the young Republic had heroes of its own, and with the advent of these—often rather ephemeral dignitaries—came a long list of towns bearing their names, or those of the Continental officers who—like Lafayette—aided the colonists in their struggle against perfidious Albion. By-and-by, local celebrities, with uneuphonious names, made still more unsonorous by having *ville* attached to them, came to the front, and the Stigginsvilles and Slocumvilles grew plentiful. After this came a reaction, and classical names—for which the early American Republicans, like the French ones, had always a hankering—covered the maps. Others, more sensible, were derived from the Indian languages. From the very first there were sensible people who considered that it would be more in keeping with the surroundings, if the new towns and villages bore the names of their Indian predecessors, where these could be got at, or when aboriginal settlements existed in the localities. It would only be a small return for much erring to thus aid in keeping alive the names of races whom the new comers had, either directly or indirectly, "civilised off the face of the earth." But this was often difficult. The early settlers—like the present Western ones—had suffered too much at the hands of the Ishmaels of the New World to be in any way inclined to perpetuate the memory of their sufferings by sentimental nomenclature. Accordingly, many of the names were lost, or were only guessed at—often erroneously, and when restored almost invariably in a corrupt and mangled fashion. For instance, to cite two modern examples of an old error, the first settlers applied the name Yazoo to the city of that name, under the idea that this was the Indian name. In reality, the Cherokees and Choctaws call the vast mounds and fortifications which are strewn along the course of the Yazoo River by that designation, and have not applied the name to any particular place. Again, when Major Savage discovered the Yosemite Valley, in California, he gave it that name, under the idea that it was the Indian word for the grizzly bear, the emblem of the Golden State. In reality, it means no such thing. Professor Schele de Vere has some amusing and instructive remarks on this subject, the gist of which may be given here, as I may not have another opportunity of referring to this subject. The numerous blunders in Indian names, real or suppositious, which the early town namers have made, are, however, in their result really not so bad as some of the absurd English ones. In fact, American town names of this class have been an inexhaustible subject of ridicule and cheap wit for foreign critics, and even for home-bred satirists. Dickens did not, however, coin a new name when he applied to the dismal Mississippi City the name of "Eden." He merely copied it from the Westover MSS. That, of course, is no reason why the proud American should defiantly adopt it to such an extent as to have at least twenty-four Edens in the Union and six New Jerusalems. "He does not much believe in ancient myths, and hence, contradicting the poet's assertion that *Ilium fuit*, he has sixteen Troys. The subject of classic names is, however, a peculiarly painful one; thanks to the preference given to this class by one of the early surveyor-generals of the State of New York—De Witt—who is responsible for the Uticas and Ithacas, the Homers and Virgils, the Romes and Athens, which abound in the Empire State, and from thence spread all over the Union. Why Demosthenes should alone be forgotten of all the great classic authors it is hard to tell; but even Shakespeare is badly treated: he has but one town named after him in Arkansas, and the two cities of Romeo and Juliet are but a sorry consolation. There is no objection to be made to well-meaning settlers who determine to identify their homes with Enterprise or Energy, with Friendship or Harmony, Liberty or Equality; but why they should ever choose Embarrassment is wholly unintelligible. Nor is it easy to understand the taste of the nine communities who chose to recognise and honour Cain by assuming his name, while not one has done the same honour to Abel." The contrast of some of these classical names is ludicrously absurd. For instance: in the State of New York you leave Carthage in the morning, dine at Leyden, and sup, if by this time your impatience has left any appetite, at Denmark; or if you choose another route, you reach Russia by noon and Norway by night. The laziness and ignorance displayed in this kind of nomenclature are surpassed, however, by the Wisconsin people, who have

VIEW OF THE JUNIATA RIVER, NEAR LEWISTOWN, PENNSYLVANIA.

absolutely transferred a whole county from the State of New York to that of Wisconsin, every name of town, village, and hamlet being faithfully reproduced. Surely the names of these entitled towns must exercise some influence over those doomed to dwell in them. The dim religious light of Piousville one could tolerate; but how any man can live in Dirttown (Georgia), Robtown (Virginia), Gin-Henry (Missouri), or Small-pox (Illinois), and yet preserve his self-respect, surpasses the understanding of the present commentator on American town nomenclature. "Can a man confess," indignantly asks an American, "that he lives at Longacoming, Fuddletown, or Buggaboo, and expect to be looked upon like any other respectable gentleman?" Kickaboo, Wegee, or Maxinkuckee would seem to be infinitely preferable names, though startling enough at first sight. There is a class of men among us who can, perhaps, best afford to connect themselves with such names; it is those who themselves enjoy peculiar designations, like Mr. Underdone Boots, of Albany, and Mr. Unfledged Hawk, of New York, to say nothing of men of historic renown, like Mr. Preserved Fish. With such an endless variety of names at our disposal, and enjoying besides—according to a most learned opinion delivered in the courts of New York—the unlimited right of 'assuming a name at will,' it is a wonder that so few new forms appear in our geographies and directories. When new territories are to be named, and new towns to be christened, nothing but repetition is thought of, and hence the multitudes of counties and post-towns which have the same designation. The same poverty of invention appears in christening children, and apparently the fatality extends even to the increased facility of changing names by means of divorce. A case in point appeared in the Court of Indianapolis. An enterprising woman, desirous of making a new experiment in matrimony, complained that she had been married four times already, without ever succeeding in obtaining a new 'start.' Born a Smith, she had first married a Smith, then a German Schmidt, next an English Smythe, after him a Smithe, and now she appeared on record as Mrs. Smythe once more. The judge was inexorable, and she may have to end as she began—a Smith for ever. A Western town had a still more grievous trial to undergo. The first settlers had called the place Grasshopper Falls, a name which the good town considered unpleasantly suggestive of a peculiarity of Kansas, and therefore applied in 1863 to the Legislature for leave to change. A wag had suggested Sauterelle, the French name of the destructive insect, and Sauterelle the lawgivers decreed it should be hereafter. The common people"—for even in America it appears that these are found—"however, unable to repeat the foreign sound, in a little while transferred it into Sowtail, and so great was the distress of certain inhabitants of the place at receiving letters addressed to them at Sowtail, that they permanently returned to the Grasshopper of their early days." Similar anecdotes might be multiplied almost indefinitely, and some instances in which the names applied to places by the early French settlers have got ludicrously corrupted have already been referred to (p. 75). For instance, Shepody Mountain, near the Bay of Fundy, is Chapeau Dieu (God's Hat) anglicised, Nancy Cousin's Bay is the sailor version of Anse des Cousins (Mosquito Bay), La Grasse Rivière, a tributary of the St. Lawrence, became in time the Grass River, Marais d'Ogée became Meredosia, Mauvaise Terre, Movister Creek, and Chenal Ecarté, Snearty. Bompare is really Bon Pas Prairie, and like Bon Peré, will probably end, as Mr. De Vere suggests, in a vulgar

Bamper. Cap à l'ail (Garlic Cape), on the Upper Mississippi, is now known as Capolite; la Rivière qui court has become Quicurre River, while a branch of the Red River of Lake Winnipeg, originally called Rivière Cheyenne—"from the Cheyenne Indians living on its borders"—had gradually changed into Rivière Chien, and was then translated as Dog River. Bozrah, in Connecticut, grew from being a mere farm homestead into a town, and got its name indirectly at the hands of the settler, whose pious zeal after Scripture names outran his knowledge. "Who is this that cometh from Edom, with dyed garments from Bozrah?" The honest man supposed Bozrah to be a prophet, and at town meetings frequently referred to the apocryphal prophet of that name. In time he was known by no other *sobriquet*; his farm also received the name, and ultimately the town which grew up was also called Bozrah. An Indian village in Connecticut was called Hammonasset, but when the whites settled there, in 1665, in memory of the place from which most of them came, they called it Kenilworth. When, however, the Assembly of the Colony, in 1705, issued a patent of incorporation, the clerk copied the name out as Killingworth, and by that name it is known to this day. Cincinnati itself—though often ridiculed for its classical name and very unclassical pursuits—narrowly escaped being called Losantiville. Was not *L* the schoolmaster who, after much labour, had produced this grandiose name, a reminder of L, the first letter of Licking River, nearly opposite the town, and did not *os* mean in Latin mouth? So here was the mouth of Licking River. And did not *anti* mean over against? while *ville* everybody knew was the genteel name for a town? Here was the whole geography of the new city in a nutshell. The people around in their envy no doubt dubbed it Mosaic Town; and jealous writers point out, with a sly hint at poetical justice, that the unlucky schoolmaster was a short time afterwards murdered on the Miami River by a single Indian.*

Not only do several towns in America bear the same name, but often whole districts, so that it has frequently happened that writers ignorant of this fact have been led widely astray on commenting on particular occurrences. For instance, in early records there is no place which occurred more frequently than the "Dark and Bloody Ground" of Kentucky —or Kain-tuckee—"at the head of the river," as the Shawnee Indians knew it by. It was, before the whites came there, the battling-ground of the native tribes. Long after great cities and fertile farms had sprung up on these plains, the remnants of the Indian tribes would come from distant localities to visit the graves of their ancestors, choosing strange, out-of-the-way paths; and shrinking from well-remembered pools and mounds, haunted, according to tradition, by the spirits of the departed warriors. Here, from 1769 to 1794, the early Kentucky settlers had to fight desperately for their lives against the united aboriginal tribes, who made a determined effort to drive the intruders back. "Surrounded by an enemy far out-numbering them, animated by deadly hatred and ferocious cruelty, wielding the same rifle with the whites, and as skilful in its use, these brave pioneers took, nevertheless, possession of the land, felled forests, laid out roads, built towns, and changed the wilderness into a garden. It is difficult to measure the greatness of their courage, more difficult still to fathom the depth and the weight of that darkness

---

* De Vere: "Romance of American History," p. 120.

in which they worked undaunted and undismayed. For nearly twenty-five years a cloud of bloodthirsty Indians was for ever hanging around them, and darkening every bright moment of their lives. No man could open his cabin door in the morning without danger of receiving a rifle bullet from a lurking enemy; no woman could go out to milk her cow without the risk of feeling the deadly scalping-knife on her forehead before she returned. Many a man returned from his hunt to find a smoking ruin where he had left a happy home, or an empty hearth where wife and children had gladdened his heart; and gratefully he blessed God if he found their remains, and was thus spared the anguish of knowing them to be in the hands of incensed brutes." Can the reader now understand the undying hatred which the early American settlers entertained for the Indian, or that conduct of the whites towards him which, looking at it now from our point of view, we are apt to think cruel and unjust? But in other parts of the Union the same struggle went on. Hence, in the upper part of Ohio, a locality memorable from the deeds of those times, became known as the Slaughter House, and afterwards as the Dark and Bloody Ground of Ohio. Again, the Dark and Bloody Ground of New York is in the Mohawk Valley. Here, either instigated or led by Sir William Johnston and Joseph Brandt—for by some it is denied that the latter was present at the massacre—to the eternal disgrace of the English name in America, the Mohawks fell on the homes of peaceful settlers. In Arkansas there is also a Dark and Bloody Ground. The Pawnees surprised Fort Mann, murdered the garrison, and departed. When the news reached the nearest settlement, a panic seized upon the people. They left the fatal region, which for some years afterwards was only spoken of with bated breath.

Pennsylvania

was originally granted to the Quaker, William Penn, in lieu of the payment of a debt of £16,000, owing to him by the Government of Charles II. In 1682 the first 2,000 settlers arrived in the colony, and in 1683 Philadelphia was fixed upon as the site for the capital. Thanks to the wisdom of Penn, the Indians and the whites maintained the most amicable relations, and the "Key-stone State" has ever since prospered. It is indeed worthy of the name. It occupied a central position among the thirteen original colonies, and its casting vote secured the unanimous adoption of the Declaration of Independence. The State is 310 miles long, 175 in extreme breadth, and contains an area of 46,000 square miles, or 29,440,000 acres. Physico-geographically, it may be divided into three natural divisions. (1) The eastern slope, from the Delaware River westwards, from seventy-five to eighty miles, to the Blue Mountains. Here the surface is slightly rolling and diversified. (2) The second region is the mountainous belt of Central Pennsylvania, comprising a tract about 100 miles in breadth. (3) The western, or Ohio River slope, sinks away gradually from the mountain summits, towards the valley of the river named. With the exception of the mountain region, most of the soil is very productive, and is well cultivated, yielding large crops of wheat and other cereals. West of the Alleghany Mountains, the soil has all the fertility characteristic of the rich valley of the Ohio. The mean annual temperature is 45° to 55°; that of the winter being 25° to 30°, and the summer 67° to 72°.

The cities are, however, hotter, the heat of Philadelphia, for instance, being often intense. One of the chief agricultural products of Pennsylvania is rye; in this it ranks first of all the States, no less than 3,283,000 bushels having been harvested in 1873. In oats it also ranked

VIEW OF "THE BRIDAL VEIL" FALLS, RAYMONDSKILL RIVER, PENNSYLVANIA.

first, and in buckwheat, potatoes, and hay, comes in the census next to New York. It also grows great crops of Indian corn, barley, and tobacco. In 1874 it contained 557,000 horses, 24,900 mules, 722,600 oxen and other cattle, 812,600 milch cows, 1,054,100 hogs, and 1,674,000 sheep. The average size of the farms was, according to the last census, 103 acres, and of the population 25½ per cent. were employed in agriculture. In 1874 there were 37,200 manufacturing establishments within State bounds. These employed 319,487 hands, used 421,197,673 dollars worth of material, valued, when manufactured, at 711,894,344 dollars. Iron, building materials, building, timber, flouring mill products, molasses and sugar refining, clothing, leather tanning, petroleum, paper, printing, &c., were among its chief industries. Mining is, however, the most important of all the Pennsylvanian industries, nearly one-half of the mining products of the United States being credited to the State by the last Federal census. The number of mining establishments was 3,086; 80,215 hands were employed; the capital invested was 84,660,276 dollars; the wages paid 38,815,276 dollars, and the value of the products 76,208,390 dollars. Coal and petroleum are the two terrestrial treasures in which Pennsylvania has no rival. Her coal mines are of that variety called anthracite, and those of the United States are so extensive, that Great Britain could almost be dropped into the hole made by excavating the coal-fields, of which the State under consideration possesses so large a share. The coal-fields of the United States have been estimated to cover an area of more than 682,485 square miles, while the productive coal region is nearly 300,000 square miles in extent. What almost illimitable prospects for the extension of this trade exist may be imagined when we state that the United States raises only 50,000,000 tons of coal annually, while Great Britain, with a coal-producing area of 9,000 miles, produces yearly more than 100,000,000 tons! The anthracite coal-field, which yielded, in 1872, 22,030,263 tons, is one of the smallest, being only about 434 square miles in area. It is now one of the greatest sources of wealth for Pennsylvania. Yet Wyoming Valley was long settled before its value was suspected. Even after black-smiths used it, in other localities its reputation as a fuel was unknown. In 1812, the Philadelphia black-smiths threatened to arrest as a swindler and imposter, who tried to palm off on them stones for fuel, the first man who brought anthracite to the city for sale. He could not dispose of his wares, and had to beat a hasty retreat to escape the wrath of those to whom he had presented his unsaleable wagon-loads of coal as a free gift. The anthracite mines are not worked without considerable peril. In 1872, one person for every 100,000 tons of coal raised was killed, and though the holocaust is not so great on the average, yet not a year passes without many accidents happening. Iron, copper, tin, plumbago, and lead are also mined, the first-named in great quantities. However, this is owing to her extraordinary wealth in fuel rather than to any pre-eminence of the State in mineral wealth over some of the others, which fall below her in the "output" of ore.

Petroleum—the familiar "ile" of a score of tales—is even more familiarly associated with Pennsylvania in British eyes than either coal or iron. It is often incorrectly termed "coal oil." It is in reality rock oil; and its production in large quantities is a business of modern origin and growth. Mr. Eaton describes the way in which it used to be obtained in former times. "A point was selected where the oil appeared to bubble up most freely, when a pit was excavated to the depth of two or three feet. Sometimes this pit was rudely walled up, sometimes not. Sometimes it was near the edge of the water, on the bank of the stream, sometimes in the bed of the stream itself, advantage being taken of a time of low water. In these pits the oil and water would collect together until a stratum of the former would form upon the surface of the latter, when a coarse blanket or a piece of flannel was thrown in. This blanket soon became saturated with oil, but rejected the water. The blanket was then taken out, wrung into a tub or barrel, and the operation was repeated." This oil was called "Seneca-oil," and used by the country people for the medication of cuts, bruises, and burns. At first the trade in oil was subject to terrible fluctuations. The first shipment to Pittsburg was made some time near the beginning of this century by a Mr. Carey, whose cargo consisted of two five-gallon kegs, slung on each side of a horse. At that time a flat-boatman or raft's man would glut the market with a barrel or two, and for a long time there would be no further demand for it. During the best year of the trade at that time, 640 dollars was the gross amount received for the petroleum yield of the United States. Then it began to be obtained by digging wells, when it came up mixed with water. Finally came the palmy days of the "ile trade," and with them the "ile fortunes," and the petroleum *parvenus* began, when the product was obtained, by sinking deep artesian wells, up through which the oil flowed. In 1864-5, the oil excitement culminated. At that time 1,100 companies for "pumping" it were formed, and the nominal capital invested in the trade was 600,000,000 dollars. Of late years many wells supposed to have run dry have been made to yield again, by means of "torpedoes." These valuable auxiliaries—into which nitro glycerine and dynamite enter—are let down into the old well, and exploded. The result frequently is that new reservoirs of oil are tapped, and the well commences to flow liquid riches again. In 1859, 82,000 barrels—each barrel containing forty gallons—of petroleum were drawn from the wells of Western Pennsylvania; in 1860, 500,000 barrels; in 1861, 2,113,000; and in 1862, 3,056,000. Then for three years there was a falling off, until in 1866 the product suddenly rose to 3,597,000 barrels. In 1869, 4,210,720 barrels was the return; in 1870, 5,673,198; and in 1873, 7,878,629, or a daily average of 21,568 barrels. In 1876 the shipments from the oil regions averaged 28,000 barrels per day. In 1874, 215,978,684 *gallons* were in all exported from the United States, while a later return shows that in 1876 no less than 32,915,786 dollars worth—or considerably more than in the previous year—was exported. It is almost unnecessary to say that petroleum wells were known before the Pennsylvanians brought the substance into notoriety. Among these, those of Burmah were the best known and the most productive, but even yet, notwithstanding the piles of print which have been devoted to the elucidation of the subject, the exact nature and origin of the substance are not well known. It seems, however, now very generally believed that it is due to the decomposition of both animal and vegetable substances. Among other evidences of wealth and prosperity, Pennsylvania has 5,020 miles of railway, 16,305 schools, 831,020 children receiving education, 19,089 teachers, eight normal schools, six universities, thirty-three colleges, fourteen schools of theology, two of law, eight of medicine, and seven of science. In 1870, there were 14,849 libraries, 5,984 religious organisations, and 510 newspapers and periodicals, including fifty-five daily ones. In 1875, the periodicals had increased to 707 (New York alone having more), and of these seventy were published daily, and 511 weekly. The chief towns are Philadelphia, situated between the Delaware and the Schuylkill, a name signifying in the Low Dutch dialect the "hidden river." Its population is now over 800,000, and its wealth and prosperity very great. It was the capital of the State until the beginning of the present century, and of the United States from 1790 to

1800. Harrisburg (27,000 inhabitants) is now the State capital, but Pittsburg, of which Alleghany city is a part (the two containing, in 1875, 208,185 people), is the second city in the State. The other towns of importance are Scranton (46,000 people), Reading

VIEW OF THE VALE OF WYOMING, WITH THE SUSQUEHANNA RIVER, PENNSYLVANIA.

(42,000), Lancaster (22,360 in 1875), Erie, and Wilkes-Barré, on the north branch of the Susquehanna, in the lovely valley of Wyoming. It derives its name conjointly from the notorious John Wilkes and Isaac Barré, and is often written Wilkesbarre, though surely no man with a proper feeling for human vanity would write a town named after Smith and Jones as Smithjones! The population of Pennsylvania was, in 1870, 3,521,951,

VIEW OF THE FALLS OF NIAGARA, WESTERN SIDE.

or 75.56 persons to the square mile. This number, which brings the inhabitants of Pennsylvania second in quantity to that of New York, had a large infusion of Scotch and Germans, and by natural multiplication and immigration must have largely increased. Indeed, during the decade between 1860–70, more people, viz., 615,737, were added to the State population than were contained in fourteen of the Sovereign States of the Union. We have said little of the scenery of Pennsylvania; but the nature of the country along the banks of the rivers is well shown in the view of the "Blue Juniata," famous among tourists at a point near Lewistown, and that of "The Bridal Veil" on the Raymondskill, near Milford, which we have engraved on pp. 205, 209.

NEW YORK,

the most populous and wealthy, if not intellectually, or even politically, the most important of the States, is in its extreme length 412 miles, and in its width from north to south 311 miles. Portions of Long Island are only eight or ten miles broad, and the south-western boundary below Lake Erie is about nineteen miles long. The ocean constitutes its boundary for 880 miles; rivers, 280; and the lakes, 350. Among the States it ranks nineteenth in area, and contains 47,000 square miles, or 30,080,000 acres. Long Island is flat and sandy; but "the highlands" on both sides of the Hudson contain a finely diversified and picturesque region, including summits which reach an elevation of 1,700 feet. North again are the Catskill Mountains, the most conspicuous peaks in which are Round Knob and High Peak, about 3,800 feet above the sea. Beyond the watershed, which turns the drainage to the north, Mr. Williams characterises the country as "rolling and diversified." In the north-east is the Adirondack Wilderness, which contains some of the loftiest peaks on the Northern Spur of the Appalachian range, with the exception of the White Mountains of New Hampshire. Among these the survey of the Adirondack range, by Mr. Colvin, shows that Mount Marcy is 5,402 feet high; Mount M'Intyre, 5,106; Haystack, 5,006; Skylight, 4,997; Clinton, 1,937; Gothic Mountain, 4,744; and Giant of the Valley, 4,530. Over this region extends a dense forest, in which bear, panther, wolf, moose, deer, and other wild animals still linger in some numbers, while the numerous lakes and streams afford extensive water communication. Here also rises the lovely Hudson River; the Saranac and Ausable empty into Lake Champlain, while other streams flow to the St. Lawrence. The Hudson has its source 4,000 feet above the sea, and flows for 300 miles. Large steamers ascend as far as Troy, 150 miles from its mouth. At one time it was called the North River, not to distinguish it, as commonly supposed, from the East River, but from the Delaware, which was known to the Dutch as the South River. Some of the lakes are very beautiful. Lake Erie we have already noticed, and Lake Ontario has more than once been mentioned. The latter is 250 feet above the sea, but its bottom must in places be lower than the Atlantic, as it is in some spots 600 feet deep. It is 190 miles long, 55 at its widest point, and about 480 in circumference. Its shores are generally flat, but the Bay of Quinte contains some more than usually attractive scenery. Burlington Bay is almost enclosed by a natural bank of sand, which forms an attractive drive for the people of Burlington, which is built on its shores. The Niagara River, which flows into it at its south-western corner, is the only outlet of the four great lakes, and at its north-east corner it issues or narrows into the St. Lawrence River, which is for some distance below known as the Lake of the Thousand Isles. The Niagara River flows over a precipice which constitutes the Great Falls of Niagara (p. 215), which are 164 feet high, and 1,100 feet wide on the American side, and 2,000 feet wide on the Canadian side. The total descent of the river is 335 feet, and its width below the falls 1,160 feet. The Falls it would be a mere work of supererogation to attempt describing. The Genesee, the Trenton, the Kaaterskill, Taghanic, the Cohoes, and Little Falls are also cataracts of some note within the State of New York. Lake Champlain (p. 216) is 130 miles long, and from half a mile to ten miles wide. Lake George (Plate XVII.) is thirty-six miles long, and is famous for its 300 islets and picturesque scenery; but the whole State is dotted with lakes and threaded with rivers. Staten Island, and Long Island (110 miles long), are among the most interesting detached portions of a State, which it would take many volumes to fitly describe, but which, in accordance with our unwilling promise, we must dismiss with the mere mention which a few lines can afford room for. There is no coal in New York State. There are, however, rich beds of marble, and salt springs, which yield immensely. There is also petroleum and natural gas in the west in such abundance that large villages can be lighted with it. Saratoga and Ballston are mineral springs, and great pleasure resorts of the fashionable world during the summer season. The climate, though mild on the coast, is in the winter extremely severe in the northern counties. But the summer temperature is high, and the fertile soil yields great crops of wheat, Indian corn, apples, peaches, melons, grapes, &c. New York is the first manufacturing State in the Union, judging from the amount of its productions, though Pennsylvania has more capital invested in industry than the "Empire State." In agriculture it is in some respects ahead of all its sisters, though in mining its results are insignificant compared with some of the other States. Its population was, in 1870, 4,382,759, or 93.25 to the square mile. In addition, there were 5,000 Indians belonging to the Six Nations settled on reservations within the State, all of them civilised, and many of them, indeed, farmers, mechanics, or professional men. In education, literature, and all the appliances of civilisation and luxury, New York stands pre-eminent. Its journals, if not the most refined in their contents, are at least the most widely circulated, and not the least enterprising. Its merchants are among the wealthiest in the world, and its enterprise, energy, generosity, goodness, and immorality have given the greatest of the American States at once a notoriety unenviable, and a reputation of which many European kingdoms might justly be proud.

New York is, of course, the greatest city in the Union. In 1875 it had 1,064,272 inhabitants, and a valuation of 1,154,029,176 dollars, which yielded a taxation of 34,620,874 dollars. It is thus the third city in the civilised world. It occupies the whole of Manhattan Island (which was bought in May, 1626, by Peter Minuit for 60 Dutch guilders, or about £5), and twenty square miles of West Chester county. Albany, on the Hudson, containing, in 1875, 85,584 inhabitants, is the capital. In all, the State contains twenty-four chartered cities, with a population, in 1870, of 1,965,650 inhabitants. Brooklyn is really a part of New York, with which it is connected by thirteen steam ferries. Buffalo, Rochester, Troy, Syracuse, may also be mentioned as "cities," which even in our Old World, not so liberal in its distribution of the civic title, would be accounted worthy of the name.

New Jersey and Delaware.

New Jersey was one of the thirteen original States—indeed, nearly all of those we have lately had to speak of are in that honourable category—and is 168 miles long, and from fifty-nine to ninety-two in breadth, its area being 8,320 square miles, or 5,324,800 acres. It is divided into twenty-one counties, and, including bays, its coast line is 510 miles. The northern portion is hilly, and even mountainous, and includes the Palisades—a wall of trap rock 200 to 500 feet high forming the western bank of the Hudson for fifteen miles. Here is some of the finest scenery on this charming river. In the centre of the State the country is rolling, and the southern and eastern portion is a sandy plain sloping to

VIEW OF LAKE CHAMPLAIN, NEW YORK STATE.

the sea, along which a strip of marsh also runs, in common with broad tracts of salt meadows. In this southern region is also found "The Pines," a belt of forest which lines the coast from end to end, and in some places widens and penetrates far across the State. "The Piners" are a peculiar people, frugal but (as a rule) not sober, "apple-jack" whiskey being one of the few luxuries for which they pay out money, and for that they are very prodigal in their disbursements. The young men are wiry, and not particularly *débonaire*; but the "Piner" girls are noted for their beauty. Most of them are tall, graceful, and dark-haired, with well-developed features and complexion, such as only ripe peaches possess, and they retain their beauty generally much longer than most of the American women. "The older men," to use the language of a correspondent who visited them and communicated to us his experience, "are awkward and angular, and stoop a great deal, for they are hard workers. They live by fishing and charcoal burning, and within the past fifty years have chopped down the entire old growth of pine-trees, so that the pines of to-day are mainly young. Twenty

VIEW OF A SPUR OF THE BLUE MOUNTAINS, DELAWARE WATER GAP, NEW JERSEY.

years ago schools, churches, and railroads were unknown among the inhabitants of 'The Pines'; but with the advent of farming, towns rose here and there, and the outside world began to send teachers to these simple people. Nevertheless, there is a low code of morality, a greater capacity for apple-jack, and a more contented state of ignorance in 'The Pines' than in any other district so near New York." Iron, copper, marl, clay, glass sand, &c., are among the mineral riches, while the usual crops raised in that latitude of America flourish. There are many manufactories, railways, and canals, which add to the prosperity of this little State, which in shape has been compared to a bean. The early colonists were Dutch, Swedes, and English, but at present out of its population (in 1870) of 906,096 only 188,943 were foreign born. At one time New York and Philadelphia drained New Jersey, so that Dr. Franklin likened it to a "cider barrel tapped at both ends." However, the overflow of these two cities is now adding to the census of this State, which in diversity of population is fourth among her sisters—in other words, having 108.91 people to the square mile.

Trenton is the State capital, with a population of 22,874 according to the last returns, but Newark, which had, in 1811, 1,838 inhabitants, cannot now have less than 125,000. Jersey City, which had, in 1870, over 85,000, Paterson 34,000, and Camden 34,000, are all larger cities. The other towns are smaller. Long Branch, Cape May, and Atlantic City are popular sea-side resorts, competing with Saratoga for the patronage of the fashionable New Yorkers in search of health and dissipation. New Jersey is, we believe, the only State of the Union in which aliens may hold land. The Act enabling them to do so was passed in 1808, so as to enable Joseph Bonaparte, ex-king of Spain, and his nephew, Prince Murat, to acquire property and settle in the State. The Pennsylvania people refused to be so accommodating, and when they found that the foreigners scattered money freely, and greatly benefited the land of their adoption, they raised the taunt that the Jersey men had imported a king to reign over them. This is said to be the origin of the humourous phrase that "New Jersey is out of the Union." The climate is generally good, though there are malarious districts. Among the most striking scenery along the North-Western Boundary are the Blue Mountains, through which the Delaware River breaks at the Water Gap, the sides of which are 1,600 feet high (p. 217).

Delaware is one of the thirteen original States, and one of the smallest, being only ninety-six miles long, and from nine to twelve miles broad in the north, and thirty-six or thirty-seven miles in the southern line, its whole area being 2,120 square miles, or 1,356,800 acres. Wilmington is the only town worthy of the name. It had, in 1870, a population of 30,841, while Dover, with a census of over 2,000, is the capital. In 1870, the total population of the State was 125,015, including 22,794 free coloured people. The northern portion of the State is elevated and healthy, but in the southern section there are many swampy places and much endemic fever. Noxious reptiles abound. Iron mines are found, and the usual crops suitable for the climate prosper. From this State immense quantities of fruit are sent to the more northern markets. From three to four million baskets of peaches are shipped annually. In 1871, 7,470,100 quarts of strawberries alone were sent. There are no good harbours, hence its commerce is not great. Delaware has the distinction of having no State prison, criminals being confined in the county jails.

Among its Old World institutions are the pillory and the whipping-post, conservators of morals, which are still in healthy vigour. Delaware was originally settled by the Swedes and Dutch, who lived on terms of the greatest friendship with the Indians, who used to style the former their own people. It is believed that during their dominion not a single life was lost in encounters with the natives. For some time William Penn governed it as a part of Pennsylvania, and it was not allowed a separate General Assembly until 1703. Though a slave State, it refused to secede, and, indeed, raised several regiments to aid the North in asserting the Federal rule over the rebellious South.

CHAPTER XII.

THE UNITED STATES: NEW ENGLAND.

THE Western, Southern, and Middle States—or by whatever name those we have most recently sketched may be called—have been left behind, and we conclude our last chapter on the topography and statistics of the commonwealths of the Great American Republic by a few words upon New England—or as the divisions in it are generally called—the Yankee States. With the exception of Virginia and most of the neighbouring regions, the extreme northern portion of the United States, owing probably to its comparative proximity to Europe, was among the earliest settled sections of the Continent. New England was essentially the home of the Puritan Fathers who fled here for peace and liberty, and carried with them much of that love of peace and hatred of despotism, with at the same time the narrow religious prejudices and even bigotry which for long distinguished the "Yankee people," and to this day are among the most marked features of some parts of Connecticut and Massachusetts. A "Deacon" from Rhode Island, or a "select man" from Martha's Vineyard, would not in most circles be accepted as the type of great liberality. To this day the New Englanders are impressed with many of the features of that Old England from whence their ancestors came two hundred or more years ago. They live in a soil not the most bounteous, and under a climate only a trifle better than that from which their forefathers fled. Hence they are frugal, prosperous, keen in business—possibly a little sharp—intelligent, educated, and generally speaking imbued with a contempt for shams, show, extravagance, or anything which savours of foreign ways and foreign wastefulness. New England really constitutes the brains of America, and Boston, though jocularly styled the "hub of the universe," has some right to be considered so from the Trans-atlantic point of view. Words peculiar to New England are often called "Americanisms." In reality most of them are Old England words in common use at the time these people left for the New World, but are now only seen in Shakespeare and the writers of the period. The very word "Yankee" now vaguely applied to all Americans of the Northern States shows the light in which the early dwellers in New England were looked on by the aborigines. It is in fact the Indians' corruption of the term *Anglais*, or Englishman, applied to the first settlers by the Frenchmen of the neighbouring country of Canada. Hence Yengees, Yenghes, Yanghis, Yankees. It was used at least as early as 1713 by one Jonathan Hastings, a farmer, who applied it to his cider and other wares, as expressive of his intention that the world should believe that they were something very superior. At least such is the legend. It is interesting, and may possibly even be true. The New England States are Connecticut, Rhode Island, Massachusetts, Vermont, New Hampshire, and Maine. Let us glance in the briefest possible manner at them, more from a statistical than a social point of view, interesting though New England is to the student of philology,* history, and the progress of mankind.

CONNECTICUT.

The Dutch from the New Netherlands, or the region of which New York is now the centre, first explored the Connecticut River, and in 1635 colonists from Massachusetts settled along its banks. Its extreme length is 100 miles, and its breadth 70 miles, while its whole area is 4,750 square miles, or 3,010,000 acres. The physical features of the State are moulded by the mountain chains of the States to the north being continued in four ranges of high hills extending through the country. There are Housatonic range, the Green Mountains, the Mount Tom and Talcott Mountains, and the Lyme range. Hence, as a whole, the soil is not rich, though the valley of the Connecticut—the Long River of the Indians who navigated it—is one of the finest agricultural regions of New England. The river overflows every spring, and leaves behind an alluvial deposit, which is a splendid fertiliser for crops. Back from the alluvial meadows are river terraces, which bear farming well, and though along the coast there are some marine alluvial flats of very deep and fertile soil, the land is, as a rule, sandy and unproductive, while the elevated and broken regions of the North-West possess a soil cold and sterile, but well adapted for grazing. At the last census, Connecticut had 25,423 farms of the average size of 95 acres, the staple crops being Indian corn, wheat, rye, oats, barley, buckwheat, potatoes, tobacco, and hay. The numerous streams which intersect the State were early turned to use by the busy, thrifty Connecticuters. Hence this section of New England is one of the busiest manufacturing centres in the United States. Its iron works, clock factories, india-rubber goods, sewing machines, and woollen and cotton goods, go to all parts of the world, as, indeed, it is said that another but not so highly-appreciated manufacture did. However, though the tale has obtained for Connecticut the *soubriquet* of the "wooden mutmeg" State, it is probably a scandal that once on a time (until found out) the 'cute inhabitants "went into" that line of industry indicated by the name given to their commonwealth! The iron mines underlie a great portion of Litchfield county; the copper mines were worked prior to the Revolution, and later still the abandoned shafts were used for the State prison. Lead, antimony, plumbago, and cobalt have also been found, and marble, white and clouded, is mined at two places. Limestone is quarried throughout the Housatonic Valley, while the freestone

---

* Perhaps the best essay on, and specimens of the New England dialect, may be found in Lowell's "Biglow Papers," where also the prejudices and ideas of the "Yankees" find full play, often unconsciously on the part of the distinguished author.

VIEW OF SALMON BROOK, GRANBY, CONNECTICUT.

quarries supply the "brown stone fronts" so dear to every aristocratic New Yorker's heart. Over 800 vessels belong to the five customs districts, while more than 1,100 persons were employed in the cod and mackerel fisheries. The State has nearly 900 miles of railroad, 1,638 public schools, 2,477 teachers, and 131,748 children, according to the latest census, under tuition. In no other State of the Union does such a large proportion of the population attend colleges and other institutions for higher education than the ordinary district school affords. Yale is the great university. It has eighty-two instructors, and more than 1,000 students, but there are other smaller institutions with a considerable matriculation roll. The population of the State was, in 1670, 15,000, and 200 years afterwards, 537,454, of whom 113,639 were of foreign birth. The density of the population (113.15 to the square mile) is greater than that of any other State, with the exception of Massachusetts and Rhode Island. There are nine incorporated cities, of which New Haven, "the Elm City," with over 51,000 inhabitants, and Hartford, the capital, are the chief. The latter has more than 38,000 inhabitants, and an immense tobacco trade. Bridport (19,000 people) is given over to the manufacture of sewing machines, carriages, and iron work. Norwich on the Thames, Waterbury, New London, famous for the whale fishery, Middletown, Meridin, and New Britain, are the other towns of note. As a pleasing specimen of the scenery of the State we have engraved a view of Salmon Brook, in Granby, the latter a town situated west of the Connecticut River (p. 221).

RHODE ISLAND

is the smallest State of the Union, being in its greatest length only 48 miles long, in breadth 39 miles, and in area 1,306 square miles, or 835,840 acres. Near the sea-coast the ground is level and sandy, and in the interior rather rolling and hilly. But the soil is nowhere very fertile, and is better adapted for grazing than tillage. Indian corn, rye, oats, barley, potatoes, and hay are the chief crops, but no wheat, tobacco, or buckwheat is reported in the census as being produced in this State. It had, in 1870, 1,850 manufacturing establishments, employing over 49,000 hands, and producing articles valued at 111,418,354 dollars. Coal exists, but has not hitherto been mined profitably, but serpentine, marble, freestone, and limestone are quarried to a very considerable amount every year. The cod and mackerel fisheries employed, in 1870, 98 vessels, and in 1874, 284 vessels were registered as belonging to island ports. Education is high. In 1874, there were in the State 43,800 children between the ages of five and fifteen, and of these 39,101 attended school. The schools numbered 752, the teachers 805, and the expenditure for school purposes 690,852 dollars. Brown University is the only college. The white population was, in 1870, 217,353, of whom nearly a fourth were of foreign birth. The population is 166.45 to the square mile, or rather more than any other State, Massachusetts excepted. About eighty of the once powerful Narraganset Indians still remain on the island, but they have long ago lost their savage habits, and are now among the most civilised and best behaved people in the State. In Rhode Island there are two cities and thirty-four towns. Providence, the second city in New England, is one of the former. It had, in 1870, 88,499 inhabitants, though it is believed that since its consolidation with North Providence it cannot have less than 100,000 within its bounds. Newport had at one time an extensive trade with the West Indies. But this has now almost entirely disappeared. It is now a fashionable resort for New Yorkers, Philadelphians, and Bostonians, who here take up their summer quarters in mansions which they modestly style cottages. How deserving they are of the name may be inferred from the fact that some of these "cottages" cost from a hundred thousand to half a million dollars. In Newport is the "old stone mill," the only thing on the Atlantic shore which, as Higginson remarks, "has had time to forget its birthday." Woonsocket, with a population of 12,000, and Warwick, with 11,000, are the only other towns of any consequence, the rest having a census below 8,000. The only other thing remarkable about Rhode Island is the fact that in 1874 a stringent prohibitory liquor law was passed here. But it is generally understood that in Newport or Providence, the thirsty soul can casuistically evade it by asking in a druggist's shop for a sight of the "striped pig," or "the baby." And never were a people so solicitous after the health of "the baby," nor so full of anxious inquiries regarding the whereabouts of the "striped pig."

It is believed that the Norsemen—wanderers from Greenland—visited Rhode Island as early as the tenth century. But it is to Roger Williams, who was driven from Massachusetts for his religious opinions, that Rhode Island owes its start as a colony. He called his new city "Providence," as a memorial of "God's merciful providence to him in his distress," and made that "liberty of conscience" which the children of the Pilgrim Fathers had omitted to put in practice in their New England home the fundamental law of his settlement. Rhode Island had, among its other eventful episodes, a long and bloody Indian war, which ended in the killing of "King Philip," near Mount Hope, in 1676. Up to 1840, the State was ruled by a constitution, the basis of which was the charter granted in 1663 by Charles II. to "the Colony of Rhode Island and Providence Plantation." By it the suffrage was limited to the holders of a certain amount of real estate, and to their eldest sons. The result was that not over one-third of the population possessed the franchise, or had directly any voice in public affairs. In 1841, an agitation was started against this state of matters, and the State was divided into, "suffrage men," and "charter" or "law and order men." The former secured the passing of a new Constitution, but the latter declared that many of the votes were fraudulent, and the whole affair seditious. Civil war was imminent, the State militia was called out, and the "suffrage party" attacked and dispersed. Dorr, the leader of the party, was tried and convicted of treason, and sentenced to imprisonment for life. The end of the Dorr rebellion was, that in 1843 the suffrage was extended, and Rhode Island has, since that date, been governed somewhat more in accordance with the political principles of the other States.

MASSACHUSETTS.

The "old Bay State" I should be inclined to claim as the greatest of the American commonwealths, as it is one of the oldest. Its soil is cold, and its climate bleak. Its people are not rich, nor its foreign commerce great. But Massachusetts is nevertheless the brain of New England, as New England is on a larger scale of America. Its example is all-powerful, and in the men which it has given to the State it ranks alongside of Virginia; only with the difference that while Virginia has produced politicians

VIEW OF BOSTON, MASSACHUSETTS, FROM BUNKER'S HILL

VIEW OF MOUNT MANSFIELD, VERMONT.

and statesmen, Massachusetts has been fertile, not only in these but in men of letters, inventors, poets, philosophers, and a brilliant galaxy of scientific discoverers. The history

VIEW ON THE CONNECTICUT RIVER, MASSACHUSETTS.

of Massachusetts is tolerably well known. But it is a mistake to suppose that the Pilgrim Fathers were the first settlers. They were not. In 1602, a company of English colonists landed on the Elizabeth Island, but soon left disheartened. It was not until December 22,

1620, that the Pilgrim Fathers landed at Plymouth, and commenced their "grievous dolours." Before spring commenced one-half of them had perished, and little more than fifty years afterwards the Indian "King Philip" destroyed a dozen towns, burned 600 houses, and killed as many colonists. In 1773, it was in Boston Harbour that the enraged colonists destroyed the taxed tea, and in 1775 it was at Lexington where the first blood in the Revolutionary War was spilled. Yet the whole State is not much bigger than some Californian counties. It is 160 miles long from east to west, and 90 miles broad on the east, and 48 miles on the west. Its area is 7,800 square miles, or 4,992,000 acres. "The south-east section is low and sandy, the northern and central hilly and rolling, the western broken and mountainous." Greylock, or Saddle Mountain, in the north-west, which attains a height of 3,600 feet, is the highest land in Massachusetts. The Connecticut (p. 225), Merrimac, and Housatonic are the chief rivers. The Merrimac rises in the White Mountains, and is navigable for eighteen miles. But its chief importance is as a motor power; it is said that no other river in the world turns so many spindles. At Lowell the mean annual flow is 5,400 cubic feet per second, and during freshets the volume will swell to 90,000 cubic feet per second. All the ports along the Massachusetts coast have a peculiarly "Yankee" flavour. Cape Cod, Buzzard Bay, Nantucket, and Martha's Vineyard, are all familiar localities to any one who has read much about the nooks and corners of New England.*

Nantucket Island, fifteen miles long, and from three to four wide, has been inhabited since 1659, but is almost destitute of trees, though the State is plentifully furnished with these. Ash, aspen, beech, birch, butternut, cedar, chestnut, elm, hickory, larch, boxwood, maple, oak, pine, spruce, sycamore, and tupelo, sourgum, or a species of *Nyssa*, are among the 1,737 species of plants recorded from the State. Much of the land is sterile, the only very rich alluvial soil being found in the valleys of the Connecticut and of the Housatonic Rivers. Careful tillage has, however, done much. In 1873, the average yield was as follows:—Indian corn, 35 bushels; wheat, 19; rye, 17; oats, 35.3; barley, 22; buckwheat, 15.6; potatoes, 125; tobacco, 1,159 pounds; hay, 1.04 tons to the acre. The climate is not good, alternating between extreme heat during the summer, and disagreeable cold during the winter. In 1870, there were 26,500 farms of the average size of 103 acres. Over sixty-three per cent. of the land was "improved," and the value of the farms, farm stock, and implements was 138,182,891 dollars. In January, 1874, the live stock was estimated to comprise 102,800 horses, 122,600 oxen and other cattle, 136,300 milch cows, 76,300 sheep, and 78,000 hogs. Massachusetts is, however, a manufacturing, not an agricultural State. Of the 579,811 people reported as being engaged in all kinds of occupations, only 77,810 were employed in agriculture. It is, however, in proportion to its population, the greatest manufacturing State of the Union. In the items of boots and shoes, cotton goods, woollens, cutlery, and chairs, it is ahead of the rest of the country. There were, by the last reports, 13,312 manufacturing establishments in the State, employing 279,380 hands, of whom 86,229 were females above the age of fifteen. In wages, 118,051,886

---

* Drake: "Historic Fields and Mansions of Middlesex." Flagg: "Birds and Seasons of New England," and "The Woods and Byways of New England." Drake: "The Nooks and Corners of the New England Coast," "Census of the State of Massachusetts," and for the more literary aspects of New England life, the works of Thoreau, Wendell Holmes, Lowell, and other lettered "Yankees."

dollars were paid, and the value of the product was 553,912,568 dollars. Building stone—such as marbles, granites, &c.—is the chief item under the head of "mines," in which the State is poor. More than one-half of the fisheries of the United States are to be credited to Massachusetts. In the cod and mackerel fisheries, 1,026 vessels, 8,993 men, and capital of 1,287,871 dollars are embarked. The annual product averages over six million dollars, while of late years the whale fisheries—chiefly of the South Sea—employed 170 vessels. In 1874 there were 2,418 miles of railroad in the State, which, during that year, carried 12,180,194 passengers, and earned 34,652,483 dollars. The Hoosac tunnel, 4¾ miles in length, is next to the Mont Cenis Tunnel, the longest in the world. In 1874 there were 5,485 schools, 8,715 teachers, and 297,025 pupils under their charge. There are five State normal schools, and seven universities, with 2,529 students, in addition to various institutions for professional and scientific training. In 1870, there were in the State 3,169 libraries, containing 3,017,183 volumes, of which number the Boston Public Library must be credited with 270,000, and the Harvard one more than 200,000. In 1874, there were 324 newspapers and periodicals in the State, and about 1,850 religious organisations. The Pilgrim Fathers numbered about 100, but in 1692 the population was estimated at 40,000. In 1870, it was 1,457,351, of whom 1,104,032 were of native birth. The density of the population is 186·84 to a square mile, no other State in the Union being so thickly populated. Nearly one-half of the people reside in cities, and 360,000 of these in the metropolis, Boston, alone. Cambridge, the seat of the celebrated Harvard University, has 40,000 people, and is the literary centre of the United States. Lowell, famous for its mills, is another flourishing town of 41,000 people; Lawrence, also a milling place, has 29,000; Haverhill, engaged in making boots and shoes, 14,000; Worcester, with its machine shops, 41,000. Springfield (27,000) is the site of the United States Armoury; while Salem (25,000, p. 228) is famous in the early annals of the colony. New Bedford devotes itself to the whale fishery; Gloucester is the head-quarters of the cod and mackerel fishermen; Lynn is famous for its shoes, while Wellfleet bears the reputation of being much prejudiced against strangers. It is in this village—though the story is also told of many of the sleepy fishing villages along the Massachusetts shore—that a stranger was hailed with the savage cry of the youthful aborigines, "Rock* him! He's got a long-tailed coat on!"

VERMONT

is not an important section of the world, though it does contain an area of 10,212 square miles, or 6,535,680 acres, extending from north to south 158 miles, and east and west between 40 and 90 miles. The Green Mountains—the *Monts Verts* of the early French travellers, from which the State derives its name—runs through its whole length, and forms the watershed between the streams that flow on one side to the Connecticut, and to Lake Champlain and the Hudson on the other. Mount Mansfield (4,430 feet, Plate XVIII.), Camel's Hump (4,088 feet), Killington's Peak (4,221 feet), and Ascutney (3,320 feet), are the chief elevations. Most of the hills are smooth and rounded, and covered with wood or grass to the very summits. The river valleys possess a deep, rich, alluvial soil, of great

* Stone.

fertility, and even on the uplands there is good loam, which bears heavy crops. The Connecticut River drains an area of 3,750 square miles, and constitutes the Eastern boundary of the State. Flowing westward are the Lamoille, Winooski, Otter, and the picturesque Missisquoi River (p. 229), all of which discharge their waters into Lake Champlain. The slopes of the hills and mountains afford good pasturage, but the valley of Lake Champlain is the centre of the agriculture of the State, owing to its protection from the cold north-east winds, and its exposure to the south. The climate is cold in winter,

VIEW OF SALEM, MASSACHUSETT.

and correspondingly warm in summer. Snow begins to fall in November, and usually lasts until the end of April. Agriculture is flourishing. Of its land nearly 67 per cent. is improved. The average size of the farms is 134 acres; the farms, farm implements, and stock being valued in 1870 at 168,506,189 dollars, and the farm produce at 34,617,027 dollars. The forest products were estimated to be worth 1,238,929 dollars, and the orchard products 682,241 dollars. The State ranked first in the production of maple sugar (Vol. I., p. 252*), making annually about 8,894,502 pounds, and in cheese next to New York and Ohio. Maize, wheat, rye, oats, barley, buckwheat, potatoes, tobacco, and hay were among its other farm crops. It had, among other live stock, 543,600 sheep, which is more than is possessed by any other New England State. Its manufactures

* See also "Science for All," Vol. I., p. 25.

VIEW ON THE MISSISQUOI RIVER, VERMONT.

employ nearly 19,000 people, lumber alone figuring for a large sum in the annual income of the State. Iron, copper, lead, and manganese ores exist, but, with the exception of the first, in no great quantity. Kaolin, or china clay, is abundant: so are marble and many good varieties of building stone. Considerable commerce is carried on with Canada through Lake Champlain; a canal also connects the lake with the Hudson. The State had, in 1870, 330,551 people, which was an increase of less than 5 per cent. over 1860. There were 32.37 persons to the square mile. Montpelier is the capital, though a mere village of about 3,000 people. Burlington is the largest town; in 1870 it had a population of over 14,000. Rutland, Bennington, Brattleborough, and Middleburg are the other chief towns, though none of them reached by the last census 10,000 inhabitants. As this book is not a gazetteer, we may leave any account of them to that kind of literature. We may, however, add that Vermont was first settled by Frenchmen in 1724, but at a later period New Hampshire claimed jurisdiction over the newly explored territory. New York also tried to have a hand in the government and settlement of the country, and even went so far as to attempt to dispossess the settlers of their lands, under the plea that one of the Merry Monarch's rather peculiar charters authorised the State of New York alone to grant them. The effort was not, however, a signal success. The "Green Mountain Boys" promptly applied the "beech seal" to the backs of the New York officers. In other words, they tied them to trees and whipped them with beechen rods, until nobody could be found willing to serve the writs. In the Revolutionary War Vermont played a prominent part, though, owing to the State being a refuge for loyalists, these voters managed for eight years to keep it out of the Union. It experienced no more of the horrors of civil strife until, on October 19th, 1864, the Confederates made a raid on St. Albans. Like Rhode Island and Massachusetts, a prohibitory liquor law is in force, and the liquor dealer is responsible for the damage done by an intoxicated person, a law which curiously enough has been introduced into Madagascar.

NEW HAMPSHIRE

was one of the original thirteen States. Its first settlers, who had very early to contend for the soil with the savages, were chiefly of Scotch and Irish origin. In 1790 the inhabitants numbered 141,885, and in 1870 the census gave 318,300 as the total number. Of these the great majority had been born in the State, and only 29,611 were of foreign birth. Even this moderate census shows a falling off of 7,773 people in the decade from 1860 to 1870. Concord, the State capital, has a population of about 13,000. Manchester (24,000 people, of whom over 7,000 are foreigners, chiefly Irish) is the principal town. It owes its prosperity to its manufactures of cotton and woollen goods; and Nashua, Dover, and Portsmouth are also thriving towns with over or upwards of 10,000 people. New Hampshire is in no degree an agricultural country. It is too much broken up by lakes and mountains to afford room for great farms. Accordingly the "Switzerland of America" depends on manufactures and the crowds of visitors which every year flock to find health and pleasant scenery among its Alpine heights. The principal district visited is the White Mountain section. It covers an area of 1,270 square miles, mostly wooded, and very thinly inhabited. "The Saco River," writes Mr. Williams, "cuts it very nearly in the centre. From Gorham to Bartlett, a distance of twenty-two miles, the main range stretches in a direction from north-east to south-west. The principal peaks, taking them in succession from the north, are Mounts Madison (5,365 feet in height), Adams (5,794), Jefferson (5,714), Clay (5,553), *Washington* (6,293), Monroe (5,384), Franklin (1,904), Pleasant (4,764), Clinton (4,320), Jackson (4,100), Webster (4,000). Mount Washington is the only one of the group which reaches an altitude of 6,000; eight are more than 5,000 feet high; fourteen more than 4,500; twenty more than 4,000; and twenty-eight equal or exceed 3,000. Mount Lafayette, at Franconia Notch, is 5,500 feet in height, and the Twin Mountains 5,000 feet. In only one other State—North Carolina—east of the Rocky Mountains are there such elevations. No ascent of Mount Washington was made by white men until the year 1642. It is a remarkable fact that while so many of the streams and lakes of New Hampshire are known by Indian names, the great mountains had no individual designation in the Indian vocabulary. It is said in explanation that the superstitious savages never reached the summits because they feared to expose themselves to the wrath of the spirits, with which their imagination people the heights. The name of Agiocochook was applied to the whole group in one dialect; in another the designation was Waumbekket Methna, signifying 'mountains with snowy foreheads.' An Indian tradition says that the whole country was once flooded, and all the inhabitants were drowned, save one Powaw* and his wife, who fled to the summit of Agiocochook, and thus survived to repeople the earth. The White Mountain Notch was discovered in 1771. It soon became a considerable thoroughfare, and long strings of teams from northern New Hampshire found their way to Portland through this avenue. The Notch, which is two miles long, is only twenty-two feet wide at 'the gate,' and through it runs the Saco River. The first horse taken through this gap, to prove that the route was feasible, was let down over the rocks by ropes. Not more than ten or twelve persons from a distance visited the mountains in 1819. On August 21st, 1820, a party spent a night at the summit. The throng of summer visitors now number 10,000 a year. The elevated railway (which goes to the summit) has a maximum grade of 1,980 feet to the mile, or 13½ inches to the yard. Among the objects of special interest to tourists are the 'Lake of the Clouds,' and the 'Old Man of the Mountains,' whose profile, elevated 1,200 feet above the lake beneath, measures thirty-six feet from the chin to the top of the head. In Coos County there are two other mountainous districts, separated from the White Hills by deep valleys. New Hampshire has an average elevation of 1,100 feet above the sea" (p. 232). It is said—though I have not taken the trouble to verify the statement —that no less than 1,500 streams are delineated on the map, and that altogether one-sixth of the whole area of the state is covered with water. Among the lakes may be mentioned the Connecticut Lake, Lake Umbagog, Lake Winnipiseogee, and the Sunapee and Ossipee, and Squam Lakes, and perched 500 feet above the sea of the Lake of the Clouds, the source of the Ammonoosuc River. All the streams are abundantly stocked with perch and salmon, the latter of which were at one time so abundant that it is said "labourers bargained that they should not be fed with salmon more than five times a week." Here is the tiresome old myth

---

* Suspiciously like *Noah*. All these so-called aboriginal traditions, in America at least, painful experience warns me ought to be received with very profound suspicion. ("Races of Mankind," Vol. I., p. 143.)

again! Of what country has not the same story been related, and of what place in Scotland is it not yet told? But I am not aware that anybody ever saw a labourer's agreement, or an apprentice's indenture in which this stipulation was entered. New Hampshire is still covered with much forest; indeed, Coos County is an almost unbroken primeval jungle of pine, oak, walnut, cedar, hemlock, fir, beech, maple, poplar, and butternut. On the White Mountain, above the line of 3,000 feet, and in sheltered places 4,000 feet, where forest ceases, alpine (or rather arctic) plants are found. From the height of the country about one-twelfth of it is above the line of successful cultivation. The Connecticut Valley is, however, fertile, and in the Valley of the Merrimac there are elevated sandy plains above

VIEW IN THE WHITE MOUNTAINS, NEW HAMPSHIRE.

the clay banks. The uplands are rocky, though possessing a quick strong soil. The climate is severe, but healthy, the inhabitants attaining a good old age. On Mount Washington, where a party of scientific men passed the winter of 1870-71, the climate was much the same as that of Disco Island, in Greenland. On the 5th of February, 1871, the temperature fell 59° below zero, and two days afterwards it rose to 62° Fahrenheit. In 1870, the number of farms in New Hampshire was 29,642, the average size being 169 acres. Only six contained over 1,000 acres. Not over one-third of the soil was improved in the 3,605,994 acres of farming land. Among the products, in addition to the usual crops, were 1,800,701 pounds of maple sugar, 16,884 gallons of maple molasses, and 2,146 gallons of wine made from the wild fox-grapes common in the woods. As a manufacturing centre for cotton and woollen goods New Hampshire ranks fourth among the States. Copper, lead, zinc, tin, and arsenic have been found. Gold was at one time mined to the value of 30,000 dollars. Soapstone is abundant, and the New Hampshire granite is extensively employed for architectural purposes. The product of the mines was, in 1870, 523,805 dollars, but of this 509,720 dollars must be credited as the value of the quarried stone. The commerce and navigation of the State are rather insignificant; in the year 1874 only 54 vessels entering and 65 clearing in the foreign trade. In 1873-4 there were 2,148 schools, 69,178 pupils, and 3,812 teachers, education having been compulsory since 1871. In 1875 nine daily newspapers and 68 weekly periodicals were published in the State. New Hampshire is thus not a very prosperous State, but the falling off in population must not be put to its discredit,

VIEW ON THE COAST OF MAINE.

this being greatly owing to the number of people who emigrate from it to the Western States. Perhaps, however, this just proves the same thing, namely, that the State is poor?

MAINE,

the most eastern of the New England States, is in its extreme length 302 miles, and at its widest portion 224, and has an area of 35,000 miles, or 22,400,000 acres, including the many islands which lie off its coast line of 278 miles, or taking account of the indentations by bays of 2,486 miles. The northern portion of the State is studded with lakes, one of which, Moosehead, is thirty-eight miles long. Most of the country is hilly, a spur of the White Mountains stretching into the State. Mount Katahdin, the highest point, is 5,385 feet high. The climate is one of extremes. Agriculture is not a leading pursuit, that distinction being reserved for cotton-spinning and lumbering, for the profitable pursuit of which the 21,000 square miles of fine woodland covering a great portion of the State affords abundant facilities. Indeed, as Thoreau has it, "a squirrel could traverse the whole length of the country on the tops of the trees." There are 1,620 lakes and 5,151 streams represented on the official map, the whole of this water surface being 3,200 square miles. In the value of its hay crop Maine ranks seventh among the States. Indeed, this was in value three and a half times that of the Indian corn, wheat, rye, oats, barley, and buckwheat grown. Peas, beans, hops, flax, wool, butter, cheese, maple sugar, honey, fruit, &c., were among the other products. Nearly all, if not all, the available lands of Maine are now taken up, and except as labourers or residents in the town, the State now receives scarcely any immigrants. Manufactures are, however, progressing, a law permitting any town to exempt from taxation manufacturing establishments for a period of ten years. The statistics for 1875 report 6,072 establishments, employing 55,614 hands, and producing an annual value of 96,209,136 dollars. There are 1,099 establishments for sawing timber; 8,506 hands are employed, and the value of the produce of their labour is 14,395,747 dollars. But cotton has now taken the place of lumber as the leading industry, Maine as a cotton manufacturing State ranking sixth among her sister Commonwealths. In fisheries it comes next to Massachusetts, and in education will compare favourably with any part of New England. In 1870 there were over 26,915 people in the State, including 1,606 designated as "coloured." Portland is the largest city. In 1870 it had 31,413 people, but Augusta (population 7,808) is the State capital. Bangor (18,289), Lewiston (13,600), Auburn (6,168), Biddeford (10,282), and Saco (5,755), are the chief other towns in Maine. Maine has been always noted for the eccentricity of its laws. The prohibitory liquor law, which takes its name from this State, is still in operation, but is not very strictly observed. In 1874 there were 276 convictions under this law. Forty-one people were sent to jail, and 30,878 dollars were collected in fines. White persons are prohibited from marrying negroes or Indians. In such circumstances we need not be surprised to learn that in one year there were 487 divorces granted! This finishes the United States of America.

CHAPTER XIII.

Mexico: Anahuac—Aztec and Spanish.

No doubt, long before that date, vague rumours of a wonderful land—very great and exceeding fertile—full of that gold and those precious stones which alone had a charm for the explorer's eyes, and peopled by a race far surpassing in intelligence and civilisation those lying to the south of them, had reached the ears of the discoverers of the New World. But it was not until 1517 that Francisco Fernando de Cordova visited Mexico. He was, therefore, the first European to set foot in that land which has been the grave of so many since. Even then he never got farther than Yucatan. Among those who heard of Don Francisco's exploits was Hernan Cortes, a military adventurer, and in all verity a swashbuckler of the most pronounced type. But he had that courage which was and is the birthright of his race. He was, moreover, greedy of gold, cruel and unscrupulous as to how he should get it, and pious to a degree that recognised no reason, no limit to the propagation of his creed, and no respect for man, woman, or child who did not profess it. Thus the reader will perceive that the future conqueror of Mexico was a typical Spaniard of the sixteenth century. On Good Friday, April 22, 1519, Cortes disembarked on that portion of the coast where Vera Cruz now stands, of which city of "the true cross" he then and there laid the foundations (p. 237). The very first day he landed he had to give battle to the warlike inhabitants; and these battles he fought with little intermission, until his career of rapine and conquest was crowned by the taking of the city of Tenochtitlan, and the capture of the young king, the last of the native monarchs of Mexico. What cruelties he inflicted on the people, how ruthlessly he destroyed their monuments and trampled out their national existence, let the history books tell. Suffice it for our purpose that he established a military government, of which he was the head, and decreed local councils, who promulgated laws, some of which are still in force in the Republic of Mexico. Meantime, fire and sword did their work against the unfortunate natives of the country. They were a race known as the Aztecs. Of their origin we know little, though much has been guessed regarding it. It is believed that they displaced a people of similar character—the Toltecs—from whom they obtained most of their arts and their religion. The Toltecs are said to have come from the north and to have gone to the south. But where that north was, or that south is—unless it be in Central America, where there are remains of great monuments—we can only conjecture (p. 241). When the Spaniards landed, the Aztecs might be said to have been a civilised race, and Anahuac* a non-barbarous country. At all events, their civilisation was far beyond anything which was found among the North American aborigines in any portion of the continent. There are some grounds for believing that even before the Toltecs there lived in Mexico a race surpassing in civilisation and culture that which so astonished the rude soldiery of Cortes, and even filled them with admiration, however little that admiration failed to restrain their iconoclastic propensities. Who these races were we can only conjecture vaguely from tradition, and from the perhaps rather mythical records preserved by their successors. The Toltec era is considered to have begun in the seventh century and ended in the thirteenth; after which the Aztecs founded (in 1325) the city of Tenochtitlan, or Mexico. We know only for certain that when the Spaniards landed they found the Aztecs in possession of the greater portion of the country, and their laws and civilisation the laws and civilisation of Mexico. The head of the State was a king, at first elective, but latterly despotic. He ruled, aided by three councils: one for revenue, one for war, and a third for the government of the provinces. The nobles and priests had the greatest influence in regulating the affairs of the State. The former were intrusted with the education of the children, and were consulted on all grave family affairs. Hence it naturally followed that their influence, socially, was almost unbounded. The fundamental principles of morality

---

* Anahuac is an Aztec word, signifying "by the waterside." At first it was applied only to the valley of Mexico, but it was afterwards used to denote the greater part of the country now comprised in the Republic of Mexico.

were greatly regarded by the ancient Mexicans. Property was respected, but security of person was the chief point they aimed at providing for. In the wilder districts, public inns were established at convenient intervals for the accommodation of travellers, and boats and bridges were also provided gratuitously for their use in crossing rivers. The roads were kept in good repair, at the public expense. An excellent and elaborate system of courts of law was in operation through the Empire. The Aztec laws were as remarkable for their moderation in civil actions as for their severity in criminal ones, though it seems that they were a little too flexible in regard to the priests and nobles to be looked upon as very impartially administered. "Creditors could imprison their debtors, and had a claim on their inheritance, but could not enslave the widows or orphans; and slaves about to be sold might free themselves by taking refuge in the royal palace. Adultery was punished by death, no matter how noble the offender might be. For treason or any crime against the sovereign, embezzlement of the taxes, &c., the offender was put to death, with all his kindred to the fourth degree. Murder, even of a slave, was always a capital crime. Drunkenness in youth was a capital offence; in persons of maturer years, though not capital, it was punished with severity; but men of seventy years, and all persons on festive occasions, were permitted the use of wine. He who lied to the prejudice of another had a portion of his lips cut off, and sometimes his ears. Finally, he who robbed in the market, altered the lawful measures, or removed the legal boundaries in the fields, was immediately put to death; and conspirators against the Prince, and those who committed adultery with the Prince's wife, were torn in pieces limb by limb. The murder of a merchant or an ambassador, or any injury or insult to the latter, was considered a sufficient cause of war. During a series of very cruel wars, all prisoners were devoured or enslaved. At one time, the laws were so few that the people knew them by heart. They were represented by paintings; and the judges were attended by clever clerks or painters, who by means of figures described the suits and the parties concerned in them. The Mexicans had two sorts of prisons, one for debtors and persons not guilty of capital crimes; the other a species of cage, in which were confined condemned criminals and prisoners taken in war, both of whom were closely guarded; those doomed to capital punishment being sparingly fed, and the others abundantly nourished, in order that they might be in good flesh when led to sacrifice. For the same reason the Mexicans in battle preferred to capture their enemies alive. Polygamy was permitted, but seldom practised, save by princes and nobles. Marriage generally required the consent of the parents of both parties; and there was a special court for divorces in which a wife might sue. Filial affection was a characteristic of the Aztecs. Except in the royal family, sons succeeded to all the rights of their fathers; and if these died without male issue, their rights reverted to their brothers; and in the absence of the latter, to their nephews. Daughters could not inherit. The government revenues were derived from crown lands, set apart in the various provinces from certain of the agricultural products, and chiefly from a tribute, consisting of provisions and manufactured articles; besides which, a contribution was received from the merchants and craftsmen every twenty or eighty days." Of all professions, that of arms was the happiest. He who died in defence of his country was deemed fortunate. Their armies were admirably organised. The priests went in front, and

VIEW OF VERA CRUZ, ON THE GULF OF MEXICO.

the signal for battle was kindling a fire and blowing a trumpet. In the performance of their religious duties they were most sincere. They were essentially monotheistic, believing in one supreme being—Taotl—assisted by many inferior deities, presiding over some special phenomenon of nature or phase of human existence. The war god, Huitzilopochtli, was the most dreadful of these Aztec gods. Thousands of human beings were yearly sacrificed to him in many pyramidal temples scattered all over the country. Quetzalcoatl was, on the other hand, a beneficent deity, who forbade human sacrifices, only permitting bread, roses, and perfumes to be offered to him. This "god of the air," as he was called, having incurred the displeasure of the other deities, was compelled to leave the country, but he always promised to return; and to the day when their natural life was extinguished this hope never died away. The most horrible feature in the Aztec religious life was human sacrifice, believed to have been introduced, with other horrible rites, by the Aztecs themselves, and not inherited from the mild Toltecs. On the most trivial occasion human beings were offered up; and, indeed, the performance of these hideous rites latterly formed the chief work of the priesthood. The Franciscan monks calculated that at least 25,000 persons were annually slaughtered on the altars of the war god in the capital and other principal towns. "Days have been observed," writes Herrera, "on which above 20,000 had thus perished, reckoning all the sacrifices in all the parts." The smell of the temple of the Aztec Mars was like that of a shamble. The description of the slaughter of the victims, selected from prisoners of war or from the population of revolted cities, is almost too horrible to quote."* The records of the country were kept in a kind of picture-writing, not unlike the Egyptian hieroglyphics. In addition to these picture-writings, and the aid of tradition, history was preserved by old men, whose duty it was to keep important events, genealogies, &c., in their memory, and to draw upon it whenever required by the exigencies of the State, or of private individuals authorised to call upon these primitive historians. They had orators and prose writers, and picture-writing ever aimed at the perpetuation of the efforts of the Aztec poets. They had also national music, and a variety of musical instruments, such as trumpets, whistles of bone and clay, horns of large sea shells, flutes, drums, and stringed instruments. Theatrical performances were given in the open air, the stage being covered with the foliage of trees. All the performers wore masks, as they still do in China and at the "medicine feasts" of the present North American aborigines, and all the theatrical performances were connected with religious rites. Occasionally the merchants would give performances in the temples, disguised as beetles, frogs, birds, butterflies, &c., the play being usually pantomimic, mingled with recitations, and the whole masquerade ending with a dance. The following is a description of their arithmetical and chronological system:—"The first twenty numbers were expressed by a corresponding number of dots. The number twenty was expressed by a flag, and larger numbers were reckoned by twenties and expressed by repeating the number of flags. The square of twenty, 400, was described by a plume; 8,000, the cube of twenty, by a purse or sack. The year was divided into eighteen months of twenty days each, and both months and days were expressed by peculiar hieroglyphics. Five complimentary days were added to make up the 365, and for the fraction over, of nearly six hours,

* See Helps: "Spanish Conquest of America." Vol. II., Book x. Chap. 4.

required to make the full year, they added thirteen days at the end of every fifty-two years, or cycle, which they called *xiuhmolpilli*—'the tying-up of years.' A month was divided into four weeks of five days each. The epoch from which the Mexicans computed their chronology corresponded with the year 1091 of the Christian era. They had no astronomical instrument, except the dial, but their skill in the science of astronomy is shown by their knowledge of the true length of the year, of the causes of eclipses, and of the periods of the solstices and equinoxes, and of the transit of the sun across the zenith of Mexico. Most of their astronomical knowledge was derived from the Toltecs." Their physicians were learned in botany and zoology, although their *modus medicandi* was mixed with superstitious ceremonies. Even the Spaniards were astonished at the skill of the Mexican surgeons in the treatment of wounds and in blood-letting. Geography was studied as far as their knowledge of their own country and the region lying in its immediate vicinity was concerned. Agriculture was tolerably advanced, but the want of the plough and beasts of draught was a great drawback, for the absence of which other simple instruments and arduous toil on the part of the farmer could not altogether compensate. Most of their cutting instruments were made of obsidian, but they had also axes of copper or bronze, alloyed with tin, so as to give it as great hardness as possible. Gardening was assiduously attended to, and many plants were cultivated for beauty and for use. The magney, or aloe plant (p. 245) then, as now, furnished many articles of food, clothing, drink, and domestic use; sugar they obtained from the stalk of the maize plant; and from the cacao they made what they called *chocolatl*, which was simply the same as our chocolate, which derives its name from the old Mexican manufacture. Mining, metallurgy, casting, engraving, chasing, and carving on wood and metal, were arts in which the Aztecs were most expert; while on looms of the simplest construction they manufactured cotton-cloth of exquisite fineness, and interwove it with rabbit-fur and feathers—which supplied the place of silk— and painted or dyed the fabric in gay colours. The garments made of this fabric were magnificent to a degree beyond anything now seen in America. Buying and selling were carried on in the public squares, there being no shops. Earthenware was manufactured extensively, and some of which still remains is painted in showy colours and designs. There being no beasts of burden in Mexico, everything was carried on men's backs, or in vessels on the lakes, where the number of boats and ships of every description is said to have been marvellous. Their maritime commerce was, however, trifling; and of countries beyond the sea, it is almost needless to say the Mexicans knew nothing. They had no seramphores; but that invention, ingenious at the time, though clumsily antiquated now, was practically forestalled by the Mexicans hundreds of years before it was known in Europe, by the use of towers for the transmission of rapid news. These towers were erected at intervals of six miles along the highways. In these were always waiting couriers, ready to start at full speed on foot with the despatches which might be brought, and they being enabled to travel their short stages rapidly, news could be carried through the empire at the rate of 300 miles per diem. This was the forerunner of the Russian, or Chinese, courier system, or of the old "Pony Riders," who used to gallop with the mails across the North American prairies (p. 22). Trade guilds were common—most frequently those pursuing one occupation united in a kind of corporation, or on a small scale what was, at the time of the European discovery of Mexico and for some time afterwards, so universal among the handicraftsmen of Europe. Women, though not occupying an inordinately high position in the social system of Mexico, were, nevertheless, far from degraded. They shared with the men on equal terms in labour and festivals, and on high occasions were dressed simply, though, as to ornaments, with an extraordinary degree of superfluity. All the proprieties of life were observed. They were courteous, and even polished, dignified in their intercourse with each other, and respectful to their superiors. Yet there is something—notwithstanding all this external civilisation—repulsive in the accounts which we read of the life and manners of the ancient Aztecs. For instance,

VIEW OF THE PYRAMID OF CHOLULA, NEAR PUEBLA.

in their banquets—often held, and always very costly—human flesh was frequently served as a delicacy, more especially when these feasts were connected with their terrible religion.

COLONIAL MEXICO.

No sooner had the Spaniards fairly crushed the Aztecs than they commenced that course of action which they had before put in force in the New World, and which afterwards became stereotyped with them. When Cortes arrived he found the throne occupied by Montezuma, an energetic prince. The kingly office was elective, the candidates, however, being always taken from the brothers of the deceased prince, or, failing these, from his nephews. On coming to the throne, Montezuma had made war on the State of Tlascala, and on Nicaragua and Honduras. But after a time he grew indolent, exacting, and arrogant, and devoted nearly all his time to the service of the temple. This alienated the affections of his subjects from him. Tales of impending changes took possession of the minds of the people, so that when Cortes landed he found comparatively little difficulty in penetrating to the capital. Montezuma even sent him envoys and presents, and in every way tried to propitiate the strangers, who, however, basely rewarded his clemency. He treated the *conquistadores* as superior beings, and when the inhabitants rose against Cortes he appeared in order to pacify them. But he was wounded accidentally by a stone thrown by one of

COLOSSAL HEAD CARVED IN STONE, IN AN AZTEC RUIN AT IZAMAL, IN YUCATAN.

the mob, and felt the indignity so keenly that he repeatedly tore the dressings from his wounds, and died June 30, 1520. One of his descendants was Viceroy of Mexico from 1697 to 1701. His last descendant—or at least the last who was recognised as of the blood of Montezuma—was Don Marsalio de Teruel. He was banished from Spain, and afterwards from Mexico, "on account of his liberal opinions," and died at New Orleans in 1836. The Indians were distributed among the *conquistadores*, to toil in their mines, to till the lands that had once been their own, to wear their fragile lives away in labours beyond their strength. Yet the aborigines of Mexico were more fortunate than those of Hayti. There the same policy of *repartimientos*—if policy it can even be called —was put in force. But the Haytians were weakly people, of low vitality and little muscle, and in a short time, as we shall see by-and-by, the Carib race was numbered among the peoples that had been. The Aztecs, on the contrary, were, if not robust, at least wiry, and survived their life of slavery longer than did the feeble folks farther to the South. And so the Aztec dug gold, and the Spaniard spent it in Spain, or bought office, titles, and vineyards with it. His plate ships were on the high seas, and his wealth the wonder and envy of the poor world that lay around his Iberian home. The "Mexican pistoles" have been celebrated in many a ballad, and in a score of tales, while the treasure-laden vessels of the *conquistadors* were the prime moving causes to many a naval enterprise, in which Raleigh, and Drake, and all that glorious band gained their fame or met their death. By-and-by—at the beginning of this century, for instance—the Mexican population was made up of four classes. First, there were the pure-blooded Indians; second, the Creoles, or pure-blooded descendants of the Spanish settlers; third, the Mestizoes, or half-breeds; and last of all, fewest in number, but unfortunately the most important, the Spaniards of European birth. The Indians were held in tutelage, and though their lot under the later Viceroys was not so hard as it had been under the *audiencias* whom they replaced, yet they had to pay tribute, and were never considered as human beings capable of exercising their judgment, or indeed of having one. Their chiefs—the caciques, or nobles—were, however, exempted from the degrading restrictions which weighed on the others; and though never treated in that manner which the slightest ideas of humanity or the art of ruling men would have dictated, were at least not so badly used as their predecessors had been under the rule of the early conquerors. The lot of the Creoles was, however, most galling to that proud race. On such as had accumulated wealth, titles of nobility or crosses of honour were conferred. But these cheap emoluments failed to make the Spaniards of Mexican birth contented. From a fear, apparently, that their patriotism would tempt them to throw off the yoke of the mother country, they were excluded from participation in the government of their native land. They could hold no office, and were even forbidden to participate in the foreign commerce which was enriching the colonies. These, and other grievances which I have partly described in another place,* rankled in the Creoles' mind, until, from being loyal subjects of His Most Catholic Majesty, they entered into a chronic condition of "veiled rebellion." They disliked the Government which could treat its children so unjustly, and they detested the Old Country Spaniards, for whose benefit these laws were made. These, office-holders they called *gachupines*, and the gachupines returned the hate of the *Creoles* with the lordly interest of contempt. Still, the inevitable rupture between Spain and Mexico might have been put off for a little longer, had not the events of 1808 occurred at home. In that year the throne of Ferdinand the Seventh was usurped by Joseph Bonaparte, a step which united the Spaniards and the Creoles for a time. Both loudly protested against this high-handed action of the French Emperor; though, as seems ever the unhappy fate of Mexico, the short-lived unanimity was endangered by intemperate dissensions regarding the new provisional form of government of which the state of matters at home called for the organisation. The excitement increased when the Viceroy, Don José de

* "Races of Mankind," Vol. II., pp. 4–5.

Iturrigaray was imprisoned on suspicion of designing to seize the crown of Mexico, in entertaining which suspicion we daresay Don José was not greatly wronged by the Mexicans. Nevertheless, he was released, but from his incarceration may be dated the beginning of that longing for independence which culminated, in 1816, in the rebellion of Don Miguel Hildago, a priest so popular among the Indians that at one time he had 10,000 soldiers—or at least troops, disciplined and undisciplined—under his command. He was a man of much talent, but he was not soldier enough to withstand the trained army which the Government brought against him. The end was that, after being defeated, he was—*more Mexicano*—betrayed to his enemies and—shot. But his blood watered the seeds of revolt which he had sown. The contest was carried on until, on the 2nd October, 1814, the first Mexican Constitution was promulgated at Apatzingam. But the end had not yet come. Morelos, the successor of Hildago, and, to the honour of his Order be it told, a priest also, was finally captured and—shot. Then followed a guerilla warfare, so far useless that by 1820 the power of Spain was again more firmly established than any Government has ever for the last fifty years been in Mexico. But in the course of that year, the news of the promulgation of a Constitution in Old Spain by Ferdinand VII. renewed the agitation, and again Don Augustin Iturbide, a rebel not of yesterday, raised the standard of revolt. The people were ripe for it: even the Viceroy, one Don Juan O'Donoju—under which Spanish sophistication it is difficult to conceal his Hibernian origin—sided with the rebels. In a few months thereafter Mexico was wrested from Ferdinand by what looks like a *coup d'état*, if the whole revolution had not been one extensive affair of the same sort. Then the leader of the revolutionaries emerged from being simple Colonel Iturbide into the magnificence of Augustin I., Emperor of Mexico.

Mexico Independent.

The farce was soon played out. My not particularly esteemed acquaintance Santa Anna proclaimed the Republic, and Iturbide, to avoid civil war, resigned, was exiled, and the Republic inaugurated. Finally Iturbide, returning next year, to show his ex-imperial countenance in Mexico, was—need we say it?—*shot*. A taste for shooting is like the appetite for absinthe: it is unpleasant at first, but in time grows agreeable, and then becomes a recognised institution. And so in time the Mexicans found it. It would require a good memory and a long chapter now, too, to record all the contests for the supreme power, half of the revolutions, or a tithe of the revolutionaries shot in the next half-century of Mexican history. For instance, in 1828, Gomez Pedraza and Guerrero, of revolutionary fame, were candidates for the Presidency. Both were generals, a soldierly crop which flourishes greatly on American soil, Latin, or Anglo-Saxon; and on the election of the former, the latter took up arms and slaughtered a great many voters on both sides, and then Pedraza, who had the worst of it, fled the land to avoid being—shot by Don Guerrero, President elect of the bayonet. In due time Spain tried to regain possession of the country, but with such indifferent success, that the invading force was disarmed and sent to Havanna. In the fulness of time, the Vice-President, Anastasio Bustamente, likewise a general, finding he had a good many soldiers raised to repel the Spaniards, like a true Mexican patriot utilised the force to unseat the President and proclaim himself. In other words, Don Bustamente "pronounced" against Don Guerrero, and Don Guerrero was, of course—shot. Then in his turn Bustamente was succeeded by Pedraza, and Pedraza in due time was ousted by Santa Anna, and Santa Anna, after a good deal of miscellaneous murder at home, abrogated the Constitution of 1824, and converted the confederation of States into a consolidated Republic, of which he constituted himself President in name, but Dictator in reality. It was even proposed to appropriate the Church property to the payment of the national debt. But that was too much. Mexico, if not very particularly honest, was above all things pious, and so, after a little murder and "further complications," these honest gentlemen agreed to cheat the Government creditors and be true to the Church; a resolution, be it added, to which they have, with singular if not exemplary fidelity, kept to the present day. This was in the year 1835. Now, among the Mexican States was Texas, peopled by several thousand American settlers, who decidedly objected to this consolidation of States, and above all to Santa Anna being Dictator of them. Accordingly, they "went in" for "a revolution," and at San Jacinto smote the President hip and thigh, annihilated his army, took himself prisoner, and proclaimed the Lone Star Republic, for which see Chapter VIII. Then back came Bustamente, and back came Santa Anna, as President for four months, to be succeeded by Nicolas Bravo, who held office for one week; after which came chaos, during which the Dictator was the *caballero* who could manage to struggle to the front. Out of the farrago of names we can pick Santa Anna, Bravo, and Canalizo, and then Santa Anna again. Indeed, my sinister host of 1863 had a wonderful capacity for escaping being shot, and, perhaps on a principle allied to the *post hoc, propter hoc* kind of reasoning, a most all-consuming capacity for shooting other people. He believed in blood-letting as a political curative agent. Never did a Spanish republican statesman—and that is saying a good deal—so speedily manage to reduce a majority into a minority, as this terrible wooden-legged general of evil fame. Accordingly we are not surprised to find him in power in 1843, and in banishment in 1844, at the hands of Canalizo, who, before the year was out, was deposed and succeeded by Herrera, who also found himself out of office, and across the border, on the last day of 1845. Though the Mexicans could not complain of political monotony, General Herrera must needs find himself at war with the United States, owing to the annexation of Texas—after a brief existence as a separate government—to the Great Republic. No good came of that. The Mexican Government gained no glory, and lost California and New Mexico. Meantime Santa Anna came back again, and again had to flee, after being for a time President. Then Herrera, and Arista, and Cevallos shared the fate of all Presidents. The arrogance and insubordination of the soldiers threatened anarchy, until, in 1853, for the fifth time, the irrepressible Santa Anna was hobbling through the Plaza de Armas in his capacity as head of the State. But this everlasting election, deposition, revolution, seizure, and election again, had grown tiresome to the *blasé* old man. He accordingly determined to have the position for life, just to save trouble and travelling expenses. But the Mexicans thought otherwise, and once more Santa Anna was deposed, and Alvarez, without the formality of popular sanction, reigned in his stead. But Alvarez, to avoid being shot, resigned in favour of Comonfort, who soon found a revolution on his hands. This was instigated by the priests, whom he had

ALOES (AGAVE AMERICANA) IN BLOOM.

incensed by his "promoting" a law, since passed, for the sale of Church lands and the freedom of religious belief. But still, though a very democratic constitution was promulgated, Mexico did not pay its debts. Indeed it repudiated an acknowledged debt to Spain; and conspiracies multiplying, Señor Comonfort had to flee before the bayonets of General Zuloaga, his quondam ally. But Zuloaga found himself confronted by Benito Juarez, who, though unsuccessful, was, strange to say, neither shot by his rival nor exiled by himself. On the contrary, he went to Vera Cruz, and commenced presidential business on his own account. He was singularly successful in this venture. In due time Zuloaga gave place to Robles, and Robles—a futility though a General—to Miramon, and Miramon—a skilful slaughterer of his fellow-citizens—to Juarez, who had the good fortune to be recognised by the United States. Very early he promulgated many popular reforms. "Among these," remarks Mr. Hawes, to whom we are indebted for most of these data, "stand most prominent the making of marriage a civil contract, the abolition of perpetual monastic vows, and of ecclesiastical tribunals, the suppression of monasteries, and the appropriation of Church property to the service of the State, the total value of which was estimated at rather more than 300,000,000 dollars, or one-half of the value of all the landed property in the country. These measures were soon followed by the complete separation of Church and State. But the Church party had resolved on the destruction of Juarez's government, although national liberty should be sacrificed for its accomplishment." An opportunity for carrying the design into execution soon occurred. Mexico, as usual, plundered subjects of Spain, France, and England, and treating all demands for reparation with supercilious neglect, was, in December, 1861, visited by a joint expedition from the three powers. An agreement was entered into—though it would be rash to say that it has ever been strictly carried out—whereby a portion of the custom's receipts should be appropriated to paying the debt due to the Spanish and English, who accordingly withdrew. But the French remained. The end of it all was, that in 1864, the Archduke Maximilian of Austria became Emperor, and in 1867, being left to his own resources on the withdrawal of the French troops under Bazaine, the expedition having cost France 48,000,000,[*] was captured, and after a farcical trial condemned and shot, along with Mejia and Miramon, two of his generals, the latter of whom had for a time tasted the bitter sweets of power himself. Juarez, who had, during the Empire, been carrying on a kind of government of a scrambling sort, was re-elected President, and once more endeavoured to set the machine of State in order. Meantime, an old acquaintance of ours had been watching all these experiments at government with that peculiar interest which had made him a marked man, even among Mexican patriots. And in due time, once more General Santa Anna, ex-Liberator, ex-Dictator, ex-President, arrived upon the scene. He had figured for a time in the ill-assumed rôle of a disinterested lover of his country, and had attempted to carry this out by offering Juarez his services in driving the foreign occupants out of the country. But Juarez, remembering the fable of the horse who asked the man to aid him against the stag, declined to make his government another example of the fate of all such bargains. Then came out Don Antonio Lopez de Santa Anna in his old colours. But the old man was not

[*] See Chevalier: "Le Mexique, Ancien et Moderne" (1863).

what he was, or Mexicans had ceased to be charmed by his serenic *pronunciamentos*. For on July 12th, 1867, he was captured, condemned to be shot, and finally banished. He died in 1873. Then followed more pronunciamentos, more insurrections, more revolutions, more disorder, and a good many more murders, and all manner of legal and illegal wickedness. Most of them were very abortive, but that of Angel Santa Anna—a son of the arch-plotter and patriot—was so far successful, that for four months he plundered and slaughtered, until he was, with the chief of his followers, captured and—shot. In 1871, Juarez was again elected, and in 1872 died, and was succeeded by Lerdo de Tejada, who had been one of his opponents. Juarez was in many respects a remarkable man. He was a pure-blooded Indian, but had received an excellent European education. By profession he was an advocate, and when first "called" to the Presidency was Chief Justice of the Supreme Court of Mexico. He is remarkable as the first President of Mexico who held power during his full term of office! Lerdo de Tejada gave place to Diaz, and Diaz is still at the time of writing (1878) in power. But how long it would be rash to vaticinate. For on the borders, big with a pronunciamento and ready for revolution, is Escobedo, the captor of the ill-fated Maximilian; and it is only a reckless man, very sanguine or very ignorant, who ventures to foretell Mexican affairs many months in advance. For some time past, forgetting nothing and not anxious to learn, the country has been acting rather high-handedly in reference to American citizens plundered by Mexican bandits. It is, therefore, just possible that by her own haughty contempt for "Gringos" generally, and those of the United States in particular, that Mexico may hasten her inevitable destiny by a few years. Meantime we cannot say that the country is greatly improving. The laws, however, are said to be better obeyed, or " rather less disregarded," and the military have got reconciled to the idea of a civilian as President, a very necessary step in good government, which it may surprise the reader had not been fundamentally understood long ago. We have devoted some space to a brief sketch of Mexican history, for the reason that the history of Mexico is, take it all in all, the history of all the Spanish ex-colonies in America, substituting one name for another, and the date of a revolution in one year for that in another. Pompey is, in Hispano-American history, much the same as Cæsar, "especially Pompey," and Don Juan José does not differ widely from Don José Maria, except that the one " pronounced " in Mexico, and the other in Peru. *Arcades ambo!* *

CHAPTER XIV.

Mexico: Its Physical Geography and Resources.

Mexico—the Mexitl of the Aztecs, the Estados Unidos de Méjico of official documents—is 1,990 miles long from the frontier of Guatemala to the extreme north-west limit. Its

* It may be added that our intercourse with Mexico is at present non-diplomatic, the States which recognised Maximilian being excluded by the Government from the community of nations whom the Republic will honour by its intercourse.

maximum breadth is 750 miles, and its minimum on the Isthmus of Tehuantepec only a very few. At present it is divided into twenty-seven states, one federal district, and one territory. According to the statistical reports of 1869 and 1873, though chiefly the former, the area, population, and capitals are as follow:—

States.	Area in Square Miles.	Population.	Capitals.
Aguas Calientes	2,216	160,630	Aguas Calientes.
Campeachy	26,084	80,366	Campeachy.
Chiapas	16,769	193,387	Chiapas.
Chihuahua	105,295	179,074	Chihuahua.
Coahuila	61,959	95,597	Saltillo.
Colima	2,334	64,453	Colima.
Durango	12,643	185,077	Durango.
Guanajuato	11,130	874,043	Guanajuato.
Guerrero	24,236	309,629	Guerrero.
Hidalgo	8,480	404,207	Pachuca.
Jalisco	48,967	924,580	Guadalajara.
Mexico	9,398	530,063	Toluca.
Michoacan	21,069	618,219	Morelia.
Morelos	1,898	147,659	Cuernavaca.
Nuevo Leon	11,563	174,000	Monterey.
Oajaca	27,389	646,725	Oajaca.
Puebla	9,308	697,788	Puebla.
Queretaro	3,120	153,286	Queretaro.
San Luis Potosi	28,889	476,500	San Luis Potosi.
Sinaloa	25,927	163,095	Culiacan.
Sonora	81,022	108,488	Ures.
Tabasco	12,716	83,707	San Juan Bautista.
Tamaulipas	28,059	108,778	Ciudad Victoria.
Tlaxcala	1,498	121,605	Tlaxcala.
Vera Cruz	27,483	459,262	Vera Cruz.
Yucatan	32,638	422,365	Merida.
Zacatecas	26,585	397,915	Zacatecas.
Federal District	85	275,396	Mexico.
Lower California (Territory)	59,053	21,645	La Paz.
Total	761,640	9,169,797	

The population may at the present time be over 9,600,000, but the tables of 1873, which gave the census as 9,100,000, undoubtedly exaggerated the number of people in some districts. The table-lands and the slopes of the Cordilleras are the most thickly inhabited districts, but in many sections of country the settlements are few and far between. In 1861 there were 8,000,000 inhabitants, distributed as follows:—Indians, 4,800,000; whites, 1,001,000; half-breeds, 1,190,000; negroes, 6,000. This general myscegenation makes up the Mexican nation. In the Republic there are eighteen cities, whose population exceeds 20,000; in twelve of them it is above 30,000, and in five more than 50,000. The position of Mexico is highly favoured, both as to climate and facilities for commerce. Lying between two seas, it has a seaboard of 6,086 miles, 1,677 of which belong to the Gulf of Mexico and the Carribean Sea, and 4,408 to the Pacific, including 2,010 which line the Gulf of California. The shores of the Mexican Gulf are low, flat, and sandy, except near the Tabasco River. Yet the voyager along the shores of this portion of the country is not apt to think it monotonous,

* See Vol. I., p. 344.

for in the background, though many leagues distant, it is relieved by the picturesque mountains of Vera Cruz. On the Pacific coast the shores are also, as a rule, low, but

TROPICAL CLIMBERS.

here and there spurs of the Cordillera extend towards the ocean, and thus vary the monotony of the flat sandy beach, backed by that unvarying fringe of rich vegetation, relieved here and there by the slender thatched cottage of some Indian, or settler, whose habits of life are not much more complex than those of the aborigines. Off the north-eastern coast of Yucatan are a few islands. The only one of any consequence is that of Cozumel—the Swallow Island of the original inhabitants—the Santa Cruz of the Conquistadores, which is 300 square miles in area, and abounds in forests of valuable timber. It is also famous as the site of the shrine to which the ancient Mexicans made pilgrimages, in order to pay adoration to idols, the temples of which can be still traced in ruins. Carmen Island, in the Bay of Campeachy, is about sixteen miles long, and two miles wide. In the Gulf of Mexico are the Islós de los Sacrificios, near Vera Cruz, and the islet on which was built the fortress of San Juan de Ulloa, which looms so largely out of Mexican history. The others are smaller. The physical geography of Mexico is extremely varied. Its surface is nowhere within thirty miles of the sea higher than 1,000 feet, except, perhaps, in Chiapas; but after this distance from the sea the country everywhere north of the Isthmus of Tehuantepec rises by a succession of terraced mountains to a table-land with a mean elevation of 8,000 feet. This table-land extends far beyond the northern limits of Mexico. The effect of this gradual rise is seen in the course of the railway journey from Vera Cruz to the City of Mexico. In the space of a few hours every variety of climate is experienced, and in rapid succession the railway train, which appears so out of keeping with the sleepy primitive life past which it darts, passes through sugar-cane and indigo plantations, patches of plantains and bananas, and almost insensibly to the pines, firs, and lichens of the north. "The interior of Mexico," to use the language of Mr. Tylor, "consists of a mass of volcanic rocks, thrust up to a great height above the sea-level. The plateau of Mexico is 8,000 feet high, and that of Puebla 9,000 feet. This central mass consists of a greyish trachytic porphyry, in some places rich in veins of silver ore. The tops of the hills are often crowned with basaltic columns, and a soft porous amygdaloid abounds on the outskirts of the Mexican Valley. Besides this, traces of more recent volcanic action abound, in the shape of numerous extinct craters, in the high plateaux, and immense "pedrigals," or fields of lava, not yet old enough for their surface to have disintegrated into soil. Though sedimentary rocks occur in Mexico, they are not the predominant features in the country. Ridges of limestone hills lie on the slopes of the great volcanic mass towards the coast; and at a still lower level, just on the rise from the flat west region, there are strata of sandstone. . . . The mountain plateaux, such as the plains of Mexico and Puebla, are hollows filled up and floored with horizontal strata of tertiary deposits, which again are covered by the layers of alluvium constantly accumulating. Of the mountains, Popocatepetl rises to the height of 17,540 feet, and Orizava to the height of 17,370. Both are volcanoes, though with Istaccihuatl and Toluca rising above the limits of perpetual snow. San Martin, in the State of Vera Cruz, belches out, day and night, smoke and flame visible far to seaward in the Gulf, though it has not "erupted" since a few years after the Conquest. Mexico is imperfectly watered, there not being many rivers, and few of them, owing to the construction of the country, being navigable for more than short distances. The Rio Santiago, 500 miles long, is broken near Guadalajara by sixty falls in the space of less than three miles, while the Rio Grande Del Norte, though winding 1,800 miles, is only navigable for sixty miles from its mouth, and even then only for small vessels. The same may be said of most of the rivers flowing into the Gulf of Mexico.

Mexico is rich in lakes of considerable size. For instance, it has Tezcuco, with an area of ninety-nine square miles; Chalco, fifty-four square miles; Xochimilico and Xaltocan, each about twenty-seven square miles; and about fifty-five other lakes and lagoons of smaller size. During the rainy season the city of Mexico is often jeopardised by the overflowing of these sheets of water. The old Aztecs devised and carried into execution almost the only system of drainage which exists. But even this has been allowed to fall into decay and ruin. For instance, they made a canal connecting Tezcuco, Xochimilico, and Chalco, which is still navigated. But as it is the receptacle for all the sewage of the capital, to its miasmatic exhalations most of the insalubrity of the city is due. Lake Chapala is also a sheet of some importance, and is traversed by steamers. The Mexican and the South American Andes, or Cordilleras, bears a striking similarity in this respect, that both are intersected by barrancas, or vast fissures, while the backs of the mountains form elevated plateaux or basins, so uniform in height that they may be regarded as one continuous table-land. The Valley of Mexico is an elliptical plain, about 940 square miles, fringed on all sides, except the north, by lift peaks, none of them active volcanoes. Indeed, the plain may be regarded "as one vast volcanic hearth, roughened at intervals by isolated hills running abruptly from the surrounding level." Among these peaks may be enumerated Popocatepetl and Istaccihuatl,* which tower over all the others. So regular, indeed, is the mountain plateau of Mexico, and so easy the slopes where depressions occur, that a wheeled carriage could be driven all the way from Mexico to Santa Fé (p. 16), a distance of 1,200 miles. Prescott's "Conquest of Mexico" has left on the minds of most readers a tolerably accurate picture of the valley as it was in the days of Montezuma. "The amphitheatre of dark hills surrounding the level plain, the two snowy mountain peaks, the five lakes covering nearly half the valley, the city rising out of the midst of the waters, miles from the shore, with which it was connected by five causeways, the straight streets of low, flat-roofed houses, the number of canals crowded with canoes of Indians going to and from the market, the floating gardens moved from place to place, on which vegetables and flowers were cultivated, the great pyramid up which the Spanish army saw their captured companions led in solemn procession and sacrificed at the top—all these are details in the natural picture." How they have been altered within the last three hundred years, how the present city of Mexico is not the city of the Aztecs, we shall have occasion to notice by-and-by, when sketching the more salient features of Mexican towns and town life.

The Climate.

Like all the countries of Spanish America lying on the slopes of the great Andean ranges, Mexico has three climates, corresponding to the three terraces into which it is divided. There is first the coast region, or *Tierras Calientes* (the hot lands). This comprises all the country lower than 3,000 feet above the sea, and has a temperature of from 75° to 80°. The second, or *Tierra Templadas* (temperate lands), extend from 3,000 feet to the

* This is not a volcano, though given as such in some physical atlases. It has not even a crater.

mean elevation of the central table-land, 6,600 feet, and has a mean temperature of from 65° to 72°. The third region, or *Tierras Frias* (cold lands), which is above this altitude, but is really only cold by comparison with the hotter regions lower down, for its temperature is from 55° to 60° in the dry season, and never higher than 80° in the wet. The Mexicans speak of their climates as hot and dry, and hot and moist; temperate and dry, and temperate and moist; and cold and dry, and cold and moist. Really there are in Mexico, as in most countries similarly situated, only two seasons: "the dry from October to May, and the rainy comprising the remaining months. The heaviest rains fall in August and September. The rain is generally excessive on all the coasts, but especially so at Guaymas, Mazatlan, and Acapulco, on the Pacific; and Vera Cruz, Merida, Sisal, and Progreso, on the Gulf." The healthiest climates are the dry ones; the most unhealthy the moist ones. In the high land the air is so rarefied that acute lung diseases are common, and disorders of the stomach and bowels are also frequent and fatal. But yellow fever is the great scourge on the coasts. It and the terrible "black vomit" usually set in at Vera Cruz about the end of May, and rage until the November colds check their fatal progress. Some of the coast towns occasionally escape for several years. But these exceptions are not found among the towns on the Peninsula of Yucatan at all, of which the mortality is usually excessive.

Vegetable Products.

In the *Tierras Calientes* the soil is usually very fertile. Maize, rice, when irrigation is practicable, bananas, pine-apples, oranges, manioc, are among the crops; and in the swampy forests along the shore the sarsaparilla, jalap, vanilla, and other tropical plants flourish (pp. 249, 253, 256). In the winter the north winds blow in this region, and the hurricanes often desolate it. In the summer the breezes also blow, but not so severely as at other seasons, and, as we have already mentioned, the yellow fever sets in. In the *Tierras Templadas* wherever rain falls an almost perpetual summer reigns, and all the grains, fruits, and vegetables of Central and Southern Europe—including maize, oranges, lemons, grapes, and olives—are produced in exuberant abundance. The *Tierra Frias* has a keener air, and generally a more arid soil than the lower lying lands. Here agriculture does not find outlet in the cultivation of a variety of crops. Barley and the agave, or American aloe (p. 245), which was to the ancient Aztecs what the vine is to the Southern Europeans, or the bamboo to the Chinese, are the chief crops. It is still extensively cultivated for the sake of its fibre, and the juice, which is fermented into *pulque*, the favourite drink of the Indians, and which is even liked by some whites, though, speaking from experience, I must pronounce it one of the abominations of the earth. A sort of brandy, or *mezcal*, which is highly intoxicating, is also distilled from it. The value of the pulque trade was, in 1862, calculated at 1,187,523 dollars, and that of mezcal at 2,576,616 dollars. But since that date, though we have no accessible statistics, the trade must have largely increased with the greater facilities of transit afforded by the opening of the Mexico and Vera Cruz Railway in 1873. A special train, known as the "pulque train," runs between the capital and Sultepec, the centre of the district where

A LAGOON IN THE TIERRAS CALIENTES.

the beverage is manufactured. In some portions of the highlands village after village is separated by plantations of aloes. In the Llanos de Apam the best pulque is made.

They are planted in long, regular lines, and no sight is more common to the traveller who looks out of the carriage window than to see the Indian "tlachiquersee," each with a pigskin on his back, and his long calabash in his hand, "milking" such plants as are in season. The cultivation of the *maguey* is almost a religious duty among the Indians. Mr. Tylor tells us, "The Indians have a great fancy for making crosses, and the aloe bends itself particularly to this kind of decoration. They have only to cut off six or eight inches of one leaf, and impale the piece on the sharp point of another, and the cross is made. Every good-sized aloe has two or three of these primitive religious emblems upon it." The juice is collected in great hollows, which are cut in the heart of the plant to receive it. This is allowed to accumulate during the night, and then regularly extracted by the Indian milkers in the morning. Here is the description which the same eminent ethnologist whom we have already quoted gives of the process:—"Getting to the top of the ravine, we found an old Indian milking an aloe which flourishes here, though a little further down the climate is too hot for it to produce pulque. This old gentleman had a long gourd, of the shape and size of a great club, but hollow inside, and very light. The small end of this gourd was pushed in among the aloe leaves, into the hollow made by scooping out the inside of the plant, and in which the sweet juice, the *aguamiel*, collects. By having a little hole at each end of the gourd, and sucking at the large end, the hollow of the plant empties itself into the *Acole* (in proper Mexican *Acotl*, water-throat), as this queer implement is called. Then the Indian stopped the hole at the end he had been sucking at with his finger, and dexterously emptied the contents of the gourd into a pigskin which he carried at his back. The pulque is taken to market in pig-skins, which, though the *pig* is taken out of them, still retain his shape very accurately; and when nearly full of liquor they roll about on their backs, and kick up the little dumpy legs that are in them in the most comical and lifelike way." In the aloe district huts are built of the stem of the shrubs which have been allowed to flower, stuck into the ground side by side, with pieces of leaves tied on outside with aloe fibre. These huts are no doubt cheap, and possibly picturesque, but in the cold nights which the *Tierras Frias* often experience cannot be said to be at all times comfortable. The manufacture of aloe fibre is a branch of industry only second to that of the fermentation and distillation of the juice of the plant. The bags, or *costales*, in which the ore is carried from the mines, are almost invariably made of this fibre. The fibre itself is made in two qualities—the coarser from the long pulque aloe, and the finer from a small species of the same genus; and the uses of both qualities seem almost endless. Coffee, tobacco, yams, capsicums, pepper, pimento, indigo, ipecacuanha, dragon's blood, copaiba, fan-palms, india-rubber trees, ebony, mahogany, rosewood, and many other vegetable products are cultivated in Mexico. The cactus is the plant which is almost universally seen in Mexico, and, indeed, gives a characteristic aspect to its scenery (p. 53). Indeed, the arms of the Republic are an eagle perched on a cactus, and holding a serpent in its mouth. The tale is that the old Mexicans, when they first came into Mexico, were a mere tribe of savages, and wandered about from place to place, now fixing themselves here for a time, and now there, just as circumstances would permit them. In time they had a revelation, that when they came to a plain where they should see an eagle with a serpent in its mouth, perched on a cactus, there they should make their permanent abode. And so the story goes that when they arrived on the spot where the present city of Mexico is built, they found what the diviner had told them to look out for. There was the cactus— there was the eagle—and best of all, for neither an eagle, nor above all a cactus, is sufficiently remarkable to call for attention in Mexico, there was the serpent in its mouth. And so they settled down. Endless species met the eye everywhere in all the regions of Mexico. Hedges of them are made of the organ cactus (*Cereus hexagonus*), which is also grown in Italy, though it does not seem to be turned to account for fences anywhere except in Mexico. In appearance its stems look like the pipes of an organ; hence the popular name. Some are so full of watery sap, that in dry weather the cattle will bite at them; just as in the north coast of Africa the camels delight in munching, regardless of the thorns, the juicy leaves or branchlets of plants of the same order. The fruit of the plant, the so-called prickly-pear, or pitahaya, is also eaten, and even made into preserves. The roots of the plant have also a curious property of rapidly breaking lava into a soil capable of supporting other plants. In Sicily, for example, the lava fields are often planted with the cactus, and in a few years the soil is sufficiently disintegrated to allow of vines being planted on it. The same effect is seen in Mexico, but it is also breaking up the ancient pyramids of porous amygaloid lava which are among the few remnants of Aztec religion which the fanaticism of the priests allowed to remain in Spanish Mexico. The cactus is also cultivated in order to feed the cochineal insect which lives upon and furnishes the well-known red dye. This is carried on chiefly in the province of Oajaca, where the nopal, or great prickly-pear—the same species which has established itself on the shores of the Mediterranean—is cultivated extensively. The grape flourishes, and in some districts wine is extensively manufactured. Cacao cultivation and chocolate manufactories have been already referred to, while the cultivation of the sugar-cane and the cotton plant are too familiar operations to require any special notice. Jalap is exported to the extent of 50,000 dollars *per annum*, though at the beginning of this century the trade amounted to 170,000,000 dollars.

Animal Products.

Among the animals may be mentioned cattle, horses, asses, mules, goats, and sheep, all of which are now plentiful, though originally introductions from the Old World. In the lower ground are buffaloes (not, of course, the wild bison, but the domestic ox of that name), while the tapir, wolf, lynx, jaguar, wild cat, skunk, brown porcupine, stag, &c., are common in the wilder parts of the country. The rivers and lakes abound in fish; turtles are taken off the coast, while another reptile, the iguana, is by some considered excellent food. A few boa-constrictors are found in the southern forests, and some other snakes, particularly the rattle and coral snakes, which are said to be exceedingly venomous. The *alacranes*, or scorpions, are almost as much dreaded, and it is said that not a year passes in which some children are not killed by their stings. Among other insectal pests, gigantic spiders, the dreaded tarantula, scolopendras, and a plague of mosquitoes abound. The country is rich in many other kinds of animals, and among these we need only name its parrots, humming-birds, trogons, and the zenzontli, or humming-bird; while eagles, hawks, and turkey-buzzards abound. The latter is, indeed, as in all tropical countries where it is found, the scavenger of the towns, and is protected by law from being injured or destroyed. They have, however, to compete for garbage with the dogs, who drive them away when anything better than usual is to be got. Mr. Tylor describes them in Vera Cruz as sitting in compact rows on parapets of houses and churches, and especially affecting the cross of the cathedral, where they perched two on each arm, and some on the top. "When some offal was thrown into the streets, they came leisurely upon it, one after another, their appearance and deportment reminding us of the undertaker's men in

GATHERING VANILLA (*Vanilla planifolia*).

England coming down from the hearse at the public-house door, when the funeral is over." The mines and other resources of Mexico we shall have occasion to notice further on, when we purpose to devote a little space to consider the commerce and the resources of the country, so far as they affect the sombre future of the Republic of Mexico, apart from that of the country itself. That no incurable ignorance, stupidity, or innate viciousness can well permanently affect.

CHAPTER XIV.

MEXICO: ITS MEN AND MANNERS.

To give a brief characterisation of the people of any country is always very difficult. Especially is this a delicate task when the Mexican population has to be described. The race is heterogeneous, and what may be true of the people of one part of the country may be

ENTRANCE TO THE CATHEDRAL OF MEXICO.

utterly untrue regarding those of another section. Of course no one will agree with what is said. Every man who ever spoke to a Mexican considers that he has a right to pro-

A MEXICAN MONK OF FORMER TIMES.

nounce *ex cathedra* opinions regarding the people; and perhaps from his own point of view he has. Still they are rather extensive inferences from very imperfect premises, and as the vast majority of mankind have never learned the most elementary rules which must be attended to in weighing evidence, the result of this multitude of councillors is a rather embarrassing series of dogmas. One traveller represents the Mexicans as a fine race, possessing all the virtues of the rest of mankind, and some peculiarly their own. Others will assure the reader on their word of honour that they have searched the vocabularies of the language in which they write without being able to pick out a series of adjectives strong enough to express the utter turpitude of these degenerate descendants of a degenerate race! Let us take a middle course, and endeavour to glean from different writers whose ability, truthfulness, and length of residence in the country have been such that, now and then checking their statements by our now slight personal acquaintance with the Mexicans, we may in the end be able to get some tolerably sound ideas regarding the people of this portion of Spanish America. And before saying one word regarding either the Indians, the Spanish Creoles, or the mixed races—before attempting to give some idea of the men or women of Mexico—let us say a few words about those whom Voltaire characterised as the third sex, namely, *the priests*. They have in times past exercised such influence, and even now are so powerful, that the character of the Mexicans can never be properly understood unless the character of their spiritual advisers be explained. In doing so we must, perforce, be severe, but severity in this case is only justice; for I should be extremely sorry to think that the ecclesiastics of any country or any creed were to be judged by the Mexican standard.

THE PRIESTS AND THE CHURCH.

In Mexico, as in every Spanish country—free or in bonds, Republican or Monarchical—the priests are all-powerful. The lordly race of Castile is like the king "most Catholic." Hence it follows that the Churchmen, as the administrators of the creed, share in the influence which that creed exercises over the actions, the thoughts, and—I am not inclined even to gainsay that—the morals of the Spanish people. Indeed, in no matter what administration or form of government, the Church "is the *imperium in imperio*, the *respublica in respublica*. It was even more so prior to the confiscation of the Church property, and the proclamation of liberty of conscience; an era, however, which does not date earlier than 1859. Take Puebla (p. 264), for example. This town of over 60,000 inhabitants may be said almost to have belonged to the clergy. It has seventy-two churches, some of them by no means small, though none of them so fine as the Cathedral of Mexico (Plate XIX.). Here the Church party focuses itself; here it plots that steady, powerful opposition to all reform, which is one of the many causes of the unhappy political condition of Mexico, with its sempiternal revolutions, and its endless changes of government, pronunciamientos, and political annunciations, which, after a time, grow wearisome in their murderous sameness. The reader must take Mr. E. B. Tylor's authority for the statement that, " as is usual in cathedral towns, the morality of the people is rather lower than usual." The revenues of the Mexican Church were enormous. Tejada estimated the income at twenty millions of dollars yearly, or more than the whole revenue of the State; but we are inclined to think that this estimate, which far exceeds that given by any other authority, is exaggerated. Tejada, however, explains it by the well-known fact that the priests have always tried as much as possible to conceal the wealth of the Church. When the property was confiscated it was found to be worth, even making allowances for the circumstances of the time, which did not permit of the full price being obtained, half the value of the real estate of the country. There is, Mr. Tylor remarks, nothing remarkable in this wealth. For two centuries the relations of the people to the Church remained much as they were in the Middle Ages in Europe. It was until comparatively recently accounted pious and extremely meritorious to leave large sums of money or good estates to the Church. To this day, Mexicans, in no way distinguished for their ecclesiastical zeal, and, indeed, all but heretics, through long habit and national feeling, will insert a clause in their will, leaving some nominal sum "to these charities, which, from time immemorial, it has been considered pious to benefit." Half the city of Mexico at one time belonged to the Church, a fact not difficult to credit when we remember that in the time of Philip V. half the freehold property of Spain was in the same plight. Yet at the time when our informant, Mr. Tylor, visited Mexico (viz., in 1856), the clergy, secular and regular, only amounted to 4,600, and the number has been steadily decreasing of late years. In 1826 it was 6,000, in 1844 it had fallen to 5,200; while, taking the census of 1856 as the year of calculation, the income of each priest and monk was on an average, at the lowest reckoning, £200 *per annum*. But we must not suppose that this was anything like the maximum. The pay of the poor country cures, as now, was only from £30 to £50, so that it necessarily follows that some of the higher dignitaries and monks were in receipt of very handsome incomes. Yet the profession was not overstocked, but on the contrary, year after year, became more and more deserted, notwithstanding the prizes which it had for its followers. It is difficult to account for this, unless on the ground that the Mexicans were growing more enlightened and liberal with the progress of years, or that the Churchmen *in posse* shrewdly looked forward to a time when the property of the Church would be confiscated. No doubt many, some will say a majority of the clergy, are and were men of extremely loose morals. Indeed, so evil is their odour, that heads of families not unfrequently prohibit any priest from crossing their threshold. "But we do not generally find Mexicans deterred by a little bad reputation from occupations where much money and influence are to be had for a little work." Every writer on Mexico holds up his hand in holy horror of the immoral lives of the Mexican clergy. This is not a something of yesterday. In 1625, Father Gage, an English friar, travelled in the country, and was so disgusted with their ways that when he came back to England he turned Protestant, and died Vicar of Deal. It may be added, by way of comment, that in 1626 the Romish faith was not a very popular or a very lucrative creed in these isles, and that "converts" were personages of no little consequence. Until the suppression of monasteries the monks (p. 257) led most unmonastic lives, and the clergy's piety generally is chiefly distinguished for its absence, and entire disregard of Church discipline. Many of them, for instance, are celibates only in this respect, that they dispense with the marriage ceremony. For a priest to be the son and grandson of a priest is not considered in most parts of Spanish America to be anything more than rather a curious circumstance, interesting but not sufficiently remarkable for note. It ought, however, to be allowed that there are many exceptions to this general rule. Some of the country curates are men of exemplary lives; and the same may be said of the priests of the Order of St. Vincent de Paulo and the Sisters of Charity with whom they are associated. But then it so happens that few of these, either priests or sisters, are Mexicans.* When Don Ignacio Comonfort came into power he conceived the laudable ambition of holding office for a little time longer than the usual eight months, which for the ten previous years had been the average Presidential term of his predecessors. He, therefore, as we have already noticed, tried hard to gain popularity by attacking the *fueros*, the extraordinary privileges of priests and soldiers which had survived the War of Independence, and the adoption of a Republican form of Government. Neither were, until lately, amenable to the civil tribunals for debt, or, indeed, any offence. They were judged by special courts, composed of members of their own body, who naturally administered peculiar justice to complainants and creditors. But the Mexican people were suspicious of politicians, and especially of politicians who tried to buy popularity by such specious bids as this, and accordingly failed to support him in his projected reforms. However, under Juarez, even this was managed, though it does not at all follow that the Legislature may not in time again slide backward to the old state of affairs. Yet, after all, the influence of the Pope is but slight, as his Holiness discovered when he attempted to interfere in a dispute about some church benefices. Nor is religion high. Notwithstanding the penal laws against monks, nuns, Jesuits, and the clerical garb, the priests' influence is still immense, but it seems to be chiefly among women of all classes, and both sexes of Indians and the uneducated whites. The higher class Mexican gentlemen—though nominally Roman Catholics—pooh pooh! the Church, and break endless jokes on it, and tell an infinitude of amusing anecdotes not to the credit of the priests. They do not attend church, though the young Mexican dandies have a habit of gathering around the church doors, and then, as the ladies pass between the rows of these exquisites, criticising their looks and their dress in a most audible tone. If they hold any opinions, these seem to be those of the French school of Freethinkers, and, indeed, in some respects, as far as the influence of the priests and the Church goes, Mexico is not unlike France. Tithes are now optional; but the priests are never weary of telling their congregations that all manner of dire things will befall here and hereafter if they neglect to contribute freely of their substance to the Church. There can, however, be no doubt about the piety of the poorer classes of the Mexicans—they are thoroughly priest-ridden, the schoolmaster notwithstanding; though whether the outward observance of the forms of religion and the festivals of the Church entitle them to that much-abused designation is doubtful. Indeed, if all tales are true, there is an amazing amount of hypocrisy prevalent even among these people. A ludicrous story is told how on one occasion a son of Belial, by dint of professions of great sanctimoniousness, managed to get into a church where some extremely ascetic people were scourging each other. The church was perfectly dark, and as the whip descended on the penitential backs of scourgers and scourged, whining cries, intended apparently for agony, resounded through the building. The new-comer watched the operations for some time, and then set to work and laid about him lustily. He instantly noticed that the cries which

---

* E. B. Tylor: "Anahuac," p. 287.

VIEW OF PUEBLA (FROM THE EAST)

VIEW OF PUEBLA (FROM THE WEST)

proceeded from those who experienced the strength of his arm and the weight of his whip were of quite a different character from the previous lamentations, and altogether more natural. The sequel to the tale is, that there were serious objections raised against the continued admission of this new scourger to the meetings of these good people—he was much too pious!

INDIANS.*

The Mexican aborigines, though divided up into many races, have still the marked common characteristics of the North American Indians. They are not usually a good-looking race, but the men are sturdy, and the women short and fat. In appearance they are usually rather melancholy-looking, but in reality they are a merry people, who, when at home, chat and jest till late in the night; "amuse each other with jokes and puns play tricks, and laugh." But to the Creoles and Mestizoes they are cold and reserved, having no confidence in them. The Indian is, however, remarkably without sensibility otherwise. He is tenacious of life, wounds which would undoubtedly kill a white person appearing hardly to inconvenience these swarthy races. He never has lockjaw, even though sustaining the injuries which would produce it in other men. He drinks freely; many of them are under the influence of brandy half of their lives, yet they never have *delirium tremens*, and seem not greatly affected by their potations. To nervous fevers the Indian, however, succumbs rapidly: "he neither rages nor becomes delirious, but all energy is wanting, and in a few days he expires of exhaustion." The Mexican Indians of the present day are by no means stupid, but they want the power of originality. They are good imitators, but are deficient in imagination. They can master details, but can rarely extend or expand these details. As to poetry they are totally deficient in it. Some of these characteristics are exhibited in the Aztec work, though we must remember that the origin of the Aztec civilisation is as yet unknown. There is no likelihood that it originated among them; the first elements may have come from without. Sartorius justly enough remarks that the two children of the sun of the Peruvian Incas, and Quetzalcoatl (p. 238), the white men of the Mexicans, may indicate the early influence of the Caucasian race. To the west of the Aztec country lay Hvitramannaland, known to the Icelanders as early as the tenth century (Vol. I., p. 144). Who can say how the threads of the civilisation of the East are attached to those of the West? The Indian is like all his race, full of distrust and suspicion. But this is exhibited not only to the whites but to the people of his own race. "Their salutations among themselves, especially of the women, are a long rigmarole of wishes and inquiries after health, which are repeated monotonously and unsympathisingly on both sides, often without looking at each other, or even stopping. The Indian who is desirous of obtaining something of another never asks for it directly, or without beating about: first he makes a small present, praises this or that, and at last brings forth his wish." If he has a request to make of any one he always prefaces it with a present.

* Sartorius: "Mexico," p. 63; Tyler's "Anahuac," pp. 69 *et seq.*, and private notes of the author, during two short visits to Mexico in 1865 and 1866.

Even an interview with the village alcalde, or magistrate, though he may be an Indian of his own tribe, is never entered upon without sending an *avant-courier*, in the shape of a fat hen or some such gift, to smooth the way to a favourable reception. They are admirable diplomats—never speaking straightforwardly, if they can manage to say their say in ambiguous language, so as to leave some outlet by which they may afterwards interpret the terms of the discussion to their own advantage. In dealing with them the other party to the bargain must be careful to define previously all the conditions of the contract, otherwise there is certain to be in the end an endeavour to wriggle out of it. Priests who speak the language perfectly will sometimes acknowledge that they find a great difficulty in understanding the confessions of a penitent, on account of the riddles, metaphors, and other ambiguous expressions which are used, and in which their native languages abound. Like most of the tribes from the North—and, indeed, all the native American races, the Eskimo included—they have a great repugnance to tell any stranger their names, lest they should be compromised. Rather than do so directly they will tell a lie. If questioned about a third person they will rarely tell anything. If they have, for instance, met him on the road, they will unhesitatingly declare that they have never seen him. From fear of being taken advantage of, they are distrustful, and accordingly always cheat in a small way if they can manage it, so as to get "even" with the knavery of the other dealer. Their plantations they generally try to conceal in the depth of the woods, or in secluded ravines, so as to be out of the track of man. They approach their huts by secluded paths, and if by chance a stray hunter comes upon one of their little settlements they are by no means pleased; they even resent it as an intrusion. From long subjection to the Spaniards they are now quite cowed and servile, and though numerically superior to the whites, they are too divided up by differences of language, tribal hatreds, and the want of any national pride—for their old traditions are now almost forgotten—that no fear need ever be entertained of their rising *en masse* against the other races in the country.

Many of them cultivate their own patches of land, collect pulque, or work as day-labourers, though living in their own villages instead of upon the estates of their employers. For instance, in the State of Vera Cruz it is very common to contract with the village authorities for a number of labourers. They receive money in advance, but the whole village is responsible for their due fulfilment of the contract, and the alcalde will send workmen till the debt is liquidated. Some branches of agriculture, such as the preparation of vanilla (p. 256) and the cochineal, are carried on solely by the Indians. Among their arts are admirable life-like models of the Mexican people in clay, models of fruit in wax, &c., as well as a brisk trade in ancient idols, which they dispose of to the British antiquaries for excellent prices. On the rivers they are fishers and ferrymen, and in the neighbourhood of cities they furnish wood and charcoal for the inhabitants. "Early at break of day, and till late after sunset," writes Mr. Sartorius, "one hears in the streets of Mexico a melancholy long-drawn 'Ousior,' and sees the collier trudging in with a tall coal sack upon his back, who with inarticulate croak designs to say, 'Carbon, Señor' (Coal, Sir). Whoever has not seen the canal of Santa Anita (*la vigas*) has not yet an adequate comprehension of Indian industry. Unnumbered crafts of all kinds come rowed along, from the clumsy flat boat to the light canoe which can hold one. The little Indian girls row lightly on with their double-bladed paddles. Their boat is filled with vegetables, the outside decorated with flowers, as if it were a bridal boat, and the young people are gaily laughing and singing, while the old frog-catcher paddles past with his booty, solemn as a baboon, and equally ugly. Everything hurries and rushes along towards the market: whole loads of wild ducks and strand-snipes come from the lakes; fowls' eggs, fruit, &c., from the villages, in order to supply the daily necessities of the great city. In the neighbourhood without the city [of Mexico] lie the vegetable fields of the Indians, and those "Chinampas" called the floating gardens. These garden-lands have been won from a marsh: a simple turf covering under which the water stands. On a strip of this land brush has been thrown, whilst at the sides deep ditches have been dug and the earth thrown up over it. As often as the ditches become filled, which is pretty often, the earth is again thrown on the bed. When the sod is a foot thick over the water it is planted, and the plants flourish well because they never lack moisture. These gardens rest upon insecure foundations, and totter beneath the tread; and we can well imagine that in earlier times, before the artificial draining, when the valley [of Mexico] was often inundated, such little islands may have been detached in a storm and floated off." Indeed, in Humboldt's day, though by that time the waters of the lake had been long drained off, there were still some of these artificial islands in the lake of Chalco, which the owners towed by a rope, or pushed about with a long pole. They originated and were resorted to at a time when the city of Mexico was built—like some huge collection of "lake dwellings"—in the midst of the water-covered valley, and the inhabitants were not strong enough to hold land on the shore; hence they were driven to these and other strange shifts to get food. Chalco is and was a fresh-water lake, but the mud of Lake Tezcuco was at first too full of salt and soda to be used for covering the rafts of brush and reeds which form the foundation of these floating gardens, with any prospect of the soil bearing a crop. However, by pouring the water of the lake into it and letting it soak through, the ingenious Aztecs dissolved out most of the salt. Then the island was fit for cultivation and bore excellent crops of vegetables. Sometimes the island was so large that the proprietor was able to build a hut on his floating farm, and live there with his family. The name *Chinampas* is still applied to the garden patches along the canal. "However, at present they all lie at anchor, and the wind is no more able to blow them adrift. They look right cheerful, surrounded with balsams and pinks and border flowers, and planted with plump cabbages, lettuce, and parsnips." A peculiar jog-trot is the ordinary gait of the Mexican aborigines. They are so accustomed to carry loads, light or heavy, on their backs, that if, say, on a journey with a letter, they will make up a parcel of stones ten or fourteen pounds in weight; to this the letter is added, and the whole fastened on their backs, so that they will not forget that they have a commission to execute. They usually carry all loads by means of a strap which passes over the forehead. Some of them are able in this manner to bear great loads. At the foundry of Arcas there was living, twenty years ago, an Indian who carried a weight of rock of 600 pounds, from the quarry to the smelting-house, a distance of three-quarters of a mile! It is common in the mines to see labourers carrying about a quarter of a ton to the surface, the only ladder being trunks of trees fixed slanting across the shaft, with notches cut in them for steps. Laboriously, in this fashion, they carry their produce to market for twenty-five or thirty miles over mountain paths of the most rugged character. Tired out and perhaps wet through, they pass the night in an open porch, in order next morning to offer their products for sale. And yet they will not earn for this

AN AZTEC RUIN AT TULOOM, IN YUCATAN.

labour as much as they could by a day's work in the city. Still nothing can turn the Mexican Indian from his old habits, or convince him that the little luxuries he values—soap, sugar, brandy, and frequent indulgence in strong drink—can be procured more advantageously in other ways.

The following graphic description I borrow from Mr. Sartorius. I give it in his own words, as I could not presume to reduce the picture by any condensation:—" By going out in the afternoon to one of the gates of Mexico, either that of Belen or San Antonio, or of San Cosme, one can observe the trains of Indians wending their way home to their villages. What a contrast with all that surrounds them! Splendid houses, magnificent carriages, elegantly-dressed ladies and gay dandies; and close by, these poor half-naked Indians, the men in front, the women in the rear, the children upon their backs, speaking another language, wearing another dress, and of another colour from their fellow-creatures who promenade the streets. They are trotting along towards their home, joking and laughing among themselves, caring little for the world about them, a distinct people within a people. Yonder, under the tall poplars, stands a pulqueria, a shop where their favourite drink is sold. There they must stop to take one drink more. Dense groups are standing round; godfathers are greeting each other with hat in hand, and making profound obeisance; the well-filled cup goes round from mouth to mouth, and the discourse grows eloquent in praise of the precious nectar. The wives sit on the ground and take their children from their backs, give the little one breast, and stop the mouth of the bigger one with a cocole (a kind of small, sweet, dark-coloured roll), but all the while keep an eye on the pulque bowl, if perchance the husband or a gallant neighbour should hand it to them. Want and toil and the long journey are forgotten. Louder and louder grow the assurances of friendship among the men; often and oftener they come to the wife for another quartilla (a quarter real, or about three cents), for she has the proceeds of the market done up in her girdle, and after every new drain upon it, she reties the lessening bundle with a sorrowful look. Now the mirth grows boisterous. In some groups the women begin to follow the example of the men. Here is a crowd making merry and dancing to the strumming of a farana (a small stringed instrument); yonder, the rising hilarity makes them tender; whole drinking circles embrace each other, lose their equilibrium, and fall, to the infinite delight of the others. The bestiality is now under full sail, and no one notices that the sun has already set. Jealousy brings the women in contact, who commence a vigorous conflict, tug at each other's hair, and scratch and bite one another; even the men get to blows, excited by their fellow-lazzaroni. The uproar becomes fearful, till at last the police are among them, who take the combatants to the watch-house, and disperse the rest, who, after many vicissitudes, at length reach their village, fully resolved upon a repetition the next time they go to the city." The Mexican Indians are, moreover, like all the races which have ever come into contact with the Spaniard, punctiliously polite. On two Indians meeting—say on the Chalco Canal—they will simultaneously burst out into a long string of complimentary phrases, often meaningless, and always high-flown. They will pour these forth without looking at each other, and often without stopping the canoe which they are paddling. Sometimes they will shout in Mexican, sometimes in Spanish, "How is your worship this morning?" "I trust that I have the happiness of seeing your worship in good health." "If there is anything I can have the honour of doing for your worship, pray dispose of me," and so forth. It is doubtful whether they learned these exaggerated bits of courtesy from the Spaniards. As they are so thoroughly Castilian in this formal politeness, and as completely without meaning as the low bows and complimentary grimaces which accompany them, it is probable that the Indians have only borrowed the lofty courtesy of a nation whose politeness has been defined not inaptly as " the diamond ring on the dirty finger." Mr. Tylor, however, seems to believe that it is possible that its origin dates further back. The Indian in his native village is a very different person from the Indian selling his wares in the market-place of a great town. At home they are talkative enough, but, as we have already noted, in the presence of the whites and half-breeds they are constrained. This is due not only to the fact that the whites and Mestizoes do not understand his language, but also owing to the distinct line of demarcation which long prejudice has drawn between the Indian and the rest of the population of Mexico. The whites call themselves *gente de razon*—people of reason—to distinguish themselves from the Indians, who are, of course, people without reason. Indeed, the distinction in ordinary parlance is carried still further. It is common to hear a white talking of his section of the community as "gente"—people—and of the aborigines as "Indios"—Indians—that is, not people at all. The Indian mode of talking, as has already been well remarked, shows how broad the gulf between them and the whites is. The Indian is full of quibbles, puns, and small jokes. On the contrary the Spaniard's talk is not about names, but almost invariably about things; hence the facility with which Spanish writings and conversation can be translated into other languages. The Indians' dress is usually very scanty, even in the valley of Mexico, where the mornings and evenings are rather chilly. "The men have a general appearance of having outgrown their clothes." The sleeve of the cotton shirt only reaches to their elbows, and the drawers, of the same material, end at the knees. A kind of blanket, a pair of sandals, a palm-leaf hat, and the Indian is dressed. The woman clothes herself in a kind of cotton sack, very short at the sleeves and open at the shoulders, in addition to some sort of petticoat. Not unfrequently she wears a cloth folded on her head like a Roman "contalina." But most commonly her head-dress is her own thick black hair, which hangs down behind in long twisted tails. The appearance of the Mexican Indian is fairly well given in the figures on pp. 269, 277. "His skin is brown, his limbs muscular, especially his legs, his lips thick, his nose Jewish, his hair coarse and black, and hanging straight down." The modern Aztec cherishes few of the traditions of the former greatness of his race. About the only signs of his caring anything about it is evinced in the flaring paintings outside the drinking shop. These represent Aztec warriors joyously drinking great bowls of pulque, a hint to their degenerate descendants to go in and do likewise. However, these daubs are mere copies of the French coloured prints depicting or supposed to depict the scenes of the Conquest, and though scattered all over the country do not seem less popular from the circumstance that they represent the followers of Cortes, with the faces and limbs of Europeans. They are, however, quite as like the real Aztecs who get drunk under them as the simpering shepherds and shepherdesses who hang on the walls of English cottages, or their brothers and sisters in stars, garters, periwigs, court swords, and diamond necklaces, who sprawl over the walls and ceilings of many a German Schloss, and perhaps in some palaces nearer home. We have already hinted that the Mexican Indians are not particularly honest. This is unfortunately a marked characteristic of the nation, and though it would be harsh, and probably unjust, to attribute it all to their religion, yet there cannot be a doubt that dishonesty has increased in Mexico since the Conquest, and is greatly due to the Roman Catholic system, which inculcates a belief, but allows their morals to shift for themselves. Theft has even ceased in some districts to be considered an offence worth punishing, and for long public opinion, at least among the Indians, has not included pilfering among those acts which bring disgrace to the perpetrator. To be found out is unfortunate; to escape is highly meritorious, and makes at once a man of superior talent. He confesses to the priest, certainly. But the priest absolves him, and tells him to sin no more, an injunction which he treats very much as a sacerdotal common-place, and proceeds forthwith to sin again. All the Indians working in a mine are searched as they come out, though unhappily this degrading thief-making scene can be witnessed at the gateway of many of our public works in England. A half-naked Indian might be supposed to be able to secrete very little. But long experience has taught the superintendent—generally a Scotchman, and Don Pedro McTavish is usually the most acute of his nation—that even this is possible. The labourers' ears, mouth, hair, and even less likely places are scanned by the doorkeeper, and as rich ore can be kept in little bulk, he is not unfrequently successful in finding considerable quantities of the company's property. Still, a great deal more escapes. This fact is very apparent from the presence in close proximity to the mine of little smelting works, the owners of which profess to be employed in smelting the ore from a few small mines in the vicinity. But in reality they are kept going solely on the ore stolen by the labourers at the bigger ones, and which ore is purchased from them at exactly one-half its value, a fact which the company robbed knows perfectly well, but must bear as best they can as one of the "institutions of the country." "If the superintendent," writes Mr. Tylor, from whom I derive these facts, "should roast the parish priest in front of the oxidising furnace, till he confessed all he knew about the thefts of his parishioners from the company, he would tell strange tales—how Juan Fernandez carried off thirty-four pennyworth of silver in each ear for a month; and how Pedro Alvarado (the Indian names have almost disappeared except in a few families, and Spanish names have been substituted) had a hammer with a hollow handle, like the stick that Sancho Panza delivered his famous judgment about, and carried away silver in it every day when he left work; and how Vasco Nuñez stole the iron key from the gate (which cost two dollars to replace), walking twenty miles and losing a day's work in order to sell it, and eventually getting but twopence for it; and plenty more stories of the same kind."

Never did a people more thoroughly exhibit all the signs of subjection than the Mexicans. They were enslaved, treated as such, and no effort made to elevate them. Hence they became dissimulators, thieves, and adepts at outwitting the Conqueror by cunning when they could not do so by force. However, it is cheering to find that the influence of foreigners upon them is contrary to the usual rule—good. Indeed, it could not be well for the worse. The Mexican miners have arrived at the lowest depth of vice, and accordingly, when they came in contact with the Cornishmen, the result was that they improved. They saw that the foreigners had a contempt for theft and lying, and for the first time in their lives learned that these were accomplishments not highly valued in the society of the Old World. Conscience is only a matter of education, and crime a mere question of skies. And so the Mexican began appreciably to tell the truth on very high occasions, though lying, as a well-worn garment, was good enough for ordinary uses, and if he stole, was not particularly anxious to tell of his exploits to the stalwart *Inglés* who was his fellow-labourer. Instead of squandering his wages he imitates the Europeans by putting them in the savings bank. Thus, in course of time, the influence of foreigners may exercise some appreciable influence on the Mexican character. And, in all verity, it will bear improvement—it is difficult to deteriorate it. When they

MEXICAN INDIANS OF THE TIERRAS CALIENTES.

cease to consider a steady course of dram-drinking the height of enjoyment, and standing in the sun doing nothing for hours a gentleman-like enjoyment—and a very pleasant one it is when the sun is not too hot and the scorpions not fond of the locality—then we may expect not great things, but something of *Los Indios*, and the fewer shades of paler colour above them. The Mexican Indians are, however, not apt to change. They are truly "people of unreason." They do the most absurd things—as we all do—and fabricate the rudest articles for household use and for sale for no other reason than simply because their fathers did so. They are Conservatives who decline to be educated into a new policy. The Indian makes unglazed pottery, without taste, and with not a great deal of utility either, and "packs" it on his back twenty or thirty miles to market, seemingly careless of the waste of labour and the loss of time. But time is nothing to all his serenely stolid race, and in Mexico at least, where the Indian is not quite so lazy as the rest of the American aborigines, the fact of beasts of burden having been in the country for three hundred years seems never to have occurred to him. "They cultivate their little patches of maize, by putting a sharp stick into the ground and dropping the seed into the hole. They carry pots of water to irrigate their ground with, instead of digging trenches. This is the more curious, as at the time of the Conquest irrigation was much practised by the Aztecs in the plains, and remains of water canals still exist, showing that they had carried the art to great perfection. They bring logs of wood over the mountain by harnessing horses or mules to them, and dragging them with immense labour over the rough ground. The idea of wheels or rollers has either not occurred to them, or is considered a pernicious novelty. It is very striking to see how, while Europeans were bringing the newest machinery, and the most advanced arts into the country, there is scarcely any symptom of improvement among the people, who still hold firmly to the wisdom of their ancestors. An American author [Mayer] quotes a story of a certain people in Italy [though the tale is a patriarchal one and has been told of many people, and even individuals] as an illustration of the feeling of the Indians in Mexico respecting improvements. In this district he says that the peasants loaded their panniers with vegetables on one side, and balanced the opposite pannier by filling it with stones; and when a traveller pointed out the advantage to be gained by loading both panniers with vegetables, he was answered that their forefathers from time immemorial had so carried their produce to market, that they were wise and good men, and that a stranger showed very little understanding or decency who interfered in the established customs of a country. I need hardly say that the Indians were utterly ignorant; and that this of course accounts to a great extent for their obstinate Conservatism." In the northern part of the country, from long association with the whites and Mestizoes, they are docile and inclined to work a little. But their ideas are no more advanced than those of their brethren in the *Tierras Calientes*. They move in the same round as they did 500 or 3,000 years ago, with this addition—that they are now frequently stupefied with brandy, which could not possibly have been the case at that date. The Mexican Indian is perfectly free. Under the republic he is now a citizen who controls his own affairs. But until he has a stimulus to do something more than eternally vote, very little good can come of him. He does not care so much to work on the haciendas as he used to do in the days of his tutelage. Why should he? He does not know what to do with the money. He cares nothing, like the Mexicans and half-breeds, for fine clothes. A few shillings' worth of cotton will amply furnish his wardrobe. He is never seen on horseback, so that a mustang, a silver-mounted saddle, or silver spurs with companalleros—little bells—are nothing to him. He accordingly spends his substance in riotous living, or if of a saving turn buries it in the earth. Having therefore so little need of money he fails to see the beauty of the Ranchero's logic that it is the duty of an Indian to labour on a farm for good wages. His little cane-walled, palm-thatched hut is enough for him in the way of house, while his garden patch will supply him with abundance of food, and the sale of the surplus with tobacco, clothes, and pulque—if, indeed, he does not have that "within himself" also. His worldly effects are few. He has—to enumerate them—a *metate* for grinding down the maize to be made into tortillas, a few calabashes for bowls, cups, and bottles; a palm-leaf mat (*petate*) or two makes a bed, and some pots of unglazed earthenware to serve as cooking utensils. A wood fire in the middle of the floor does duty as a stove, while a chimney is a superfluity in a house which is a mass of openings surrounded by sticks. If the family possess any surplus clothing, one box certainly can contain it. But probably they carry their wardrobe on their backs. In addition, the Indian farmer has a wooden spade, a hoe, some sharp stakes to make drills with, the machette, or iron bill-hook, which serves as axe, pruning knife, and if needs be other purposes also. "A knife," quoth the Spaniard, "is good for chipping bread and killing a man," and in Mexico the division of labour is about the same. The women weave cotton cloth in the simplest form of loom, using *malacates*, or spindles, much the same as those which their fathers did, and which again were almost identical with those of the lake men of Switzerland. In many parts of the country the Indians live under their old caciques—descendants of noble families who did not "come in with the Conqueror," but were in Mexico when he came. The authority of these men the Spaniards found it convenient to keep up, and to employ them as collectors of the revenue, and as agents for the oppression of their fellow tribesmen, an office for which, in accordance with the universal rule in such cases, they showed great aptitude. One other institution of the Mexican Indians and—as this is not an ethnological work—I have done. That is the *Temazcalli*, or vapour bath. It is identical with the bath of the same name used in Russia. It is a kind of oven, into which the bather creeps and lies down. Hot stones are placed in it and water poured on them. Then the Indian steams. After he feels sufficiently flaccid he creeps out, and either springs into the adjoining stream or lake, or has cold water poured on him. This institution extends as far north as British Columbia, and is found through all the intermediate country, though not to the west of the Cascade range, so far as my observations go. Some of the Indians make good soldiers, especially the Pintos, or "Painted" Indians of the *Tierras Calientes*, bordering the Gulf of Mexico. Women and children generally travel with the army, and the slaughter of these wretched creatures during the endless Mexican wars is something frightful.

Mexico is not thickly populated. The *Tierras Calientes* we can understand having only fifty people to the square league, for, from its physical characteristics, it is unhealthy. But why the lower temperate region should support only 100 people to the league seems difficult to comprehend. Here Nature seems to put forth all her resources for the support of man. The banana grows plentifully and almost without cultivation. The Indians can and do live upon it, and an acre of bananas will support twenty times the number of people an acre of wheat will. Then there is the *Yuca* (which yields cassava), rice, the sweet potato, yams, and maize. Yet even here, contrary to that theory of political economists generally accepted, there is far from a teeming population. How is this? Mr. Tylor tries to solve the problem. At the time of the Conquest the population seems to have been greater than now, and before that date even still more numerous. No doubt many were destroyed in the Aztec wars, but the Spaniards killed few; and though the wars of Independent Mexico have injured the inhabitants of the plateau, they have not greatly affected the people in this part of the country. Finally, after exhausting all explanations which might seemingly account for it, we must come back to Sartorius's solution, if solution it be. He considers the evil to be principally in the diet and habits of the people. The children are not weaned until late, and then allowed to feed all day without restriction on boiled beans, maize, or whatever vegetable may be eaten by the family. Little exercise being taken, the children get pot-bellied and scrofulous, and many die young, while those who grow up have their constitutions impaired. They lived in communities, and intermarried, so that diseases and weakness became hereditary. Besides this, habitual intemperance still further debilitates their constitutions. We need therefore be in no way surprised to find that the ordinary epidemics of the country—cholera, small-pox, and dysentery—carry them off by thousands. As is the case among all the North American Indians, whole villages have been depopulated by these diseases. In the last century, sometimes as many as 10,000 or 20,000 would be carried off at once.*

Mestizoes.†

In Mexican official classifications three classes of the population are mentioned, viz., whites, Indians, and *castas*, or castes, though the law recognises no distinction of colour whatever. Yet custom, which is everywhere more powerful than law, has ever recognised an aristocracy of race, just as in older countries there is an autocracy of birth. The *castas* comprise all the coloured population, from almost white to the darkest brown; the negroes even are included under this designation, but not the Indians. Hence the classification which we have given. The negroes, it may be added, are few in number, and almost exclusively confined to the sea-coast districts. There they maintain themselves by marriage with other pure-blooded negroes, and by the immigration of free blacks from Louisiana, and the Southern States generally, and from Cuba, and other West Indian islands. The mulattoes, originally the descendants of a white father and a black mother, are now made up of the various interminglings of mulattoes with negroes, Indians, Mestizoes, or mulattoes among themselves, so that the original name no longer suffices to distinguish them. Roughly speaking, when a Mexican is seen with crisp woolly hair, flat nose, thick lips, coarse hands and feet, and other marked African characteristics, he is at once classed as a mulatto, china, or woolly head. Like the negroes, they are chiefly found near the coast, and are a class of little consequence in the community. Very different is it with the Mestizoes, or Mestins (p. 273). They are the

* "Races of Mankind," Vol. I. p. 220; Vol. III. p. 207. Pimentel: "Cuadro descriptivo y comparativo de las lenguas indígenas de Mexico" (1862). Orozco y Berra: "Geografía de las lenguas de Mexico" (1864). Kingsborough: "Mexican Antiquities" (1831–16). Mayer: "Mexico," &c. 1852.

† Sartorius: "Mexico" (1859); Tylor: "Anahuac" (1861); Geiger: "A Peep at Mexico" (1874); &c.

offspring of white fathers and Indian mothers. They began to appear soon after the Conquest, and in the course of three hundred years have increased to such an extent as to form a very considerable section of the population, and have figured very extensively in the endless revolutions of the country. No doubt most of these revolutions have been instigated by the whites, but the fighting men have generally been the Mestizoes, for the

MEXICAN MESTIZO LADY AND MAID.

Indians never take sides in politics, if they can possibly avoid that disagreeable contingency. Naturally the Mestizoes have got very much mixed among themselves, and there are all shades of colour, from the brown-skinned youth, who is more Indian than white, to the swarthy lad who is more Spaniard than Indian. Like most hybrid peoples, they fail to partake of the good qualities of either race. They have a great deal of the absurd pomposity and meaningless pride of their fathers', and little of the dignity and patience of their mothers' people, but they have all the frivolousness and love of amusement characteristic of both races. A Mestizo will at any time desert the most important work for pleasure of any description. He is, moreover, the true Mexican. The Creole is fond of his ancestors, the Indian goes on as did his father before him, but among the Mestizoes whatever is original and peculiar in Mexico is seen. "Among the traders," around the country churches on the Sunday, writes Sartorius, "one distinguishes at the first glance the Mestizoes from the Indians, by their dress, complexion, and language. The Mestizo is also clad differently from the Creole, who imitates the European fashions, but the garb of the former is not ill-looking, and is particularly becoming to the horseman. In the warmer regions the Mestizo wears on Sunday a white shirt, carefully plaited or embroidered, long wide trousers of drilling or various-coloured cotton stuff, fastened round the hips by a gay silken girdle, brown leather gaiters, a broad-brimmed felt hat, and a parti-coloured covering thrown over the shoulders for a mantle. He never cumbers his neck with a cravat, his shirt is generally open, and upon his naked breast a small rosary or a scapular is seen. The peasants, or rancheros who live upon farms, usually distinguish themselves from the villagers by means of the calzoneras, a sort of trousers, left open on the outside from above the knee, and through the opening the broad white linen drawers appear. They also gird themselves with a straight hanger, which is in no case omitted. The village artisans, tradesmen, and mule-owners (arrieros) wear upon holidays a calico or tick jacket, but dress like the rest on other days. The Mestizoes of the table-lands wear everywhere the calzoneras, of cloth or brown buckskin, and set off with many buttons on the sides. A sort of riding-boot of pressed leather, fastened above the knee by a coloured knee-band, protects the leg from thorns. The colder climate demands also warmer covering for the body: a cloth or leathern jacket is consequently worn, the latter often richly ornamented in front with silver buttons, needlework of bright-coloured leather, and the like. A coloured silk handkerchief, loosely tied about the neck, and the woollen mantle [serape], *i.e.*, the great carpet-like garment, resembling a herald's mantle, hanging down behind and before, complete the defence against cold. An embroidered band of pearls, fur, gold, or silver-work about the hat is never wanting." The small landowners, scattered peasants, and shepherds are nearly all Mestizoes, and for this reason, as well as from the fact that they are more numerous than the white Creoles, they constitute the "heart of the Mexican nation." Still, to be a Mestizo is not honourable, and accordingly, whenever they possibly can, they are fond of passing themselves off as Creoles. In the town this is especially the case. Thus, though many of the citizens are Creoles, yet the Mestizoes have had ambition and energy enough to compete manfully with them in arts and even trade, which was at one time entirely in the hands of the whites. The arrieros, or muleteers, who constitute such an immense body in Mexico, are also usually Mestizoes, and among a nation so little distinguished for honesty as the Mexican, it speaks much for their character when a merchant will commit into the charge of these arrieros gold and silver ingots, and rich ores and coin, without exacting any other security than a bill of lading. The Mexican muleteer, a familiar figure to any one who has ever travelled in the region from Panama to British Columbia, leads a toilsome life. He has to be travelling over the rugged mountain paths, or over the torrid plains, at all seasons of the year. His first duty in the morning is to load his beasts; his last at night to unload them. He almost invariably sleeps in the open air, and has to prepare his frugal repast by the camp fire where he bivouacs. Yet he is the most cheerful of mortals. The tinkling of the "bell-mare" which leads his "train" may be heard in the quiet valleys or among the mountains mingling with his cheery song, or his not unmusical objurgations to his animals. For, it may be remarked, that long association with brevet-asses has not improved the manners of the arriero, whose conversation, especially to his charge, is usually more forcible than polite. No sooner has he eaten his supper and fed his beasts, than he begins dancing around the fire to the sound of the jarana, accompanying the dance with lively songs. The labourers in the mines, and the male and female domestics in towns and villages, are also usually Mestizoes. Indeed, it may be safely affirmed that in all employments which require intelligence, the Mestizo has the preference, the Indian only performing the mechanical drudgery which the half-caste superintends or plans out. They have also in some cases attained to good positions in the public employ, and in professional life Mestizoes may be seen on the Bench, at the Bar, in the officer's uniform, and until late years in the monk's cowl. His position is socially fixed a good deal by the degree to which his blood and colour separate him from the Indian, and the tendency of the Mestizo is always to cling to his father's race. This is natural. The white is the superior being, and the most honourable in the State. The Indian is the lowest and least intelligent. Moreover, the Indian never intermarries with the white, and rarely with the Mestizo, whose ambition it is to contract an alliance with the Creoles. The Indian also detests the Mestizo as the living witness of his daughter's shame, and the hate is returned with interest. Hence, though the Mestizo class will, owing to the numerical superiority of the Indians in the country, and their immense preponderance in some parts, increase, yet it is not likely, as is the case in some other parts of North America, that the Mestizoes will ever sink lower than they are. They will not become Indians, but on the contrary will in successive generations attain nearer and nearer to the social position of the Creoles. Like most of the Spanish race, the Mestizo, in towns especially, is a most determined gambler. He can scarcely wait for his wages in his eagerness to venture all at the gaming-table; and if on Saturday his wife does not manage to secure part, the Sunday may be a hungry day for the "punter's" family. Experience has no effect on him. Time after time he will lose all. Saturday after Saturday he will skulk home in the twilight perhaps without shoes or trousers. But on Sunday he will sing and dance as if nothing had happened, and will cheerfully go to work on Monday, with the cheap luxury of hope, and sanguine of again retrieving his fortune at the gaming-table when pay day next comes round. But though he has many of the worst vices of both races whose blood runs in his veins, the Mestizo is not without many good qualities. He is less effeminate than the Creole, and possesses more decision of character, elasticity, and capacity for toil. Hence he easily accommodates himself to every privation, and overcomes difficulties at the sight of which the Creole would wince, and most likely shrink. In his family he has also something of the old patriarchal disposition of the Spaniard of Don Quixote's day. He is hospitable to all, respectful to old age, and kind to his servants. He learns quickly, is the most practically-minded of mortals, and in private life conducts himself with "ease and propriety." This at least is the opinion of Mr. Sartorius, whose long residence in Mexico as a planter entitles his opinion to every respect. He is certainly passionate, ambitious, and apt to be troubled with passions which allows nothing to stand in the way of their gratification. But his anger is short, and when it disappears it leaves no dregs of sullen vengeance behind. In this respect he is unlike either father or mother, in whom anger will sleep for years until it can be cooled on the unfortunate object of it by the most cowardly vengeance. Gambling is indeed the bane of Mexico. But it is also about the only stimulus of the lower classes to work. "Let them play," was the remark of the owner of a large estate to one who was lamenting this disposition in his labourers. "Let them play, for it is only by this means that we have labourers. If these men were to save their earnings by leading a regular life they would soon become independent." The "fandango," or dance, is another of the Mestizo's passions. A few rockets thrown up is the signal that this amusement is about to commence. Then no business will suffice to restrain the dancers: all gather—old and young—to share in the merriment. The dance is always accompanied by singing, and these songs, which all improvise with remarkable facility, afford excellent opportunities for love-making. Indeed, "love and jealousy, assaults and evasion, anger and reconciliation," are their theme, while the dances afford about the only opportunity which custom allows for the young Mestizo to court. A young girl is always watched; and Spanish etiquette—not strictly obeyed, it may be added—forbids a young man to speak to her when she is alone. Meetings are generally arranged on these occasions in a few hurried words as the couples in the dance face each other, and if meetings in secret are not practicable, elopements are planned, and carried into execution when a convenient opportunity and a dark night supply the fitting place and time. All classes are exceedingly lax upon this point, and since the priests or monks not unfrequently set the example of levanting from their cure or monastery with the wife of one of their parishioners, it cannot be expected that the scholars will be very mindful of the duties of morality. Cock-fighting is among all the Hispano-Americans a passion. Clerics and laymen alike indulge in the amusement, on week-day and Sunday, but especially on the latter; and large sums are staked, won and lost on the result of the fowls' prowess.

It is not difficult for the traveller to become acquainted with the Mexicans. For though in the large towns there are hotels, and in the smaller ones caravanserais where accommodation for man and beast can be had, yet it is usually of the most primitive description, and in the villages he is dependent on the hospitality of the Creole or Mestizo, a hospitality which is not only never denied, but which is usually extended so frankly that the visitor to Mexico cannot fail, in spite of all their weaknesses and even vices, to remember the Mexicans with something akin to affection. Next to his ladylove the Mestizo thinks most of his horse; perhaps now and then he thinks a little more, supposing the quadruped should be of a very superior description, and the biped of a little less than average merit. No expense is spared by the wealthy Mexican on his horse and its trappings. In his *ménage* the Mestizo resembles to a great extent the Indian. His house, though somewhat more comfortable, is almost as simple. He usually takes his meals in the kitchen, detached from the rest of the dwelling, in order that he may have the maize bread hot from the fire. He does not sit at a table, but takes his plate on his knee, and in eating uses no knife or fork, though meat constitutes a great portion of his diet; the food of the

Mestizo thus differing from that of the Indian, who lives almost entirely on vegetables. The other manners of the richer Mestizoes are simply those of the Spaniards, while the poorer ones live very much as do the Indians. This might be expected, since the half-caste constitutes the link between *y gente de razon* and *y gente sin razon*, though

INDIANS DRESSING ORE FOR THE CORRALITOS SMELTING WORKS.

much bitterness and heart-burnings are the result of the struggle to keep from sinking into the *Indio* on the one hand, or inability to rise into the *gente* on the other.

THE MINERS.

Though the miners are drawn from all classes, yet, as their habits are rather peculiar, I may give a few particulars about them derived from the same sources as those to which I am indebted for the preceding account of the Mestizoes. In the mining districts mining is the all-important occupation of life. The miner knows nothing save mining, and cares for nothing else. Agriculture he looks upon with contempt, and prefers a life in darkness with the chance of gain to one in sunshine with the certainty of competence. All classes take a chance in the mines. The exaggerated stories of the fortunes disinterred influence the minds of the people in the mining towns, and the news of a lucky strike acts like an electric shock upon the excitable inhabitants, who calculate the value of money in proportion to the immediate enjoyments it can command. There are in Mexico a class of people who occupy themselves solely in seeking out new mines. They are known as *cateadores*, openers, or ore-seekers, and correspond somewhat to the "prospectors" in the Californian, British Columbian, or Australian diggings. The occupation of looking after mines becomes in the end a sort of passion. All regular occupations are neglected in the pursuit of this attractive species of gambling. Yet the cases in which the *cateador* is successful are very few indeed. Not that he ever confesses this fact, the *cateador* having always a mine to sell, just as the "prospector" in California knows that he could get "250,000 dollars sure, sir," if he only could get that ledge in Coyote Cañon brought under the notice of a sufficiently verdant capitalist in New York. In due time the *cateador* comes into the nearest town provided with a bag full of ore. He has at last discovered the mine that is to make the fortune of everybody who chooses to invest in it. But he is not too generous. He will give his friend, the keeper of the wine-shop, or the thriving grocer over the way, the first chance. The prospects look bright. They provide a few hundred dollars and some provisions, and the *cateador* "denounces" the locality and registers it. But the frijoles and maize meal are soon eaten; and money, which proverbially hath wings, is, in the development of a silver mine, particularly strong of flight. The partners of the *cateador* tire of this drain on their resources, and bitter with disappointed hopes decline to supply more funds. Other parties are more sanguine, but in the end have no more reason for it, and at last even the *cateador* gives in, certain, however, that but for the want of a little coin one of the most promising mines in all the Estados Unidos de Méjico has been lost to a *peso* loving world. On the other hand, there have been lucky hits, and few though they be, the history of some of them affords a curious commentary on the ups and downs of mining life. For instance, some forty years ago, there lived in the town of Tasco an "amalgamator" of ores named Patino. He was an adept at his business, but only able to make a moderate livelihood. One day an Indian—Miguel Jose—who supplied him with charcoal, laid a bag of silver ore in a corner of the smelting shed, and asked him to analyse it. The ore had been picked off an out-cropping ledge near the Indian's hut in the woods. As Indians were always bringing him worthless rock he paid little attention to the charcoal burner's request. At last, pestered by his importunity, he agreed to assay it. To his astonishment he found the ore, if not very rich, at least of a fair quality. The end was that Patino and Miguel Jose raised money to work the mine in partnership. It proved wonderfully rich—so rich, indeed, that the two proprietors in a short space of time were worth 3,000,000 dollars. This wealth they squandered in an absurdly wasteful manner, for the miner can bear ill-fortune better than good luck. Patino lived in the most extravagant manner; while Miguel Jose was not backward in his eagerness to spend his sudden riches. He built a fine house, stocked it with fine furniture, and though he had probably never hitherto mounted a horse, bought several splendid ones, and shod them with shoes of the metal which was "making the mare to go" at such a satisfactory speed. But he was not altogether without foresight. He had a kind of presentiment that some day fortune might turn. He, accordingly, in the midst of his riotous living, established a retail business, which he conducted profitably. And true enough, in due time came the wreck. The mine gave out, and as the proprietors had saved nothing, their wealth gave out at the same time. The end of it was, that Patino died in great poverty, while the extravagant, though not altogether improvident Miguel Jose, lived in comfort, surrounded by the remains of his former splendour. A similar tale might be told of a score of other mines. Everywhere throughout Mexico can be seen deserted mines which were once very rich, and close by the ruins of a palatial mansion, formerly the home of the owners of the vanished El Dorado. As are the masters, so are the workmen—they rarely save anything. They labour to gain money to toss about recklessly at the gaming-table, in the cockpit, at the wine-shop, at the tailor's, in the dance-house, or in buying presents for the swarthy Dulcineas of Silver Land. In Mexico all the old stories of Californian or Australian extravagance might be duplicated. For instance, a miner "flush" with a few weeks' wages, rolled into a shop in one of the cities. Being told that a piece of gold lace which he had prized was too dear for him, in bravado he bought it, and, cutting it in pieces, tossed the fragments into the street. The manners of the miner also smacks of his rough trade. The agriculturists are usually polite, if not polished. The miner is neither. He is almost invariably insolent and arrogant, quarrelsome with his "mates," and fond of squabbles about his wages with his employers. Seclusion seems to act as a hardener upon his character. Like his fellow craftsmen all over the world who live under ground, he is full of superstition. He believes in mountain sprites, pixies, and presentiments. Though not inimical to the fair sex, he will on no account allow them into the mine, even as visitors, lest their presence should bring "ill-luck," and cause the ore to disappear. His faith in charms is firm, and his belief in the saints scarcely less so. He is always vowing tapers to Santa this, or our Lady of that, and is a steady patron of all Church festivals. Nor is he illiberal to the priests. Every Sunday he bestows half a real (3d.) on the clergy as an acknowledgment of the mass said for his benefit, and a softener to the dispenser of absolutions and sin-forgiveness. After having done so, he goes and promptly sins again. When we add that he is jealous to a degree, even remarkable in a Mexican, and not always careful of keeping himself clear of the sin he so ruthlessly avenges in others, it will be evident that the Mexican *minero* has good need to keep on good terms with the Church. As always happens in these cases, the morals of the women in the mining villages are very lax. Education is neglected; the boys take to the fathers' business as soon as they are able to do any work, while the girls follow only too closely in their mothers' footsteps. We have already more than once reiterated our belief that notwithstanding the long period during which Mexico has been settled, the mining resources of the country are not by any means exhausted or even all known. To this day new and paying mines are frequently being discovered. The "stampede" to the fresh mining locality is of the same nature as those "mining rushes" with which all readers of the accounts of gold countries must be familiar. If the find proves worth anything, a village soon springs up, and a scene of riot and debauchery ensues. If it fails, then as quickly the disappointed *cateadores* and their camp followers disappear.

THE RANCHEROS.

"De labrador a minéro, gran majadero; de minéro a labrador, gran Señor"—from farmer to miner, a poor devil; from miner to farmer, a noble gentleman. This familiar Mexican proverb expresses the national appreciation of farming. Agriculture is, indeed, looked on as the profession most to be desired. The flower of the country are engaged in farming. The ranchero, or farmer, is proud of his calling, and happy if his son follows it. He can scarcely be called a peasant, even in the sense of a peasant being a small landed proprietor, for no peasant in France, Germany, or Denmark is half so independent of lord or seigneur as the Mexican agriculturist. He is ignorant and superstitious, because he has few opportunities of learning; he is hospitable, frugal, attached to old customs and old habits, simple in his ways of life, not often a gambler, but apt to be jealous, and when drinking and dancing have inflamed him not very particular as to the course he may take to soothe these by no means "gentle passions." He has many of the characteristics of the Andalusian from whom he sprang. He loves his wife and his children, and he loves his own horse in particular, and all the equine race in general, to a degree which shows that even in America the Arab characteristics of his Moorish forefathers are not eradicated. When the *conquistadores* came to Mexico they parcelled out the land among themselves just as the Conqueror did in England. The Indians remained on the land as serfs, but were allowed, however, to cultivate the soil which they held by payment of rent to those who had received the allotment of it. After a time 600 yards round every church were decreed to be public ground. Hence to this day the fairs and markets are held in such localities. By-and-by, when immigrants arrived in the country, the holders of these great estates sold parcels to the new arrivals, or when the lands happened to have been entailed, they let them on inheritable leases; hence to this day some magnificent farms in Mexico are held at a merely nominal rent. Some of these estates—such as those seized by Cortes—were as large as a German Dukedom. But even the smaller grants were often so large that only a small portion could be cultivated by the proprietor, who, to obtain some revenue, let the rest to small farmers, who paid their rent chiefly in produce. Then there were estates belonging to the Church, which have since been confiscated, and which were managed on much the same principle. To this day much the same system prevails, though year by year the great holdings are getting further and further reduced. These large estates are called *haciendas*—if for agriculture, *haciendas de labor*, and when used for cattle-breeding, *haciendas de ganado*. The latter are often fifty miles square, and are usually divided off into several small establishments, or *estancias*, under one general management. The smaller farms are called *ranchos* (in Spain *cortijos*), and their holders *rancheros*, a general term applied to landowners and farmers. The owner of a hacienda does not necessarily reside on his estate. He is usually a wealthy man, with a town house, and periodically visits his property in an ancient kind of carriage, filled inside with himself and family, and laden exteriorly with beds, bedding, chairs, chests, and other household furniture, and usually accompanied by a troop of horsemen. Most of the older farmhouses look as if they had been built soon after the Conquest. They are castellated, and have high walls, turrets, and battlements capable of defence. All the windows are firmly secured with iron gratings, and the gate—of the farm-houses of the plateaux—is secured with iron bars. Under the high walls lives the master with his confidential servants. Outside there is usually a little village composed of the Indian labourers' huts. The book-keeper manages the accounts of all business which requires writing. The major-domo, an old and trusty servant, takes the general superintendence of everything. To him the

INTERIOR OF SMELTING WORKS AT CHIHUAHUA.

labourers must look, and after them he looks very sharply. He rings the bell in the morning as the signal for them all to assemble. The roll is then called, and a hymn sung; after which work proceeds. At night he reports progress to his master, instructs the clerk as to what should be written down, and receives his orders for the next day. At every farm there is a chapel, where mass is read every Sunday, and a shop where the labourers can supply all their material wants, except that of a dram, though liquor can be bought in large quantities. The Mexican field "hands" are invariably men, except during the maize season, when the women are actively employed also. The *rinda* is, indeed, the signal for holiday. This name is applied to the last ear of maize that comes from the field. It is dressed up with ribbons and flowers, and conveyed in a triumphant procession to the master's house, as a signal that the winning of the crop is over. Of course a dance or some bottles of brandy are the sequel to this agricultural "harvest home." The labourers are paid wages and board themselves, though they usually receive a ration of maize or pulse. It is in the *Tierras Calientes* that the small ranchero is usually met with. He cultivates his land with the aid of his family or a labourer or two, and when he has harvested his own crop will occasionally assist the larger *hacendados* with theirs. These people, according to Sartorius, who, however, wrote some years ago, can seldom read or write, and receive no instruction whatever. When they intend marrying, they must know part of the catechism by heart, and are, therefore, when the time comes, examined up to that mark. "They are not fond of hard work; nor have they any need of it, as they have plenty to live upon, if they devote only a few hours a day to agricultural labour. They are good hunters, know the haunts of the deer and wild boars, and track the wild turkey. The men tan the deerskins remarkably well, dye them, and make their clothes of them; the women spin and weave cotton. During half the year, there is little or nothing to be done in the field. The chase is then attended to, or the fibres of the long-leaved *Bromelia pita*, or of the maguey, are prepared, or cordage or ropes made of it and sent to market. In other localities they collect copal, storax, and Peruvian balsam, the fruits of the oil palm, pimento, or vanilla in the forests. Many days, however, are passed extended on the mat, playing the guitar, sleeping, or staring up at the blue sky. The report of a festival in the neighbourhood, however, electrifies them; the prospect of a fandango makes dandies of them. They bathe and anoint themselves, and are then indefatigable in song and dance."

The Cattle Breeders and Herdsmen.

In the rocky parts of the haciendas, and to a great extent in the northern parts of the country, and in the warm coast regions, cattle rearing is greatly followed. This is necessary, owing to the scanty population or the unfitness of the soil for agriculture. But the Mexican also likes the work of rearing and tending cattle. The cattle cost little trouble, and obtain their food without distressing their owners' muscles. Moreover, herding enables the lazy ranchero to obtain abundance of food without toil, and to indulge without stint in his favourite amusement of being on horseback all day long. In agricultural districts where cattle are kept they do much damage to the crops, and though in most cases the farmer is compelled at great expense to fence around his land, even this is insufficient to prevent cattle and even pigs from breaking in. No doubt the cattle owner is held responsible, but in the thinly-peopled districts this is impossible. He is there expected to look after his property. Each herd of cattle on a hacienda is under the charge of a mounted vaquero, whose business is to attend to the 500 or 600 cattle under his charge. He has often to pursue stragglers, and lasso (p. 145) one which has been wounded by a jaguar or a thorn, and extract from its sides the eggs which the flesh-fly has laid in the wounds. "At full gallop he pursues the flying animal, casts the noose about its neck, quickly turns his horse's head, and drags the struggling prisoner to the nearest tree, to which it is soon bound. In a moment he has dismounted, has cast a second noose about the hind feet, and with one jerk the heaviest beast is extended on the ground. The head and fore feet are quickly tied together, and now the surgical operation can be performed at leisure." The vaqueros are all magnificent horsemen. It is to a fresh arrival something wonderful to see how they will gallop at full speed up or down the rockiest path, or under low trees through dense scrub lying on the horse's neck, and yet at the same time cast the lasso with the greatest precision. They are very proud of their skill, and when excited by emulation or the applause of onlookers will perform on horseback feats which could not be surpassed on the docile steeds ambling round the sawdusted path of a circus. All his cattle know the vaquero, and when he appears in the pasture and cries, "*Toma, toma!*" (take), they run after him in crowds. The bait which he uses to attract them is chiefly salt, a bag of which he invariably carries at his saddle. Some of this he occasionally strews on a large stone, or drops some on the tongue of a cow. His reputation as a salsiferous individual is soon secured, and henceforward he is master of his herd. Except in the vicinity of the large towns butter or cheese is rarely made on the haciendas. No cow, moreover, will allow itself to be milked without the calf. Hence, when a freshly-arrived European is told at a farm that "no calf is tied up," he need trouble himself no more with requests for a draught of fresh milk. Some of the coast estates will possess as many as from 10,000 to 20,000 horned cattle. There is in these districts considerable difficulty in getting a sufficiency of vaqueros; accordingly the cattle frequently run wild. At the sight of a man they will career over the plain like deer, and as they have not been accustomed to get salt, considerable stratagem is necessary before they can be approached. In order to be able to obtain possession of them, a number of tame cattle are kept on every hacienda. These cattle are known as *cabestros*, and are used as decoys to the wild ones, which are easily secured. But the great event on a hacienda is the *herradero*, or annual branding of the young beasts. This is a season of rejoicing, and is looked forward to by the country people for months in advance. Few of the herdsmen can write; hence they kept stock of the calves born with a kind of tally. This consists of a broad untanned strap, on one side of which the male calves, and on the other side the females, are registered by notches. Another strap keeps account of those which have died or been sold. These primitive ledgers, which are kept with a surprising degree of accuracy, are handed over in the autumn, when the hacendado makes up his accounts. Meat is the cheapest of all food in those parts of the country where cattle abound. Hence it is extensively eaten, chiefly in the form of *sexina*, or *tasajo*. This favourite Mexican comestible is prepared as follows:—The flesh is cut in long strips, finely powdered with salt, and sprinkled with the juice of lemons. It is then wrapped up in a hide over night. Next morning it is taken out, and as soon as the sun is high enough is hung in lines and thoroughly dried by the hot air. It is afterwards packed in bags and sent to market. It is easily prepared, and is, moreover, savoury and wholesome, and vast quantities of it are consumed in the country. "The herdsman is a confidential man, and his position much esteemed among the country people. The chief herdsman, who has to superintend several vaqueros, is termed mayoral, or caporal. He must be the proprietor's right hand. He attends to the sale of the cattle, supplies the herdsmen with salt, visits the different *hatos*, and controls the herdsmen. He is consulted in all matters relating to the herd; he knows whether certain operations are to be performed during the crescent or decrescent moon; he is the only one who knows how to cure the diseases of the animals; he is an excellent horseman; he selects and breaks in the young horses; he is not only minutely acquainted with the

PLAZA OF GUADALAJARA, IN THE STATE OF JALISCO.

theory of the saddle, but can also alter those which are faulty, so that they may not press; he shoes horses extremely well, and speaks about the good qualities of a horse, about the best races in the country, and other distinguishing characteristics, more learnedly than a professor. It is amusing to hear one of these mayorals speak, for he is the living chronicle of the whole neighbourhood, and is acquainted with the genealogy of the biped and quadruped races from the most remote periods. The reader must imagine a tall, broad-shouldered man, with sinewy arms, bare breast, and sunburnt face, but evidently a white Creole. The grey beard leads us to suppose him about sixty years of age; but the eye is full of life, and he manages a restless young horse with the same indifference as if he were seated on a block of wood. The vaquero's conversation smacks—not exactly of the stable, for that is a building with which neither he nor his horses are acquainted—but decidedly of horses. His similes are all derived from it, and his phraseology is so full of expressions derived from throwing the lasso, the neighing of colts, and the running of horses, that it is sometimes difficult for one less familiar with his favourite animal

AN EVANGELISTA AND HIS CLIENTS.

than he to understand his meaning. Still, there is not much democracy on the 'runs': the Spanish character is too dignified for that. The vaquero is not only the herd-man, but the faithful servant of the house. When his master travels abroad he accompanies him as an escort, at the same time consulting with him on business affairs as earnestly as if he were one of the family. If the master visits the herdsman at his hut he is entertained with the best that the establishment affords. There comes forth fresh milk, fruits, and wild honey; a kid is slaughtered; and should there be many guests, a calf. The best pieces are roasted on little spits at the fire; while the women crush maize and bake bread. The meal is taken seated on a mat, without knife or fork, the bread being handed round in little baskets of palm leaves. However, on these occasions the herdsmen do not eat with their masters, but respectfully wait on him. After the meal, business is spoken of; the number of milch kine, the fat oxen, and the state of the pasture commented on. The master examines the list of births and deaths, and has the certificate of death handed in, namely, a piece of hide with the owner's brand, or a slit ear. The herdsman has a story to relate about every death; how, guided by the flight of the vultures, he found the carcase, which had been bitten by a palanca (*Trigonocephalus*, a highly poisonous snake), or bore the marks of the sharp teeth of a jaguar. He then describes the chase of the wild beast, praises the speed of the horse, and the boldness of the dogs, who wag their tails on hearing themselves named. The master often remains all night with his vaquero, who then prepares a couch for him of mats covered with deer-skins and soft sheep-skins. The gay serape, which he wears on festive occasions only, is spread over all, and the chinks of the wooden wall are hung with bullocks' hides, in order that the draught may not inconvenience the honoured guest." *

CHAPTER XVI.

MEXICAN CITIES AND CITY LIFE.

THE visitor who for the first time sees a Mexican town can by no chance mistake its Romaic physiognomy. If he has ever been in Cadiz, in Madrid, in Naples, or even in Rome, he will see the impress of the Southern nations in the town he is approaching. There are the straight streets, the open plaza, or square, the heavy stone houses, with flat roofs, the endless churches, with glittering cupolas, "far-extending, citadel-like cloisters, mounts of Calvary, magnificent aquaducts, like those of ancient Rome—splendour and luxury on the one hand, filth and nakedness on the other." Here, as in the two Castiles from which the Mexicans obtained their modern architecture, we notice the same absence of trees, and the same lack of beautiful parks and gardens, the place of which is ill supplied by dirty and unpleasant environs.

THE TOWNS OF OLD SPAIN AND NEW SPAIN.

The cities of "New Spain," however, are in other respects widely different from those of Old Spain. In the latter every town has a far-stretching history; almost every building, every archway, and all the gates, walls, and citadels are monuments in stone of the land they are in, and of the people who walk among them. In Mexico this is not the case. The ancient people, whom the conquerors displaced, are all but dead out of the land, so far as the traditions of the life and manners extend. Their monuments left are few, and the history which they tell scant. No one in Mexico seems to care for much dating prior to the

* Sartorius: "Mexico," p. 185.

Revolution; and even then the *laudatores temporis acti* in even that circumscribed sense are comparatively few. Scores of educated people, who have lived all their lives in the capital, could not tell you where Cortes' house was situated; hardly one could point out " where the armed Alvarado leaped over the broad canal;"* and none could point out, and few even care to guess, where Montezuma fell, or where the statue of Tlaloc was worshipped. There is another difference between the cities of Northern European and Southern European origin. Approach one of the former, and you walk past gardens and villas. Approach one of the latter—and the Mexican ones are of that type—and the road lies through dirty slums, miserable huts, the homes of ragged vagabonds, or half-naked Indians. An exception to this general rule must be made in favour of some of the cities of this Eastern coast, such as Jalapa, Orizava, and Cordova, where the suburbs are a labyrinth of flower and fruit gardens, from among which the red-tiled houses peep in pleasant contrast to the mass of verdure they are embosomed among. It is not until one is into the heart of the city that any pleasant features of Mexican architecture are seen. Then one finds, to his agreeable surprise, that the *pave* is raised a little above the street, and that the well-fitted basalt flags afford a pleasant footing to the pedestrian; while the streets are also paved. Most of them are straight and broad, crossing each other at right angles. The houses of the better classes are usually two or three storeys in height, though the prevalence of earthquakes acts as a deterrent to lofty architecture; while the dwellings of the humbler residents are usually but one storey high. The architecture is Spanish, but most of the numerous churches are in the French and Italian style of the seventeenth century.

LIFE ON THE PLAZA.

The Plaza (p. 284) is to a Spanish city what the Groote Markt is to a Dutch one. It is the heart of the place. One side is invariably occupied by the chief church, while on the other side, as invariably, are the city halls, or in the chief town of the State, the Capitol. The other sides are occupied with the houses of the richer people, the under storey being invariably a colonnade in which are the finest shops, wine and coffee houses, clubs, or buildings of a similar description. The Plaza is also the lounging place of the idle, and the market of the busy. It, indeed, corresponds very much to what the forum was in Rome. Here are the head-quarters of all news, and of gossipers and scandal-mongers innumerable. It is the fashionable promenade, and often the pleasantest part of the city. It is also in the Plaza that the wondrous motley population of Mexico can be seen to advantage. It is a motley crowd from all classes—high and low, rich and poor, priest and laymen, labourers on foot, and lordly dons and donas in sumptuous equipages.

* On the *noche triste*, when Cortes was routed by the Aztecs, Alvarado is said to have cleared a wide trench which then existed by one tremendous bound of his horse. The Puente de Alvarado, now one of the most aristocratic quarters of the city of Mexico, is said to be the site of the celebrated feat. At Popotla, a small village a mile and a half out of town, is the "Arbol de la noche triste"—the tree of the sad night. This cypress is believed to have afforded a place of concealment to Cortes and one of his followers on their flight, until their pursuers had passed by. It stands in front of a church built to commemorate the disaster, and, like it, is styled "del de la noche triste"— the church of the sad night. A few years ago some fanatical priests set fire to the tree, so that all that now remains of it are a hollow trunk, and a few live twigs.

In the Plaza are also the drills of the citizen soldiery, the firework displays on great anniversaries of the Republic, and the stately processions of the Church. Here the pious ladies of Mexico walk before and after mass, and here the impious young gentlemen of the land come to meet them. If even disinclined to gallantry it is quite *en regle*— "chic," as the Parisians would say—to once a day smoke a cigar on the Plaza, hear the news, and altogether keep oneself *en rapport* with the world. In the Plaza the

VIEW OF THE CITY OF MEXICO.

principal lawyers and notaries have their offices; and here also lounge in the sun when it is not too hot, and in the shade when it is not too cool, the lazzaroni of the city, ready to run a message when the commission is agreeable, and prompt to beg at all times, though seemingly, from their languid manner, perfectly indifferent as to the result of their mendicant whine. Here is also the *evangelista* (p. 285), or professional letter writer. No figure is more familiar in a Mexican town. He is generally an intensely respectable-looking person, who, like Dogberry, "hath had losses." In his suit of rusty black he has the appearance of a bankrupt undertaker, while the way he leers out of the corner of his cunning eyes irresistibly suggests to the student of physiognomy that the *evangelista's* function might have been that of chaplain to a gambling house. His keen glance has always an interrogatory look in it. He eyes the passenger solely with a view to

NEGRO HUTS, JAMAICA.

business, and knows at once whether the brown señorita who is timidly approaching wishes to send a *billet-doux* to her lover or a threatening missive to her rival. Notwithstanding his professional title, he has nothing to do with religion, and though occasionally poaching on the notary's preserves, has as little connection with the law. He is simply the public penman, whose functions will by-and-by be unnecessary, as education spreads more and more among the people. At present his *clientèle* is composed of domestic servants, small shop-keepers, labourers, artisans, sometimes Indians, and all who do not include caligraphy among their accomplishments. For these he will write letters on business, congratulations to friends on good fortune, or condolences on ill luck, begging letters, notes requesting favours—and

FOUNTAIN AND AQUEDUCT, CITY OF MEXICO.

particularly the favour of paying the recipient's account—threats of "putting the matter into the hands of my solicitor," invitations to dances and festivals generally, requests to become *compadre* or *comadre*—godfather or godmother—and above all love letters. It is curious to watch a *belle* of the "lower orders," closely wrapped in her *rebozo tapado*, so as to avoid recognition, pouring her woes into the *evangelista's* ear, now, from the passionate, but withal tender sparkle in her eye, evidently informing Jose Maria of her sentiments towards him, now, from the fierce jealousy which every gesture bespeaks, as undoubtedly informing Maria Annunciata of the sentiments she entertains towards *her*. Here is an Indian evidently unable to get the settlement of his little bill from the small dealer, whom in a credulous moment he trusted. He has, before going to law, invested a few reals in an *evangelista's* letter, and given place to a Mestizo, in a broad-brimmed hat, who is evidently intent on a milder errand. As a matter of course the *evangelista's* epistles are rather stereotyped, and very much on the model of those in the Mexican "Complete Letter Writer and Ladies' and

Gentleman's Companion," should such a work exist in that part of the world. His charges are by no means very moderate, and are regulated not only by the length of the letter, the fervour and general style of the language, but also by the mechanical manner in which it is executed—whether bordered or plain—with or without caligraphic devices—such as two hearts pierced by an arrow, or a pair of turtle-doves cooing in the midst of a garland of flowers. If it is a threatening letter that he is inditing, and his patron a man who can afford to pay for bad language, he will sketch a hand holding a stiletto, and for a trifle extra charge, make it drop the most elegant pears of blood imaginable. But these things cost money, as they require talent to execute; hence common wickedness has to trust to mere language for its conveyance to the object of it, and to the actual knife, should the *evangelista's* orthography not meet the demerits of the case. Sometimes the *evangelista* sits on a doorstep or the plinth of a church pillar, with a rude bit of board across his knees; or when he grows a little more prosperous he sets up a regular desk, and perhaps in addition to a comfortable chair for himself, one or two extra for the accommodation of his customers. After a time he will add better paper, pens, envelopes, and other little articles of stationery to his business. But his labours are always pursued in the open air. He has no fear of being "run in," or told to "move on," for the *evangelista* is a necessary institution of the country, and knows it well. It may be added that he also reads letters for the illiterate, and as his profession gives him possession of many secrets, there are those who hint that the professional scribe of Mexico adds to his other occupation the ugly trade of *alcahuete*, or "go between."* In the Plaza also assemble the trinket seller, tempting with her wares the *China*, *Mestiza*, or *Poblana*, for by all these names is the half-blood girl known. The "loafing" idle youth of Creole descent, too lazy and too proud to work, also infest the Plaza on the look out for simple country people in search of a pettifogging advocate, and under guise of attending to their business, fleeces his victims most unmercifully. The scarcely more dishonest pickpockets haunt, naturally, the place where people with something in their pockets to pick most do congregate, while the Plaza also forms the rendezvous of the hard-worked muleteers, in search of a return freight for their train of *burros*, or pack-animals. In every Mexican town the best streets open into the Plaza, the grandees seeming never to be disturbed by the traffic which is surging past their doors on its way to the market-place. Yet to a stranger this traffic is always picturesque. Here is a train of mules laden with pulque, and yonder is a crowd of asses drinking it. A solemn little file of Indians are trudging with their pots and their vegetables from the villages, reverentially doffing their hats—if they have any—to the priests of many shades who are ever passing. Students in long gowns and caps hurry to the university, and ladies in black silk dresses and laced mantillas—to mass. Very stately and very grave are these Señoras and Señoritas; but never were a pair of black eyes quicker at detecting and returning the glances which are thrown at the possessor from under the sombrero of the much-decorated Don who is ambling past. A twist of the fan, or a glance, conveys a world of meaning, so that the lady on her way to mass or to matins is quite as busily occupied in the ancient art of flirtation as her languid sister who is actively engaged in the same business behind the iron bars of the *reja*, or casement, or the *poblana*—

* "Illustrated Travels," Vol. III., p. 264.

the *manola* of Cadiz or Madrid—who in studied *deshabille* is standing at the open doorway of her single-storeyed house freely chatting with her many admirers. Sellers of every article under the sun of Mexico abound. A baker's boy is crying "*Pan fresco, pan tostado, per medio!*" another, "*Patos frios, patos grandes!*"—roast ducks, great ducks. Indian women are selling fruit and vegetables, shoes, clothes, newspapers, pamphlets. Another man is shouting "*Nieve, nieve!*" (ice), which passers by are stopping to refresh themselves with, while another who sells *agua fresca* (fresh water) skilfully balances a number of glasses on a tray. The confectioner with tarts is sure never to be far off, and at the open booths tradesmen are busy at work. Tailors especially are fond of pursuing their occupation with open doors, and even on the pavement when it is not sufficiently light within. Shoemakers and saddlers, tin, copper, and silversmiths, in like manner work exposed to the gaze of whomsoever thinks it worth his while to look at them. Jewellers are also plentiful, though generally all on a small scale; for the Mexican distrusts the great shops, lest the articles he buys shall be "factory made." It is one of his peculiarities that he likes everything genuine of its kind. Heavy gold watches, for instance, meet with a ready sale in Mexico, while cheap silver ones find no buyers. If a Mexican can't get the best of anything, he will rather go without. Hence woollens find a market, but the coarser clothes are unsaleable. The poor labourer when out for a holiday would infinitely prefer to wear the shreds of a silk girdle rather than a new cotton one. The Mexican towns are full of churches; hence the continuous ringing of bells is deafening. Priests are also numerous, and at one time so were monks, but, as we have already intimated, by no means of a very reputable description. Every now and again the papers used to relate how a monk or a priest had been found in a gambling-house or a worse place, and so long as these peccadilloes did not come before the public, their ecclesiastical superiors cared very little. Woe be to the man, however, who told tales out of school. By all accounts the discipline of the old convents was something tyrannical, so far as punishment for offences against the "Order" was concerned. They were nests of intrigue, envy, and heartburnings; and though the monk who might be brought by the police to his cloister in a state of drunkenness would never receive a word of admonition, should the same individual have in his cups been imprudent enough to have told a funny story about the superior, and the way he passed his time, the ingenuity of the convent would have been exhausted in devising punishment for the unhappy ecclesiastic. Mr. Sartorius, whose description I have abstracted, tells us that one of their punishments consisted in the offender being compelled to lie dinnerless outside the refectory door, and after enduring the indignity of all the Order stepping over his prostrate body, to be tortured for the next hour by hearing the clatter of plates, and the jink of glasses, while from under the door and through the keyhole came the maddening odour of roast and boil, beans and pepper—*frijoles* and *tortillas!* Another bit of grim discipline was to tie up a refractory monk to a stall in the stable, and allow him no food for a week or so, except the barley in the mule's manger.

If the Plaza is that of a provincial town, amid the crowd of cits and dandies the stranger cannot fail to observe certain portly gentlemen, obviously easy as to this world's goods, and from their sunburnt faces, and garments cut by the local artist, evidently from the country. They are invariably conservatives, believers in the bad old times, and croakers of evil things to come. They hate railways, "because they will hurt the trade of the carriers," an argument which the reader, whose memory extends back thirty or forty years, or whose reading of old newspapers has comprised that period, will perceive is not new or peculiar to Mexico. They also prophesy no good of gas, any more than of the iron road from Vera Cruz, for they themselves are cultivators of olives, or keep sheep for tallow, which, as all the world knows, is the final purpose of candles. The military is always an important element in Mexico. At one time it absorbed four-fifths of the entire revenue of the country, but, though still a prime moving cause of revolutions, has of late years had its power for evil considerably abridged.

From these sketches it will be seen that the Mexicans have much of the politeness of their Castilian ancestors, and though possessing some faults peculiarly their own, are also free from vices only too characteristic of old Spain. To a stranger they are exceedingly polite. The most casual acquaintances will volunteer to show him the sights of their city, and an invitation to dinner is rarely omitted among other courtesies. One reason for this is that the Mexicans are Chauvinistic to a fault, and are anxious that their country should stand well in the eyes of foreigners, though, unfortunately, by forgetting to pay their debts, they neglect the best way of impressing their virtues on a large section who refuse to be propitiated in any other way. Still, all visitors to Mexico do not invest in its bonds; and altogether the Mexicans are very discreet in their behaviour to strangers. They are not fond of speaking about the faults of their country, though ready enough to discuss these among themselves. They could say much not greatly to the credit of Anahuac, but they don't. They might, for instance, hint that the judges are not always incorruptible, that the virtue which consists in keeping the patriot's hand out of the public pocket does not exist, that the customs duties are so absurdly high that to "arrange" with the *douaniers* has become a recognised institution, and that while styling the country a republic, it is in reality a prey to every ruffian who can, by fair means or foul, obtain an armed majority.

The dinner-hour in the cities is usually between one and four. Then everything is very quiet. The Mexican world takes a *siesta*, or nap. Even the labourers seek repose after their midday meals. All the cities abound with eating-houses of a humble type. Here a man of the most modest means may indulge in red pepper, beans, boiled and fried, with pork, &c., and fiery ragouts. The poorest Mexican dislikes wheaten bread. Hence the rather leathery maize tortillas so characteristic of the country are the almost universal substitute for it. Knives and forks are rarely used, while the napkins, after a week's use, will make the visitor, unused to capsicum as a condiment, sneeze, so impregnated are they with the red pepper which has formed so important an ingredient in the food smeared on them by a score or two of diners. After the meal is over, something sweet—be it only a lump of sugar—is always taken, and then a large glass of water is swallowed. Finally, the well-eaten and drunken cross themselves, and say, "Bendito y alabado sea Dios," and, *horribile dictu*, open their mouths to the fullest capacity, and relieve their stomach of any accumulation of the digestive gases. This "habit" is indulged in by all classes, and is considered a very wholesome one. The poorer people do it openly, the richer and more polite only *en famille*—but all do it, nevertheless, unless they have been sadly misrepresented. Life in the warm climate of Mexico is, however, passed very much out of doors. Accordingly, labourers dine generally *al fresco*. As in England, their

MEXICAN SERENOS (NIGHT WATCHMEN).

wives bring their food to them. They then collect in little knots on the steps of the churches, or in other localities where they are not likely to be disturbed, and proceed to eat. They are very polite to each other, and usually, by sharing each other's food, manage to have a meal much more varied than they would otherwise have commanded.

"What a lovely ragout this of yours is, Doña Camilla," will remark a ragged bricklayer's labourer to the wife of a fellow-workman. "Ah! you are too good, Don Pedro," will be the response. "My husband frequently remarks that I am unable to prepare such admirable dishes as those of your good lady, Doña Mariquita;" and so the little empty, but still pleasant, talk goes on while the meal lasts. The natives of Mexico, however, are in general very moderate eaters. One stout English farm labourer could consume at a sitting as much as a whole family in the *Tierras Calientes*.

The Mexican streets are built at right angles to each other, so that the towns are divided into square blocks of houses, which have the appearance of so many compound masses of dwellings. Each side of one of these blocks measures 200 varas—or 600 feet—long, and the square of 40,000 square varas is a mansana. Each mansana chooses annually a justice of the peace and a police inspector. A number of mansanas constitutes, ecclesiastically, a parish, or politically, a quarter represented by a member of the town council, or *ayuntamiento*.

Evening is the time for driving. Then the air is cool, and the promenaders crowd the streets and public gardens, and the world of Mexico is in full swing. Talking, flirting, and all the gay whirl are in progress, when suddenly, just as the sun begins, the strokes of a bell strikes the ear. It is the *oracion*, which proclaims that the day is done. Then all progress is arrested, all noise hushed. Each promenader reverently bends his head, and whispers an *Ave Maria*, while the bell is ringing out his solemn tones. At the last stroke they again uncover, and the world whose life was arrested again resumes its gaiety. But for that day it is only the world "out of society" that goes abroad. Friends bid each other good evening, and prepare for returning home again. The same ceremonies are observed in the house, should the family be there when the *oracion* strikes the ear. Even the servant who brings in the lights when the bell has ceased will wish the family good evening. *Buenos dios*, it may be remarked, is in Mexico (as in Spain) used as the salutation from early morning to noon. From noon to night it is *buenos tardes*, while for all hours on to daylight *buenas noche* is the correct good evening, day and night being always in the plural. But it may be another tinkle that is heard. It is *nuestro amo*—the priest carrying the host to some dying person. He is in a carriage drawn by two white horses, driven by a jarvey who belongs to "the Guild of the Coachmen of our Lord." Chorister boys precede the carriage with lanterns, and as it passes all in the street or in the balconies kneel at the sound of the bell. If the time at which the carriage passes be night, then lights will be exhibited in the windows of the faithful. That is to say, such was the practice in former times, and perhaps is still in country towns. But in the capital and seaports the people have sadly fallen into disgrace. It is now rather difficult to find recruits for the Coachmen's Sacred Guild, and it is even rumoured that the people in the street, unless they be Indians or others of the *hoi polloi*, have begun to get rather hard of hearing when the white horses and their holy burden pass by.

An earthquake still, however, suffices to rouse the Mexicans from their apathy. At the first sickening oscillation the people rush out of the house, singing the *Libra nos Señor*, which, whatever may be their previous neglect of what the Scotch call "church privileges," seems to be instinctively repeated, on the principle that in such an emergency "*something* must be done."

But we have left the people quietly going home after the vesper bell had ceased tolling. The bachelors go to their chocolate, or coffee, the family men to their own houses. The merchants still work on to nine o'clock, when the last post closes; but the artizans generally break off at an earlier hour. Most visits are paid in the evening. Then "tertullas" dancing and making love—which is always an important element in Mexican society—form the evening's amusements. Tea, confectionery, wine, or fruit are the refreshments. At most of such little parties the only liquor served is a kind of sweet liquid, the name of which I do not profess to know, or a *sangria* (negus), and lemonade. The Creole is a temperate individual, and does require the use of strong waters to be excited. The men often leave the ladies to themselves, and play for the next few hours at malilla, tresillo, or other games of cards. The young dandies of course remain with the ladies, and though they are ostentatious gallants, yet the utmost decorum is preserved, in spite of a wearisome amount of word play. Indeed, the demonstrative manner of the cold Northerners is not required in their torrid regions. A glance of the eye, a twirl of the fan, or a gesture which to the stolid European visitor would mean nothing, conveys to the sighing youth more than a flood of words. By ten o'clock sedately-living people ought in Mexico to be asleep, or at least commencing "an affair of outposts" with the mosquitoes behind the curtains of his stretcher bed. By eleven the streets are quiet. A few guitars twang from an azotea, or the flirtation of a coquettish señorita, with living *cocuyos*, or fire-flies* in her hair, are the only signs that the world is not a-bed. The "serenos" have it then all to themselves. These are the Mexican night watchmen (p. 293), and derive their name from calling out the hours periodically, and generally dwelling on the last words, "Las doce ye medio-tiempo ser-e-no!" The sereno is a fitting type of the backward state of Mexico. His long spear, his clumsy, antique lantern, and old-fashioned rattle, are all bits of a time gone by. He is practically useless, for though he occasionally captures a very docile thief, he more frequently dozes away the hours of night in some convent or church portico, only waking up by some roystering dandies returning home from a late supper or an early breakfast taking liberties with his person, his lantern, or his rattle. Or sometimes he will shake himself out of an uneasy sleep, roused by the great clock of the cathedral striking the last stroke of some unknown hour. Then he starts up with unaccustomed energy, and vociferates in his long-drawn treble the "ser-e-no" with quadruple drawl, just as the London watchman of former days, when suddenly awoke from his slumbers, alarmed the sleepy citizens with his "'clock and a fi-n-e night!"

Tacubaya is a place where rich Mexicans have their country houses, in the middle of great, badly-kept gardens, and not far from Tacubaya is Chapultepec, or the "Grasshopper

---

* A beetle (*Elater noctilucus*). So bright is the light emitted that two or three will enable print to be easily read at night. Southey refers to it when he describes Madoc using them as lamps—

"And by the light two fire-flies gave,
Revealed the beauteous features of his guide."

Mountain." At Chapultepec the great lion is the *Ahuehuetes*,* or cypress grove, great trees, their branches all hanging with the grey Spanish moss † (p. 297), the remains of likenesses of two Montezumas, sculptured by the Aztecs on the side of the solid porphyry hill, and the palace which the Viceroy Bernardo de Galvez built at such enormous expense, intending it either for a palace or a fortress, or rather for both. The Montezuma sculpture, some Vandal of a viceroy, or the priests in his train, thought proper to destroy as far as he could by blasting it out with gunpowder; but the cypress grove, the favourite haunt of the voluptuous court of the Aztec monarch, who had a palace here, on the site of the present one, still ran on in all its glory. Some of the trees are of great size. One, known as Montezuma's tree, is forty-eight feet in circumference, while many others are of great size, though not equal to the *Sequoia* of California.

Mexican society is rather stationary; yet even it is changing, owing to the advent of railways, and other modern improvements, and to the fact that many of the young men now go to Europe, or to the United States, for their education, or visit that world which is so much older, but at the same time so much newer than that in which they live. One Mexican town is very much the same as another. Hence the sketch of the Plaza of Guadalajara (p. 284) may serve as the model for all of them, the life on which we have noted in the preceding pages. As in other Mexican towns, it forms the centre of out-door life. It is a square of about a hundred yards each way, with broad walks round, the promenades lined with orange-trees and provided with seats. A fountain ornaments the centre, and, as usual, the cathedral occupies one length of the whole Plaza; while the *Palacio*, or State House, takes up the side opposite the cathedral. The other two sides are occupied by the *portales*, as the Guadalajarens call the covered way, formed by pillar-supported arches over the pavement.‡

The city of Mexico, of which we gave a view on p. 288, occupies the same site, but is not the same city as that which Cortes first saw with astonishment. There have been frequent attempts to make out that the conquistadores' accounts of the populous state of Mexico was, owing to a desire to magnify their victories in the eyes of their countrymen, exaggerated. But travellers who have carefully investigated the subject have very generally come to a different conclusion. Solis and Gomara were doubtless liars, as were a few others in both ancient and modern times. But the real "authorities" have, we believe, in most cases, told a fairly average truthful story. When Cortes first saw the city (p. 254), it was his intention to have preserved it. But finally to overcome the desperate obstinacy with which the Aztecs defended their capital, he resolved upon its utter demolition. He was grieved to destroy it, for, as he said, "it was the most beautiful thing in the whole world." Slowly he moved on the great teocalli the 50,000 Tlascalan allies, throwing down every house, and filling the canals with them. When the destruction was done, only one poor quarter and a few pyramids, too large for immediate demolition, were left standing. All the rest was ruins and dead bodies. Though the site had many disadvantages, yet Cortes determined, in spite of them all, to erect his capital on the spot which the old Mexicans had selected more

---

* Or Ahochoctis (*Tapolloua distacha*). † *Tillandsia usneoides*.
‡ Geiger: "A Peep at Mexico" (1874), p. 126.

CYPRESS GROVE OF CHAPULTEPEC, NEAR THE CITY OF MEXICO.

from necessity than choice, for he remarked that "the city of Tenochlitlan had become celebrated, its position was wonderful, and in all times it had been considered as the capital and mother of all these provinces." In four years, by the forced labours of the conquered

Mexicans, the fine new Spanish city arose on the fragments of the old Aztec one. After the Lake of Zumpango was drained off through the Desague de Huchuetoca—a low pass between the hills—Mexico, which in Aztec times was on some swampy islands in the middle of a lake, stood in time on dry land, or at least on land which is at the surface dry, but swampy underneath. There is no doubt that Mexico is a grand city; but neither our space nor our inclination will enable us to say more about it than that it is a Spanish town in the New World, which has been described a hundred times in as many accessible volumes, to all or any of which the reader is referred.

The "Common People."

The Mestizoes we have already spoken of. "The actual Indians form separate communities in the suburbs, and differ but little from their fellows in the villages, where they depend on agriculture. They are independent after their fashion. In the capital itself, the Indians at the extremity of the city carry on the same occupations as did their forefathers at the time of the Conquest. They seek a subsistence in the swamps and lakes, and on their sterile shores. Like the herons, they are seen wading in the ditches which transect the swamp, catching with their little nets white fish, frogs, and axolotls, that strange proteus species* forming the connecting link between the fish and the lizard. In small canoes they row along the broad canals and sedgy lakes, collecting fish and frog spawn, fowl's eggs, water-cresses or water-lilies, or chase the aquatic birds and sandpipers which cover the lakes and these shores in incredible numbers. In addition they plait reed mats, boil salt from the water of the salt lakes, increasing their store by repeatedly washing saline earths, or collecting *salron tequesquite* in the fields, which flourishes after the rainy season. All these articles of commerce, and many more— maize variously prepared, fowls dressed, young sandpipers, humming-birds in little cages of grass, earthenware, baskets, toys of wood or feathers, gourds, string, cups, &c.—they convey to the market of Tlaltelolco, which three centuries since, when Cortes first marched into the capital of the Aztecs, was so large that 30,000 persons traded there daily."

Creoles.

The word "Creole" literally means native, though in England it is often applied to people with a slight dash of negro blood, and in the West Indies they talk of a Creole negro, a Creole brown man, a Creole white, Creole mahogany, or Creole pickles. In Mexico it used to designate a white, or European born in the country. In Mexico the Creoles form about one-seventh of the population, and constitute the real intelligence of the country, as well as its aristocracy. In appearance the Creole apes the Spaniard of the Old World, but at once his physiognomy betrays to the stranger who has ever visited Spain that though a Spaniard he may be, he is one of the New World. He is quick, animated, usually not very fresh in complexion, early arrives at maturity, but is not muscular, and

---

* *Siredon piscifera*. It preserves the gills of early life throughout its entire existence, but in addition possesses lungs. In confinement it will often lose the gills. It is about eight or nine inches long, and has been eaten in Mexico from the earliest date.

is unfitted for a long continuance of labour. His speech has all the quickness of the Southern nations, and is accompanied by an endless dumb show, which is peculiarly affected by the Mexicans, sometimes as an emphasis to what is said, sometimes as a commentary. Like the Mestizo, the Creole is passionate, and easily moved, but, unlike him, he is able to govern his emotions, and soon regains his equanimity. Unlike the Italians or Portuguese, treachery is foreign to his nature. If he desires vengeance, he executes it openly with his own hand, and would never dream of hiring bravoes or bandits to murder his enemy. He is not so apt to use the knife as the Mestizo, and then only as the result of passion, or in what he considers the regular course of a kind of irregular duel. He is vain, thoughtless, and fond of enjoyment, and apt to value himself rather more highly than the rest of the world does; but at the same time, even when quite uneducated, is endowed with a natural refinement and politeness which favourably impress the stranger who first comes into contact with him. He has all the faults and most of the virtues of the Spaniard, and to these he has superadded a few vices peculiarly his own. But as our space is too short to speak of his merits, we shall leave his failings to the reader's imagination or to other narrators. "A fertile country, producing, almost unasked, a clear sky, a mild climate, where the hardships of winter are unknown, have spoiled this Creole, and rendered him more indolent and thoughtless than his trans-marine relations; but he has retained the liveliness, the excitability, and the romantic sentiments of the latter. The Spaniard is essentially conservative; the Mexican Creole is for progress, he is liberal and tolerant even in religious matters, while the Spaniard never quits the established forms in Church and State. The Spaniard labours perseveringly, seeks also to profit, and saves what he has earned for old age. The Mexican earns with facility, and just as easily lets it slip through his fingers. He seeks to enjoy the fleeting moment, and leaves Providence to care for the future."

THE PROLETARIANS.

Mexico, long before it became independent, had developed many of the characteristics of Europe's most doubtful side. Beggars, idlers, and rogues of every type flourished there exceedingly two centuries ago, and from the rankness of their growth at the present time seem likely to go on prospering. The lazzaroni of Italy are represented by the leperos, or pelados, who, so far from being the victims of misfortune, choose this calling, and look upon themselves almost in the light of "professional persons." In a thinly populated country like Mexico, endowed with a rich soil and a fine climate, there ought to be no idlers, and no men in lack of bread. But it is just these causes which produce the Mexican lazzaroni. He can get a bounteous living by a little labour; but he can exist on still less, and accordingly prefers the latter alternative. With him it is better to walk than to run, better to stand still than to walk, better yet to sit than to stand, and best of all to lie asleep in the shade. And this the lepero does. He is, after all, if not a good citizen, a happy man, and casuists are not wanting who declare the individual's first duty is to himself, his second to the State. The lepero has no spare wardrobe, hence he requires to lock nothing up, and is not troubled with luggage to impede the movement of his quarters when the last gets socially too hot for him. Occasionally he will own a spare shirt—in order to pawn or stake it. Shoes he has none—they are only impediments to flight when he is caught picking a caballero's pocket—but he has, unless in very bad luck, a *frazada*, or coarse cloth, which serves as blanket all the night, and as a garment all the day. Finally, when he dies, naturally or by the knife of an irritated acquaintance, the *frazada* serves as winding-sheet: a cord is tied round the rag-shrouded corpse, and it is buried forthwith.

But the lepero is outwardly a pious man: most Mexican rogues are. He wears a rosary with a cross or a scapula round his neck, and invariably attends church festivals, and never neglects mass, for there the speculator on the contents of other people's pockets finds his great reward. It is amusing to watch a skilful pickpocket of this type on his knees at mass. He is seemingly the most devout of vagabonds. He neither looks to the right nor left: his eyes are solely for the prayer-book. Nevertheless he is intent on business, and, as the ladies or gentlemen on either side of him may afterwards discover to their loss, the hand holding the missal is skilfully made of wax, while the legitimate members are busily employed under the shield of the ample cloak in ransacking their pockets. If married, the lepero lives in some den or other in the suburbs; if single, he passes the night under a church porch, until the pulque shop is open in the morning. If in funds, he treats himself to liquor; if not, he tries to induce others to do so; if both fail, he pawns his spare shirt or the pocket-handkerchief which he never uses, but always likes to keep in reserve, as a good pawning article, against emergencies. Then he buys a cigar, manages to get some breakfast, and holds counsel with himself as to the district which for that day he shall honour with his presence. The leperos do not all steal—directly. Some lead blind men about the streets, and in more senses than one take care of the funds. Others beg at the doors of churches, or hang about the tavern or coffee-house doors, or pick up the stumps of the cast-away cigars. These they manufacture into cigarettes, and as the State monopolises the sale of tobacco, dispose of them secretly to impecunious smokers. But the majority are *chevaliers d'industrie*. The market-place is the spot where they most do congregate, and a crowd their market. At the day's end they dispose of their plunder to certain "fences," who keep a *bacatillo*, or bazaar, for the purchase and disposal of such ill-gotten goods. These are the lowest order of leperos. But above the *sans culottes* are others, who wear shoes, sometimes a jacket, and possibly even now and then a clean shirt; but otherwise they are much the same. They are hawkers ostensibly, thieves in reality, and are frequently, in addition, spies of the brigands, employed to ascertain the time travellers are likely to leave, and what is the probable value of the property they carry with them. Last of all, there is a class—small it must be allowed—who absolutely work, and unless under strong provocation do not steal. They are proud of the latter fact, though at the same time diffident as to their powers of continuing in the way of *hombres de bien*, or morality. "God be thanked," they will say, "hitherto we have not wandered from the right path; who can tell whether we shall continue so?" and he would be a rash man who essayed the task. These men are often porters (cargadores), messengers, water-carriers (aguadores), and so on, and now and then they even end by becoming wealthy peasants or shopkeepers. But as a general rule the street porter is, it must be said, an arrant rogue—lazy, dishonest, and a gambler to his very last coin. He is,

A RURAL KITCHEN IN THE TIERRAS CALIENTES (MAKING FRIJOLES, OR MAIZE CAKES).

moreover, an impudent scamp; he will make love to the nursery-maids, and even kiss her hand in mock gallantry to their mistresses in the balconies. The water-carrier is a better type. He is more industrious and held in greater consideration. He knows the secrets of every house, is the cook's gossip and the chambermaid's message-carrier, while even the Señora herself will consult him when she is about to change her servants, as she would do the baker's or butcher's man in more northern latitudes. When the towns are flooded, as frequently happens during the tropical rains, the porters are employed to carry people across them. This is their harvest-time and the harvest-time also of their confederates. They will secrete themselves on balconies and roofs, and with a hook attached to a line skilfully detach the mounted victim's hat or cloak, and disappear before the plundered man can reach dry land and look out for the police. The wandering cobblers (remendones) are also Proletarians, while a worse class still are the pettifogging lawyers, croupiers in gambling-houses, assistants at horse-races, cock-fights, and rascality generally. They are usually Creoles who have by vice gradually sunk down to the lowest depths, illustrating the Mexican proverb, which says of these youths, *El padre comerciante, el hijo paseante, el nieto mendicante*—"The father a merchant, the son a walking gentleman, the grandson a beggar." Their shifts, roguery, and character need not be further described, as, to all intents and purposes, they are simply those with which the reader must be familiar if he has ever been unfortunate enough to come into contact with a similar class in Europe. They are all the same.*

There are a good number of foreigners in Mexico, chiefly French, American, English, German, Italian, and Spanish, mostly employed in commerce or mining. The French are best liked, the English less so, and the Americans least of all. Some of the Austrians and Belgians, who came in Maximilian's train, are now numbered among the street beggars of the large cities. Such is the fate of the last *conquistadores*.

CHAPTER XVII.

Mexico: Its Commerce and its Prospects.

That Mexico is surpassingly rich in everything that can make a country wealthy is stating a mere truism. It is far richer than India, which is a poor country, and in resources it can compete with the best portions of the United States. There is nothing which the land cannot be made to yield, while its position—between two seas—ought to attract to it the commerce of the world. Its mines are rich but not half wrought; coal even is believed to exist. In addition to every product of warm countries its crops comprise wheat in the temperate or sub-Alpine regions, at an elevation of 3,000 feet. Here is a great table-land, enjoying the most perfect of climates, and composed, for the most part,

---

* In some works the Jarochos (pronounced Tcharotcho-s) are described as the Mexican gypsies. This is an error; the name is simply that applied to the natives of the East Coast generally. Sartorius: "Mexico," pp. 6 and 141—154.

of the most fruitful of soils. The *Tierra Fria* has fine timber; the *Tierra Caliente* all kinds of tropical crops.

Some of the products derived from the vegetable kingdom we have already mentioned. Among them maize must take a front rank, as three, and even four good crops can be obtained annually in many districts, while in all parts of the country it thrives to such an extent that the yield is sometimes five hundred-fold. It constitutes, with beans and chilies, the almost exclusive food of the Indians. Wheat yields sixty-fold, and rice about forty-five. In 1873, the cotton crop of Sinaloa comprised 550,000 lbs., and in 1872 that of the district of San Juan Evangelista 1,312,104 lbs. The coffee of Colima, which yields about 31,000 lbs. annually, is reported to be as good as that of Costa Rica, while that of Vera Cruz is also held in high esteem. The great cacao centre is Oajaca, where the plant yields three yearly crops, and this renders its cultivation the most profitable in the State. Sugar is made in large quantities at Vera Cruz, while the Vera Cruz and Tabasco tobacco is quite equal to the finest of the Cuban brands. Altogether, the annual value of the food crops of Mexico may be estimated at about 58,000,000 dollars, and all the agricultural products at 110,000,000 dollars. But the mineral resources of Mexico have gained for it the widest reputation. The Cerro del Mercado, in Durango, is said to be one vast mass of iron. Its other metalliferous deposits are believed, with the exception of those of Peru, to be the richest in the world. About 500 tons of silver and 1½ tons of gold are annually exported. From 1521 to 1803, 2,027,952,000 dollars' worth of the precious metals were exported from Mexico; from 1803 to 1810, 161,000,000 dollars; from 1810 to 1826—the year of independence—180,000,000, in all, 2,368,952,000. As yet, the mineral resources of the country are not half developed. At the time of the revolution, many of the best mines were deserted, notwithstanding the efforts of foreign capitalists to reclaim them. It has been estimated that up to 1870, altogether some 1,200,000,000 dollars' worth of gold and silver have been extracted from the mines of Mexico. The seven principal mines of San Luis Potosi alone produced in 1868 silver to the value of 2,176,899 dollars 28 cents. The State of Sinaloa is dotted with silver mines, and in most other parts of the country there are rich deposits either being worked, or which, when the country gets more settled, will be developed. During the colonial days, the Mexican mines were Crown property, let out to private individuals on the agreement that those who worked them would pay one-fifth of the yield to the Crown. When the country became independent, the mines were declared public property, and the owners were only required to pay a small royalty to the Treasury. Even this was afterwards abolished, and accordingly at the present time any one can obtain permission to work a certain amount of mining ground by right of "denouncing" and recording his discovery. In addition to gold, silver, and iron, tin, lead, mercury, copper, various precious stones, marbles, porphyry, jasper, alabaster, sulphur, and other mineral products are found in greater or less abundance.

Among the animals found in the neighbouring seas may be mentioned the pearl oyster (*Avicula margaratifera*), which is extensively fished along the shores of the Gulf of California. In 1873, 636 divers were employed in this business, and the value of the shells obtained was 112,030 dollars, and of the pearls 64,300 dollars.

The staple articles of export are silver and gold (coined), silver and copper ores, cochineal, indigo, and other dye-stuffs, cabinet and timber woods, sisal hemp, &c., while cotton, linen, woollen, and silk fabrics, unwrought-iron, machinery, hardware, provisions, &c., are imported. It receives most of its imports from Great Britain; indeed, more than twice those from the United States and France, which come next. In 1876 the imports from the United Kingdom were estimated to be worth £576,814, and the exports to us £662,132. Systematic smuggling is, however, so extensively carried on, that accuracy in the commercial returns cannot be expected. In 1872-73 the revenue was 14,333,926 dollars 50 cents, and the expenditure 20,959,363 dollars 89 cents. At present both the revenue and expenditure are greater, though we have no very accurate returns, but the above may be taken as a fair average of one year with another. The national debt must be large, but there is no official return of it. In 1871 it was estimated at £81,000,000. The liabilities contracted by the Imperial Government have been repudiated *in toto*, and, practically, it would in the end be no great loss to any one were a similar course adopted in regard to the rest of the national indebtedness. The creditors could then know for certain what had become of the money they had already practically lost, while national honour, an intangible entity which has rarely troubled Mexico, need present no obstacle to this not altogether novel method of paying old debts and new. There is a standing army of over 22,000 men, which cost in 1872-73 10,252,522 dollars. Education is in a tolerably advanced state, and is year by year, through the energy of the Government and of private individuals, getting more widely spread; but female instruction is still much neglected.

Its Prospects.

Being unendowed with the gift of prophecy, and, moreover, not wishing to give unnecessary offence, the writer of these lines prefers not to forecast the Mexican horoscope. Still, it is no rashness to venture the belief that eventually Mexico will form part of the United States. Endless political disturbances, ignorance of the first elements of political economy, an absence of public spirit, venal officials, and ruffianly politicians, all war against the unhappy land. A nation industrious at home, developing all its internal resources, might, could such a thing be possible, exist and even progress in spite of such ideas. But when the majority of the Mexicans neither themselves cultivate commerce, nor care for others developing their riches for them, then, in this working-day world, it is impossible to believe in the existence of Mexico very long as an independent country.

CHAPTER XVIII.

The West Indies: A General Sketch.

If the reader casts his eye over the map of North America, he will be struck with the crescent-like indentation in the southern part of the Continent called the Gulf of Mexico. Stretching across the mouth of this gulf, from near the southern point of Cuba, is a rough curve on to near the Gulf of Paria, in Venezuela; and then westward and northward along the shore to the Paraguana Peninsula, an immense number of broken islands and islets familiar to every person as the "West Indies." They were so called from a belief of Columbus that he had lit upon a portion of India when he first sighted them; but they are also (with the exception of the Bahamas) styled the Antilles, from another similar error. Long before the time of Columbus, a land called Antilla was fabled to lie to the west of the Azores; and when Columbus returned the cosmographers affirmed that these must be the Antilla that the world had so long dreamt of;

VIEW OF CHARLOTTE AMALIE, ST. THOMAS, WEST INDIES.

and it is certain that long before any link on the Caribbean chain of islands was discovered, Cuba and Hayti, the largest among the earliest known of them, were so called. They are usually divided into four groups: (1) the Bahamas, or Lucayos, on the north; (2) the Greater Antilles, or Leeward Islands, so called from the prevailing directions of the winds, comprising Cuba, Hayti, Jamaica, and Porto Rico; (3) the Lower Antilles, Caribbean, or Windward Islands—all small isles, with the exception of Trinidad, not far from the mouth of the Orinoco; and (4) the Leeward Isles of the Spaniards, which are a series of small ones lying off the Venezuelan coast. In all there are about forty-five largish isles, and an immense number which are mere rocky islets, coral reefs, or sandbanks. The entire area is 91,765 square miles, and the population 3,855,000. All of the larger northern islands, and some of the smaller ones, are traversed by a mountain range, running in the direction of its length, and sending prolongations down to the shore on either side. Trinidad is crossed from east to west by two chains, which are prolongations of the Andes of Venezuela, and the group from Granada to St. Eustatius exhibit volcanic craters. Indeed, several of the isles in this region have, according to the late Dr. Bryce, been in eruption since the middle of the last century; while the entire group is subject to earthquakes. It is probable that the West Indian islands are merely broken remnants of a great mass of land at one time connected with America as far north as the Bermudas, down to the most southern of the present isles, and which got disrupted and sunk. Coral reefs encircle the shores of most of the islands, and raised shell beds are also seen, leading to the belief that in comparatively recent geological periods there has been a gradual raising of the whole group, and that if this elevation had continued, the West Indies might have again closed in the Gulf of Mexico, and formed a broad mountainous division between the Atlantic and Pacific, instead of the present narrow and comparatively flat Isthmus of Panama. Many of the islands are encircled with a low malarious belt, but the ground generally rises inland, increasing in healthiness with its elevation, until, on the highest points, as on the Blue Mountains of Jamaica, the climate is almost European. The regular alternation of the sea breezes during the day, and the land wind at night, gives a pleasant variety, and maintains a healthy equilibrium. The summer is from May till October, when the temperature is about 80° Fahrenheit. Heavy rains fall from October to December, and in August there are often hurricanes, which, however, seldom touch Cuba or the more Southern isles. "The trade winds blow from the north-east and east from December to March, when they decline south-east, and are succeeded by calms in the height of summer." All the tropical products abound, while the woods swarm with animal life, and the sea with fish, and, above all, with that favourite reptile, the turtle. The Carib aborigines* are now extinct on all the isles except Trinidad, where a few still exist. Politically, the West Indies may be divided as follows:—(1) One island, Hispaniola, or Hayti, is independent, and divided between two republics. (2) The following are British:—Jamaica, Trinidad, Barbadoes, Antigua, St. Vincent, Bahamas, Tobago, Grenada, Dominica, St. Lucia, Montserrat, St. Kitts, Nevis, Tortola (Virgin Isles), and Bermuda. There are also a great number of smaller isles, which may be classed as follows:—The Bahama group—Abaco (Great and Little), Bahama, Eleuthera, Andros, New Providence, Cat Island, Watling, Guanahani, or San Salvador, Long Island, Crooked Island, Caicos, Turks Island, Exuma, Aicklin, Inagua, Marignana, in addition to coral reefs and rocks; the Virgin group, comprising Culebra, Bieque, Normand, St. Pieter, Virgin Gorda, Anegada, Sombrero, and numerous smaller ones, such as Salt, Cooper's, Ginger, Guana, &c., Seal Island, Anguila, Anguileta, Barbuda; the Grenadines, such as Bequia, Canaguan, Cariacou; and Cayman (Great and Little), dependencies of Jamaica. The entire area of the British isles is 13,103 square miles, and the population is under 1,000,000. (3) The Spanish islands are Cuba, Porto Rico, and the Isle of Pines, south of Cuba, in area 46,250 square miles, and with a population of 2,060,870. (4) The French have Guadaloupe, Martinique, the northern part

* "Races of Mankind," Vol. I. p. 265.

of St. Martin, Marie Galante, St. Bartholomew, which they bought from Sweden, and the small isles, called Desirade and Saintes group, in all 902 square miles, and 231,000 people. (5) The Danes own St. Thomas (the principal town of which, Charlotte Amalie, is figured on p. 305), Santa Cruz, St. John, and Crab Island, comprising 190 square miles, and a population numbering 46,000. (6) The Dutch have St. Eustatius, Saba, Curaçoa, Oruba, Buen-Ayre, and the south part of St. Martin Island, an area of 591 square miles, and a population of 36,000; while (7) the Venezuelan Government claim Margarita, Coche, Tortuga, and a number of smaller ones, which are mere rocklets. To these notes, for which we owe our obligations to our esteemed friend the late Dr. Bryce, it may be added that the division into the Greater and Lesser Antilles is not only a geographical but a geological one. Most of the Greater Antilles present lofty granitic mountains, though in Jamaica there are many calcareous hills; while the Lesser Antilles, as a whole, are chiefly of coral formation or volcanic origin. All of the West Indies, owing to the richness of the tropical vegetation, have a pleasing appearance; but it is in the larger ones, like Jamaica, Hayti, and Cuba, that are found really fine landscapes. Here the steep, rugged mountains, broken by gorges, and closed in by magnificent forest, strike the eye of the voyager, wearied with the endless sea, as something perhaps even finer than they are; but it is only when we wander through these umbrageous tropical woods that all the richness and glory of the Antilles enter fully into our minds.

To describe all these isles separately would be a tedious, and, indeed, a useless task. Nor shall we attempt it. We shall simply devote the space at our disposal to a brief description of Jamaica, Hayti, and Cuba, as respectively the type of the British, independent, and Spanish isles, the French, Dutch, and Danish ones having many features common to the others, while the Venezuelan group is to all intents and purposes a part of Venezuela. Habits and ways of thought are, moreover, much the same all over the Antilles, though local prejudices and interests may magnify the wants, resources, and grievances of particular islands into an importance which the world, viewing them from another standpoint, cannot be expected to appreciate.

JAMAICA.*

Once on a time Jamaica, or Xayamaca—"the land of wood and water," of which the common name is a corruption—was the richest and most prosperous of all the English colonies. The land is still among the most fertile on the earth—that no social changes can affect—but the people have fallen away from their former opulence. It is the largest of the British Indian islands, being nearly 150 miles long, by 15 miles broad, and contains an area of 4,256 square miles, and a population numbering, at the last census, 506,154. It was one

---

* "Parliamentary Reports on the Colonial Possessions," Sessions 1875, 1876, 1877. Sewell: "Ordeal of Free Labour in the West Indies." Kingsley: "At Last: A Christmas in the West Indies." "The West India Sketch Book." Gosse: "Natural History of Jamaica." Hans Sloane and Browne's works on the same subject. "Letters from Jamaica," an anonymous work, so admirably graphic that the learned judge, whose work it is, need not have no hesitation in putting his name on the title page. (First [?] Skildring af Naturen [?] Jamaica." Turner: "Geographical Magazine," 1874, &c. I have also in [?] [?] [?] [?] which appeared in the Standard newspaper, describing the present political condition of the colony.

of the earliest outlying portions of America discovered by Columbus in 1494. He called it St. Jago, and under that name it was taken possession of by the Spaniards in 1509; but in 1655 it was captured by the English, and in 1670 formally ceded to us. It is intersected by several ranges of mountains, reaching in places an altitude of between 7,000 and 8,000 feet, their general trend being from east to west. Roughly speaking, these mountains may be said to divide the island into two halves, known respectively as the north side and south side. Extensive plains, lagoons, and marshes characterise the latter division, while the north is distinguished by the great number of rivers and streams—none of which, with the exception of the Black River, are navigable, even for small craft—and by the wild grandeur of its mountain torrents, rugged cliffs, and bays embosomed in rich fertile vegetation.

The climate, on account of the varied altitude of the surface, is not the same in every part of the island. The north is, take it all in all, more salubrious than the south. In the plains the heat is intense, while in the mountain districts, such as at the military station of Newcastle, in the Blue Mountains (p. 309), the temperature occasionally sinks as low as 60° or 70°. There is, however, little variation between one season and another, unless, indeed, it is during the spring and autumn "rains" in May and October. "The advent of 'the seasons,' as they are called, is, to the experienced eye, easily foretold by the increased number of fire-flies and mosquitoes, which seem 'to smell the rain afar off,' and by the appearance in the sky, often some time before, of light, cirrus clouds, which the negroes not inaptly nor unpoetically denominate 'rain seeds.' The sky grows dense with visible vapours for some days before the showers fall. As the clouds gather the coruscations of lightning become more constant and vivid at nightfall. The rains then set in every day, and continue for a succession of days, falling at regular periods in the twenty-four hours. The vernal showers descend amid lightning and thunder, and those of the autumn come with heavy gusts of wind and storm. In the mountains the rains are earlier and heavier than in the lowland country."* Many of the rivers are almost empty during the dry season, while, as in the case of the Rio Mino, during the rains they are fierce mountain torrents which carry everything before them. During "the seasons" the rains in the mountains "bring the gullies down," as it is called, and the wild avalanche of water bursts its banks and floods the plains, breaking up roads, washing away bridges and houses, drowning horses and cattle on their pastures, sweeping away fields of cane and Indian corn, and, in a word, carrying ruin and desolation in their path. Such a flood was that of 1868. On the other hand, 1869 was a year of drought, while 1870 was almost as disastrous as that of 1868. Between Lucea and Green Island all the bridges were swept away, and 150 acres of mountain land "came down with a run." In other places the accumulation of water was enough to float a frigate. Landslips were of common occurrence, houses were caused to slide down from the top to the bottom of hills, and trees travelled in the same way rooted in the moving mass of soil.

Everything in Jamaica bears an aspect of decayed grandeur. Kingston is the chief town, but a town which has not inaptly been described as one which has lost its self respect.

---

* Introductory remarks to a "Catalogue of the Exhibition of the Jamaica Society of Arts," 1855, p. 27.

"It looks what it is" remarked an American who visited it some years ago; "a place where money has been made, but can be made no more. It is used up, and cast aside as useless." St. Jago de la Vega, or Spanish Town, was for long the sleepy—the all but dead—capital of the island. Once it was the seat of a gay Viceregal Court, the home of wealth, learning, and wisdom. Now it is a stranger to all these things. "Long-tailed pigs wander about the streets; carrion crows pick up garbage in its once thronged thoroughfares, and

VIEW OF NEWCASTLE, IN THE BLUE MOUNTAINS, JAMAICA.

at the back of the handsome square, where King's House is situated, the negroes have built their shingled huts." Everything connected with it has been long dull and languid, and is still duller now, when even the dispirited officials are gone, and with them the occasional balls which, like angels' visits, few and far between, used to waken the echoes in the old ball-room of Government House, only to render the contrast more marked, and the solitude more appalling. "Yet, wandering through its deserted streets, one cannot but feel that after all there was a time when Spanish Town was indeed a city. The very houses, albeit they are tumbling to pieces, have an air of aristocracy about them to which those in Kingston have no pretension; and what we seek in vain in every other part of the colony, viz., traces of its ancient grandeur, we find in St. Jago de la Vega. Looking at the antiquated mansions, with the numbers still on their doors, we can imagine the days when governors, and bishops, and judges held high festival within. What visions of jerked hog, and black crab, and turtle soup, and old madeira does the sight of these produce! What pictures do they conjure up of those wicked old times when *aides-de-camp* used to ride alligators through the streets, when admirals used to give balls to the brown girls, when vice in every shape was more reputable than it is at present." *

Spanish Town in the past, and Spanish Town at present, gives no very inaccurate picture of what Jamaica once was and now is. Like all the great settlements—Seville Nueva excepted—it is on the south side of the island. Once it had its abbey, its chapels, and its convents; now a magnificent avenue of tamarind-trees, which led up to the vanished house of the Spanish Governor, and an old bridge, are about the only tangible remnants of that period of the colony's history. Then came the days of the English. Jamaica was, a few years after the Conquest, at the height of its glory. It was a tropical England. In the streets of its towns were fought out, as at home, the battles of the Roundheads and the Cavaliers; and "Oh, my king!" and "Oh, my Father!" were often heard, and to this day are common phrases of exclamation among the negroes. Then also was the era of the buccaneers, one of whom, Sir Thomas Morgan, was Governor; and tradition still relates odd, if not edifying tales of the high jinks which used to go on in King's House during the reign of this doughty individual. Now and then the negroes tried an insurrection, and were as invariably put down. Money was plentiful, luxury profuse, hospitality unbounded. But the government of the island was, as it has ever been, sometimes as bad as bad could be, at other times at a stage when it might possibly have been a little worse. But it always left a huge space for improvement, though improvement was slow to take advantage of the vacancy. Great fortunes were made and spent by Scottish and English adventurers; shameless political jobs were rife, and, to use a modern author's phrase, "the unhappy island was alternately governed by a knot of needy lawyers, or ignorant, purse-proud planters, just as the one party or the other happened to be in power." The House of Assembly was one of the vilest of legislative bodies; and though there are still people who regret it, as there are always people to regret everything that happens to be gone, we have never yet met any one who would like to see it again in the full blast of its incompetent oligarchy. In the words of Hector Mitchell, the first Mayor of Kingston, "One half of the members could not afford to be in, nor the other half to be out of it." The Assembly was a free institution, however. Black men, brown men, and white men, all sat together, wrangled, legislated, spoke nonsense, and did mischief. It was difficult to say who was worst. Perhaps the natural loquacity of the brown men caused, on the whole, most mischief. They were nearly all lawyers, all poor, and all greedy; and their vanity and hope of either wearying or persuading the Government to buy them up, tempted them to drown all efforts at real work in floods of loquacity and verbiage. The black members were also loquacious; but they were really more amusing than mischievous. One of them—still celebrated in the negro songs—used to ride up to the House on a dray, clad in a green coat and brass buttons, and a white hat, though with bare

* "Letters from Jamaica," p. 45.

feet. But old Vickers was not the worst of the garrulous members who brought things to such a pass that after the fortunes of the colony had collapsed with the emancipation of the negroes, and the abolition of the monopoly of sugar, they brought it to the last stage of ruin by the disturbances of 1865. Then the Assembly did the wisest thing it ever did. Feeling its utter impotence, except for mischief, it laid down its authority, the negro rebellion was crushed, and Jamaica became a Crown colony under a new constitution. The island seems, however, to be little better than ever it was. Indeed, from the letters of the correspondent already referred to, it seems gradually drifting from bad to worse, and unless the all-healing effects of a long course of wise rule succeeds in bringing about a better state of affairs, will sink into a paradise of semi-barbarous negroes. The following is a summary of the account we have gleaned from observations made by a correspondent on a journey across the colony to the north side of the island:—The parishes on the north side bear the reputation of being the most fertile and prosperous in Jamaica. The road lies across the great mountain range which forms the backbone of the island, amidst scenery of striking and diversified beauty, rich in all the forms of tropical vegetation, to which is added the peculiar charm, rare in these latitudes, of running streams of clear water. At Spring-hill, some sixteen miles from Kingston, is established in a lovely spot the Botanic Garden, which is kept up with greater care than is usual among the State institutions of Jamaica, and testifies, by the variety of its trees and plants, to the extraordinary resources of this climate. One cannot help being struck, however, here as elsewhere in the West Indies, by the scantiness of the modern results in the work of acclimatising to any useful end the products of the Eastern tropics and other similar countries; though most of the fruits, &c., we so familiarly associate with the West Indies are not native to it, but introduced from the mainland, or even from the Pacific Islands. The mango, it is true, is universal in every garden and hill-side, and flourishes in wonderful luxuriance. Most of the varieties, however, are very inferior to those of India, and even the boasted "No. 11," which is said to have been introduced from Martinique, is not to be named with the product of Mazagan or Malda. Many of the common fruits of India and China, such as the leetchee and the loquat, seem to be unknown in Jamaica. The *Eucalyptus globulus* flourishes exceedingly at a certain height above the sea, but there are many other species of that genus equally valuable for their febrifugal and economic properties, which would perhaps be better suited to these latitudes. Nature has done everything for this island, and the efforts of man to supplement her gifts have hitherto been of infinitesimal worth. In travelling by way of the roads into the interior it is impossible not to be struck with the fact that the advance once made by man's industry has not been maintained. On every side are evidences of retrogression and decay. Cultivated lands have relapsed into wilderness. The fields which once grew sugar-cane are now overrun with jungle. Roofless houses, dilapidated works, and rotting fences testify to the general defeat which the powers of man have sustained in the struggle for existence. Along the whole forty miles of road between Kingston and Annotto Bay, until one reaches the rich low lands near the sea, there are hardly any signs of cultivation visible.

Yet that the soil is of exuberant fertility, capable of yielding all the products of the tropics, as well as an extraordinary number of those proper to the temperate zone, is proved by the wonderful luxuriance of the vegetation, even up to the hill-tops. The whole of this northern coast of Jamaica, from Port Antonio to Montego Bay, is a paradise for richness and beauty, a land of unexampled fertility, capable of supporting an industrial population as large as that on any equal area of the earth's surface. The few prosperous sugar plantations—alas! every year growing fewer and less prosperous—are here situated. Besides sugar, rum, and pimento, there might be grown fruits here sufficient in quantity to supply all the Atlantic States of America. The oranges are equal in size to those of Brazil, and in flavour to those of Sydney, and being naturally furnished with exceptionally thick skins, like some other natives of this soil, would bear exportation better than any other. The cocoa-nut trees equal in size, healthiness, and fruitfulness any seen elsewhere in the tropics. To the question, why cocoa-nuts are not more cultivated, the stereotyped answer is, that "it is no use to grow them, for they will be stolen before they are ripe." Yet a single cocoa-nut tree in Trinidad or in Honduras is reckoned to yield produce to the value of fifteen shillings a year; and elsewhere in the West Indies a "cocal," or cocoa-nut plantation, is regarded as one of the most profitable of estates. The answer as to the cocoa-nuts applies to almost every other kind of cultivation in Jamaica. The curse of this island—the blight which spoils every local industry, and is rapidly eating the heart out of the community—is the habit of thieving among the negro population, a habit which assumes here the form of a national calamity. No country in the world is so weighted and pressed down by larceny as this. The universal prevalence of this vice among the negroes, to an extent which baffles all the powers of the law, is perhaps the most fruitful cause of the decadence of this once flourishing island, the pearl—or one of them, for there are several—of the Antilles, by nature fitted to take its place among the most valuable of British possessions. The one fact that this population of a little more than half a million imports foodstuffs, including corn-meal, grain, vegetables, and dried fruit to the extent of nearly £600,000 a year in value, bears unmistakable testimony to the blighting influence upon all the springs of local industry of the favourite negro sin. There is scarcely any portion of these imports which might not be furnished by the island itself under a healthy condition of industry. The people, however—so at least they affirm—prefer to import the necessaries of life rather than run the risk of growing them in the midst of their larcenous neighbours.

The slaves were freed, and no man whose *morale* is such as to make his opinions worth quoting would wish to see them again in bonds. But nevertheless this act finally ruined the island. The planters, no doubt, had themselves greatly to blame, for they seemed to wilt under what they looked upon as an affliction, and neglected the measures which have enabled the people of Demerara and other tropical colonies to partially, at least, replace the labour they were deprived of by the Act of Emancipation. But still the facts stand as we have stated. In 1805, during the height of the slavery system, Jamaica exported 150,000 hogsheads of sugar, besides other produce. In 1830 the export of sugar declined to 100,000 hogsheads; in 1839, to 49,213 hogsheads; in 1850, to 36,630; and in 1875, to 28,000. It is true that there has been a slight increase of late years in the product of coffee and one or two other articles, but it is still very far from being such as might be expected from the resources of the island, and by no means sufficient to supply the loss caused by the great falling-off in the staple of sugar. In 1817, with a population less than two-thirds of the present, the total exports of Jamaica were valued at more than seven millions sterling. At the present time they do not much exceed one million and a quarter. The total value of the property, movable and immovable, in the island was once reckoned at £50,000,000. Now it would be difficult

A PLANTER'S HOUSE, JAMAICA.

to estimate what is the value of property in Jamaica, seeing that most the estates, with the exception of those devoted to the breeding of cattle, are wholly unsaleable. Some three hundred sugar estates, and almost as many coffee, have been abandoned since the emancipation. At present nothing can be conceived more unhealthy than the state of Jamaica, judged by every test applicable to such a country. The white population, which must ever constitute the chief strength of the island, if it is to remain a British colony and not to sink back into savagery, has diminished and continues to diminish. At the time when Bryan Edwards wrote his ponderous but valuable work on the West Indies, in 1793, the whites of Jamaica were estimated to number 25,000, exclusive of the military. At the census of 1871 they had fallen to 13,816, which is some 3,000 less than the white population of Barbadoes, an island one-fiftieth of the size of Jamaica. Yet nature has not decreed the white man's banishment from this tropical island. No other island in the West Indies is, in fact, so well fitted by nature to be a home for the British race, for no other can boast of a temperate climate under tropical latitudes. Whoever has been to Newcastle, among the Blue Mountains, has seen British soldiers living in perfect health and comfort, even though debarred by the peculiar character of the country about the cantonment from their usual sports or exercise.

The change which was made in 1865, in consequence of the disturbances, was perhaps inevitable. Representative institutions had sunk to the lowest depth of degradation. The abolition of the Assembly was, in fact, unanimously voted, amidst the applause of all classes of the people, black as well as white. The franchise had been made so low as the negro population, a preponderating voice in it, and to exclude the majority of the respectable white and coloured inhabitants from any share in the government. The comparatively small number of whites in Jamaica must ever render a return to the experiment undesirable, for the idea of parliamentary institutions worked entirely by negroes is at present not to be thought of. Yet it is generally agreed that never were things worse than they are at present; and that unless taxation is lightened, and good government increased, there will be before long some fearful catastrophe. The friends of the negro are dissatisfied, and the planters are discontented with a state of things in which their interests are systematically, even offensively ignored. An absolute government has in Jamaica proved an utter failure, as may be proved by very potent figures.

In 1863-64, the year before the rebellion, the annual expenses of the island were £319,322. In 1876-77 the estimates provided for an expenditure of £510,571.* There has thus been an increase of nearly £200,000 a year in thirteen years, though without a corresponding increase in the colonial prospects, nor even from such public works as the Government has thought necessary, has the return been equal to the expenditure. The production of the island has remained stationary during sixteen years, and shows a tendency to decline rather than to increase; yet the Government Establishment has been greatly increased, the expenditure on officials having trebled since 1865.

If we compare Jamaica with either of the two colonies to which it comes nearest in size and importance—namely, British Guiana and Trinidad, either of which it at least equals in fertility and natural resources—we shall be struck with the miserable result of a pure Crown Government for such an island. With a population of 506,154 the exports of Jamaica in 1875† were of the value of £1,410,485, and the gross expenditure £586,520. British Guiana, in the same year, with a population of 220,000, exported to the value of £2,338,121,

---

* In 1876 the amount of public revenue was £529,731, and the expenditure £486,879. The public debt in 1876 was £501,415; the imports from the United Kingdom £967,952, from all countries £1,700,253; and the total exports from all countries £1,547,015.

† I have taken the year 1875, as it is the one which the *Standard* correspondent has selected for comparison. It ought, however, in justice, to be mentioned that, as may be seen from the figures just quoted, the expenditure in 1876 was less, and the exports rather more. The argument, however, remains the same.

and expended £355,979. Trinidad, with a population of 120,000, exported to the value of £1,625,082, and expended £282,294. Thus the exports of Jamaica were, per head of the population, in round numbers, £3, in British Guiana over £10, and in Trinidad £12. From this the reader may not unjustly draw the conclusion that Jamaica, if a pleasant land, is not a prosperous one.

Hayti.*

Four centuries ago, this now little-visited island was the converging point of Western adventurers, and was intended by the Spaniards to be the metropolis of the New World which Columbus had discovered, and which they were exploring. The island, first trodden by European foot on the 6th December, 1492, is about 155 statute miles in its greatest breadth, and 460 miles in its greatest length. The superficies may be approximately taken at 30,000 square miles, or nearly the same as that of Ireland; its coast-line, measuring indentations, at 1,500 miles. The surface is essentially mountainous, the mountain system consisting of three parallel chains running in a general direction from east-south-east to west-north-west. Some of the peaks attain a height of from 8,000 to 9,000 feet, but the culminating point is Pico del Yakee, or El Rucillo, so called from its crown of silvery clouds. It is nearly in the centre of the island, and is assigned by Schomburgk a height of 9,620 feet. Except the very highest elevations these mountains are everywhere covered with a deep rich soil from which spring exuberant vegetation and forests of valuable trees. A steady supply of moisture, which descends in constant streams to the plains and valleys below, diffuses fruitfulness and verdure through this rich but unhappy island of the black man. None of the rivers are available for inland navigation, but they all supply abundant means of irrigation, and even by the strength of the currents and rapids an immense water-power, as yet waiting utilisation for industrial and mechanical purposes. The island is at present divided between two Republics: that of Hayti, comprising that part of the island formerly owned by the French, and now ruled by the descendants of their former negro slaves; and that of San or Santo Domingo, which is the section once held by Spain.

At one time we used to hear a good deal of Hayti and the Haytian and Dominican Republics. But of late years they have all but dropped out of history, and it is only when they are cutting each other's throats, or swindling their citizens with brass buttons for money, that the world's attention is particularly called to them. Yet Hayti is a magnificent island, contesting with Cuba the honour of being the Pearl of the Antilles, yielding all the vegetable products of the tropics in the richest profusion, and abounding in mines of all the useful and many of the precious metals. It was the Espagnola of Columbus, whose house they still show in San Domingo, the Hispaniola, or Little Spain, of its later historians. But the Indians call it Haïté, or the mountainous land, and this name the island yet bears. The French blacks, who in 1804 drove out their masters and have ever since maintained their independence, call their Republic by the same name, while the Eastern Republic of

---

* See Major Stuart's exhaustive report on Hayti in "Report of Her Majesty's Secretaries of Embassy and Legation," Part II. ("Parliamentary Documents," 1877). Hazard: "Santo Domingo," 1873, &c.

the Spaniards and Spanish negroes is known as Santo Domingo. It is thus an almost thoroughly black island. There the negro dominates, and the few white men who live amongst them are treated as inferior beings. The European who lands there for the first time is apt to remember that land where the horses were masters and the riders serfs, and to experience feelings much the same as those which fill the mind of the American planter whose lot is now cast in the State of South Carolina. The Carib Indians have disappeared. They were a feeble race, of languid temperament, phlegmatic, and melancholy. They troubled them-

VIEW IN HAYTI.

selves about nothing, and in due time they were enslaved and died. In 1514, twenty-two years after their first acquaintance with the Europeans, they were reduced in number from 2,000,000 to 14,000. They had perished in the gold mines, men, women, and children; and their murder is one of the foulest blots on the indifferent reputation which the Spanish rule of "the Indies" bears. The destruction of the Indians was the ruin of their destroyers. To replace them Africans had to be imported, and the Africans in their turn ousted both the Spanish settlers and the French ones, and are now masters of the island, though the two black Republics hate each other with a hatred surpassing that of most men who write "Liberty, Equality, and Fraternity" on their flags. The Haytian Republic is tolerably prosperous, though afflicted with that New World mania for revolutions which so sadly afflicts their Spanish neighbours. Port-au-Prince is the capital of the

Republic, which altogether, including the islands lying off the coast, includes an area of 9,232 miles and a population of about 550,000. Its principal foreign trade is with Great Britain, France, the United States, Holland, and Germany. Its revenue is about £850,000, and owing to the difficulty experienced in borrowing, its expenditure is necessarily about the same. The population is not increasing much, and, indeed, it is affirmed in the island that it is really decreasing. The Haytians are a very favourable specimen of what the black man can become when he has a fair chance, education, and the self-respect which freedom

VIEW OF THE CITY OF SAN DOMINGO, HAYTI, FROM THE HARBOUR.

inspires. Certainly, though Hayti under its present government is by no means so prosperous an island as it used to be, these curly-headed Frenchmen, with their polished manners, and, in some cases, even high education and accomplishments, could scarcely be supposed to be of the same species as the Jamaica negro. Still the superstitions of the Obeah men and the Ananey stories which, like the other West Indian negroes, they brought from Africa, flourish in Hayti as in Jamaica, and form a dark feature in the character of even the best of the people. The people have a good deal of the vivacity of the French, and a great amount of vitality under misfortunes that would have crushed a less elastic race. They have, like their semi-ancestors over the Channel, tried most forms of government. They commenced with a Republic—formed by men such as Agé, Toussaint l'Ouverture, and Dessalines, once themselves slaves—and, after trying Monarchy in various forms, are at present believed to be living ostensibly under the former subjection. During the empire there used to be a prodigious number of dukes, marquises, and viscounts among the negroes. But they have all disappeared. Dukes are draymen in Port-au-Prince, marquises do white-washing, and among some of the best coal-whippers on board the steamers are some of the minor nobles of the Emperor Soulouque. As for his Majesty, he is dead, and one of his august widows was for some time the most prosperous washerwoman at Kingston, in Jamaica.

The Republicans of Santo Domingo were at one time united with the Haytians, but they had to separate from their black Gallic allies, of whom they now speak with intense bitterness, and perhaps not unjustly. The Dominicans are, however, always hankering after union with a stronger State. With a territory of some 20,000 square miles, and a population rated as high as 250,000, they still consider that they are numerically too weak for a separate nationality, and live in yearly terror of attempts on the part of the Haytians to conquer them. Forgetting old difficulties, in 1861 they united with Spain, but her rule being simply intolerable—for the Spaniard is unteachable—they broke loose in 1863, and again became independent. They are still talking of annexation, and more than once have made overtures to the United States to take them into its bosom. Hitherto, however, the want of a *quid pro quo* has caused Uncle Sam to harden his heart to Dominican woes. Bananas are as fruit excellent. They are mealy, digestible, nutritious, and wholesome: but as a source of revenue they are decidedly open to objections. Santo Domingo is essentially Spanish in language, habits, and mode of life, and had Spain shown the slightest sympathy with the aspirations of the people, her flag might yet have been flying on the island. At present it is about as badly governed as it could well be. Of the revenue of £170,000, nearly one-half is spent on the Ministry of War and Marine, and yet the Dominicans, if allowed, would gladly live in peace among themselves and with all others. The capital, San Domingo (p. 317), is the oldest city of European origin in America, having been founded by Bartholomew Columbus in 1494. But at present it answers but poorly to the descriptions of Oviedo and other writers of his time. The site, plan, and area are still the same; but in vain the visitor looks for the monuments which made it the rival of the first cities of Spain. Many public buildings were left unfinished when, owing to the excitement created by the Mexican discoveries of Cortez and his followers, the exodus to the mainland began, and in this condition many still remain. Siege and war have done the rest, while the position of the harbour, within a bar at the mouth of the Ozama River, is an obstruction to shipping and commerce. We have called Santo Domingo a Black Republic, and so, in reality, it is, though the people would be very indignant were they called anything else than Spaniards. Yet, truth is, the *sangre azul* is in a very small proportion to the *sangre negro*. But the white man can own property and hold office in San Domingo. In Hayti, by law at least, he cannot. The end of Santo Domingo will undoubtedly be annexation to the United States, and when annexation comes to Santo Domingo no civilised Power will long tolerate the bluster of the gallant but rather bumptious Haytian negroes.

CUBA.*

This is the largest and richest of the West Indian Islands. It was discovered by Columbus in 1492, and has successively been called Juana, Fernandina, and Santiago, the present name, now universally applied, being the native one at the time the whites landed. Its extreme length is 730 miles, and its average breadth 80 miles. Its area is 43,319 square miles, the neighbouring isle of Pinos 1,214 square miles, and the smaller islands lying off the coast 1,350 square miles—in all, 45,883 square miles. The coast is generally

VIEW IN CUBA.

low and flat, surrounded by islands and reefs, which render navigation close to land difficult and dangerous. The highest part of the island is the Pico de Tarquino (7,670 feet). One of the peaks of the mountain range, stretching in the south-east from Punta de Maysi to Cape Cruz, is known as the Sierra, or Montaños de Maestra or Cobre. Hence, though the coast-lands are tropical in climate, the elevated interior of Cuba enjoys an almost temperate atmosphere. Some of the scenery is pleasing, and even not without grandeur. The western section is the smallest portion of the island, but is the most level. Here are nearly all the great sugar factories and tobacco plantations which have given the island a commercial celebrity, and supply most of its wealth. In the

---

* Hazard: "Cuba with Pen and Pencil" (1873). Gallenga: "The Pearl of the Antilles." Peron: "L' Isle de Cuba" (1876); and Mr. Keith Johnston's admirable summary of the state of our geographical knowledge regarding the island in the *Encyclopædia Britannica* (9th edition, 1877).

central department are Havanna, the luxurious capital of the colony, many smaller towns, and, it follows, most of the population. Outside of the towns are forests and unpeopled savannahs, though at one time in the eastern department were well-cultivated districts now fallen into ruin, and, owing to the richness of the soil, becoming rapidly overrun with wild vegetation.

Sugar, tobacco, coffee, rice, and cotton are the staple products; and though of late years the whole trade of the island has been disturbed and in some cases paralysed and even destroyed by the civil war which has raged since 1868, the amount of sugar exported in 1872 was worth over £20,000,000. Honey, rum, and wax are also among the articles of commerce which enrich the Cubans; while their imports comprise rice, olive oil, flour, jerked beef, staves for casks and sugar-boxes, lard, and coals. In the western division of the island there are over 1,000 miles of railway. What the future of the country is to be it is difficult to say. For ten years the colonists have been at war with Spain. But even then they are divided among themselves. The peninsulars, or emigrants from Spain, wish for the continuance of slavery; the Creoles desire its abolition; and as the former comprise most of the great sugar-planters, and have a volunteer force of 60,000 men, they have hitherto been able to dictate their will. The Moreb law, which declares free every slave after reaching sixty years of age, and every one born after 1870, is practically a dead letter. The insurgents carry on a cruel guerilla warfare in the interior, while the Spaniards make reprisals of an equally atrocious character. From 1868 to 1873 it is calculated that the war has cost 150,000 lives, but it has now (1878) all but died out. Nevertheless, the future of Cuba can never be what its past has been. It is difficult to obtain anything like an accurate census of the island; but it is believed that the whole population is something like 1,359,000, including 200,000 slaves, and about 90,000 free "coloured people."

We must conclude this volume with the West Indies, otherwise we might have spent many chapters over these interesting land-spots in the Caribbean Sea. Their history is stirring, as it embraces much of the heroic period of Spanish and English naval adventure and the times of the buccaneers. Their natural history is equally rich, but that we may have occasion to speak of when we travel in the neighbouring portions of Central America; while the manners and customs of the inhabitants are much the same as prevail among the other Spanish inhabitants of America, or in the neighbouring English colony of British Guiana, which we have yet to visit. Let us, therefore, cross to the "Spanish Main."

www.ingramcontent.com/pod-product-compliance
Lightning Source LLC
Chambersburg PA
CBHW031850220426
43663CB00006B/563